CONTENTS

ENGINEERING GRAPHICS

ENGINEERING GRAPHICS

FRANK M. CROFT, JR., THE OHIO STATE UNIVERSITY

FREDERICK D. MEYERS, THE OHIO STATE UNIVERSITY

EDWIN T. BOYER, THE OHIO STATE UNIVERSITY

MICHAEL J. MILLER, THE OHIO STATE UNIVERSITY

JOHN T. DEMEL, THE OHIO STATE UNIVERSITY

WILEY

John Wiley & Sons

New York   Chichester   Brisbane   Toronto   Singapore

Cover design by Lee Goldstein
Cover art by Marjory Dressler
Production Supervised by Katharine Rubin
Copyediting Supervised by Richard Koreto

Library of Congress Cataloging in Publication Data:

Engineering graphics.

  Bibliography: p.
  Includes index.
  1. Engineering graphics. I. Croft, Frank M.
T353.E63 1989   604.2′4   88-27700
ISBN 0-471-85788-2
ISBN 0-471-623202

Printed in the United States of America

10 9 8 7 6 5 4 3 2 1

WE DEDICATE THIS TEXT TO THE FACULTY OF THE DEPARTMENT OF ENGINEERING GRAPHICS OF THE OHIO STATE UNIVERSITY FOR THEIR NATIONAL LEADERSHIP IN THE FIELD AND THEIR COMMITMENT TO EXCELLENCE.

ABOUT THE AUTHORS

Frank M. Croft, Jr. P.E., is an Associate Professor in the
Department of Engineering Graphics at The Ohio State
University. He has been at Ohio State since 1984. Before
joining the faculty at Ohio State, Dr. Croft had taught 11 years
on the college level. He started his academic career at West
Virginia Institute of Technology where he spent three years
and then moved on to the Speed Scientific School, the
Engineering College of the University of Louisville, where he
spent eight years. He taught engineering graphics at West
Virginia Tech and was involved in graphics and transportation
at the University of Louisville. During this period with the
University of Louisville, he completed a Ph.D. in engineering
at Clemson University in South Carolina. In addition to his
Ph.D., Dr. Croft has a Master of Science in Engineering from
West Virginia College of Graduate Studies and a Bachelor of
Science in Aerospace Engineering from Indiana Institute of
Technology.

Dr. Croft has been very active with the Engineering Design
Graphics Division (EDGD) of the American Society for
Engineering Education (ASEE) since 1974. He has served on
the publications staff of the *EDGD Journal* as Associate Editor
and as Advertising Manager. He has served as Program
Chairman for both the Mid-Year Meeting of the Division and
the Annual Conference of ASEE. He will serve the division as
Vice Chairman during 1988–89 and will be Chairman during
1989–90.

Frederick D. Meyers, P.E., has been Associate Professor of Engineering Graphics at The Ohio State University since 1982. Prior to his appointment there he served as a project engineer, engineering manager and technical director with Owens-Corning FIBERGLAS Corporation over a period of 30 years. He also served as an engineering officer in the U.S. Air Force. His early teaching experience was as an Instructor in Engineering at the A. & M. College of Texas. He holds B.S. and M.B.A. degrees from Ohio State.

Professor Meyers is an active member of the American Society for Engineering Education and the American Society of Mechanical Engineers. He was named Ohio Engineer of the Year in 1972. He is a member of Pi Tau Sigma and Tau Beta Pi, engineering honorary societies.

Edwin T. Boyer has been an Associate Professor of Engineering Graphics at The Ohio State University since 1984. He earned his B.Ed. degree at Keene State College in 1962 and his M.S. degree at Indiana State University in 1963. In 1976 he earned the D.Ed. degree at Texas A&M University.

Dr. Boyer taught drafting and graphic arts at the high school level and then founded his own business, Visual Grafix. His company produced engineering, architectural, and land contour scale models for industry, as well as other graphics and illustrations for specific client needs. He taught for a year at Texas A&M University before teaching Industrial Design and Drafting at Northern Illinois University for eight years.

Dr. Boyer is an active member of professional organizations including the Engineering Design Graphics division of the American Society for Engineering Education, The International Technology Education Association, and he is a Life Member of The Council on Technology Teacher Education. Dr. Boyer organized and supervised the team that prepared all the camera-ready graphics for this book.

Michael J. Miller, P.E., began teaching full-time in 1985. Prior to that he had served as an Adjunct Professor while working full-time as an engineering research administrator at The Ohio State University. He also spent five years in the aerospace and electronics industries as a practicing engineer. Professor Miller holds B.S. and M.B.A. degrees from The University of Michigan and The Ohio State University, respectively, and is an Assistant Professor of Engineering Graphics at Ohio State. He is a member of the American Society for Engineering Education and the American Institute of Industrial Engineers.

John T. Demel, P.E., is Professor and Chairman of the Engineering Graphics Department at The Ohio State

University. He holds the degrees of Ph.D. and M.S. in
Metallurgy from Iowa State University and a B.S. in
Mechanical Engineering from The University of Nebraska-
Lincoln. Previously he taught at Texas A & M University and
Savannah State College. He also served as an officer in the
Army Ordnance Corps.

Professor Demel is a national leader in implementing
computer graphics in engineering. He has co-authored texts in
computer graphics with Professor Miller and authored or co-
authored over 20 publications on engineering graphics. He is
active in the American Society for Engineering Education,
American Society of Mechanical Engineers, and National
Computer Graphics Association.

ACKNOWLEDGMENTS

The authors wish to thank all the members of their immediate families for their patience and understanding during the many hours required to prepare this text. Their support during this period was greatly appreciated.

The effort and support of the editorial and production staff of John Wiley & Sons is gratefully acknowledged. We would especially like to thank Charity Robey, the College Editor, for her support and encouragement during this project.

Most of the finished illustrations in this book were produced by a team of illustrators. There were many people involved with the illustrations for this project and the authors would like to acknowledge their efforts. Those illustrators who made the most significant contributions were: Casey D. Doyle, Jefferey S. Wallace, and Marilyn R. Bahney. Other contributors were: R.E. Spica, D.P. O'Donovan, R.L. Long, T.L. Myers, M.E. Edwards, K.J. Bomlitz, and W.J. Kleman, Jr. Their dedication to this project was greatly appreciated.

Faculty members of the Department of Engineering Graphics of The Ohio State University who contributed to this text were: Professors L.O. Nasman, R.I. Hang, R.A. Wilke, G. Bertoline, and R. Herrera. One graduate student who made a significant contribution in problem development was M.S. Stephenoff.

Finally, the authors wish to express their gratitude and sincere appreciation to the reviewers of the manuscript. Without their input this text would not have been possible. The reviewers were: Ronald Barr, University of Texas at Austin; Larry Genalo, Iowa State University; Jacques Simoneau, The Royal Military College of Canada; W. George Devens, Virginia Tech; Klaus Kroner, University of Massachusetts; Joe Tuholsky, West Virginia Tech; Homer L. Bosserman, Monterey Peninsula College; J. David Holloway, Ohio Northern University; Jon E. Freckleton, Rochester Institute of Technology; and George W. Eggeman, Kansas State University.

PREFACE

This book is designed to meet the needs of modern engineering
design graphics curricula and employs the current American
National Standards Institute (ANSI) graphics standards.
Instructors today have less time available to teach, and
students have less time available to learn more material than
ever before. The number of credit hours allocated for teaching
engineering design graphics has steadily shrunk while the field
of engineering design has steadily become more complicated.
In recent years we have seen the traditional steel and cast iron
technology be replaced by a wealth of new materials. These
new materials often require radically different fabrication
methods than the older materials. At the same time, a great
variety of new design, drafting, and analytical tools have
become available. Many of these are computer based. There is
much more material to cover, and fewer hours to do the job.

This textbook meets the needs of today's graphics programs,
by addressing the new technologies. The traditional graphics
topics have been streamlined so that they can be presented in a
shorter period of time. We have eliminated obsolete material.

Our approach to this text has been to critically examine
every paragraph of material to assure that it is up-to-date,
needed in the text, and presented in a forthright manner. All
material is written in clear, uncomplicated language. The
examples used throughout this book are of objects and
assemblies that should be familiar to the student. For instance,
a bicycle chain and sprocket are used to illustrate a line tangent
to one or two circles.

Computer aided design and drafting (CADD) methods are
included where appropriate. The student who is using a
computer graphics system to do the drawing exercises will be
able to complete the exercises without having to refer to an
appendix or a special chapter. Instruction in CADD will be set
in a special typeface so that the student will be able to tell

which portions of the text are dealing with manual methods and which are for the computer aided methods.

The book does not assume that the student has had previous graphics experience. It begins with the development of basic skills that the student will need. Geometric constructions, lettering, visualization and sketching techniques, orthographic projection, and pictorial drawing are described in detail. Illustrations and pictures are used at every step of the way to illustrate the points being made by the written material.

The material proceeds logically to develop the student's ability to describe the shape and size of objects. Drawing conventions that are important to the student are described and illustrated. We describe the use of sectional views to show the inside parts of objects. Conventional methods of drawing sections are presented along with illustrations of each convention.

We cover size description, adding dimensions to the drawing, in detail. How to specify tolerances, the allowed variation in dimensions, is presented as the student begins to understand methods of making objects and reasons that variations in size must be allowed.

As the student gains in ability to draw a single object, the idea of joining two or more objects to create an assembly is introduced. We discuss fastening and joining methods and describe various types of fasteners. Welding and use of adhesives, two very popular methods of joining objects, are covered in detail.

Design tools that are available to the student are covered next. Descriptive geometry techniques are presented using real-world examples whenever possible. The pattern for a three dimensional model is included inside the back cover of this text. When assembled, the model can provide tactile as well as visual input for thinking in three dimensions. We present graphical mathematics as an analytical tool that can provide quick approximations as well as a technique that increases the student's understanding of the significance of the numerical analysis techniques that are in common use.

The design process is described so that the student is aware of the many steps that are necessary before an idea becomes a completed design. With this description, we show how graphics is used to document the steps of the design process.

In every chapter, care is taken to provide instruction on the latest technologies including CADD and drawing reproduction techniques. Old techniques that are not often used are not presented unless they will help the student to understand the new techniques. Since there are many CADD software packages available, the CADD discussion is generic and the instructor will have to provide specific information on hardware and software systems. This text is available packaged with a CADD program designed specifically for educational applications that is easy to learn and use. It has many of the functions of the commercial CADD packages. The student will own the software if it is purchased with this text; this ownership gives the student the right to use the software on any available computer.

Graphics, whether completed manually or by computer, is heavily dependent on the application of geometric principles. For this reason, it is important that the student have a good mathematics background in plane, solid, and analytic geometry as well as trigonometry. If graphical mathematics is presented as part of the course, it is important that the student have a basic understanding of differential and integral calculus.

Students who are going to use a computer and CADD software must know or learn how to perform basic computer operations. These include booting the hardware system, starting the CADD program, disk initialization, file saving and retrieving, making backup copies of disks, and recovering in case the program is halted by an error.

Typical CADD computer system components are identified in Chapter 1. It is desirable, but not necessary, that the students have some understanding of the inner workings of computer system components. It will give them more confidence when they are using the system and may help them to know what to do if an error occurs. Programming experience is not needed.

When students complete their education and enter the work force, they will probably have access to or be required to use engineering workstations. They also may be required to select the components for a computer graphics workstation or to choose computer graphics software to be purchased. Instruction in system component selection is not within the scope of this book, but might be provided as an additional topic.

The instructor's manual contains suggestions for organizing courses or course sequences of various types and lengths as well as problem solutions and transparency masters. Two workbooks of additional problems are also available from the publisher as well as solutions for the workbook problems. The book is available with or without the software packaged inside.

FRANK M. CROFT, JR.
FREDERICK D. MEYERS
EDWIN T. BOYER
MICHAEL J. MILLER
JOHN T. DEMEL

CONTENTS

TERMS YOU WILL SEE IN THIS CHAPTER

DESIGN Process of planning and creating a solution to a problem.

SIMULATION To create a mathematical description or a pictorial representation of an object or a process.

SOFTWARE Programs that make a computer run. System software provides the computer's essential general functions. Application software performs specific tasks such as CADD.

MODELING Design technique where the characteristics of an object or process are stored in a computer and used to produce and work with an image of the design on the screen.

CADD Computer-aided design and drafting. Refers to computer software and hardware for creating engineering drawings on a computer screen.

DIMENSIONS Information added to a drawing to show the sizes of the objects the drawing represents.

.25

ORTHOGRAPHIC PROJECTION Method for describing objects using multiple views. Each view is taken at right angles to the adjacent views.

SECTION VIEWS Cutaway views of objects showing internal features.

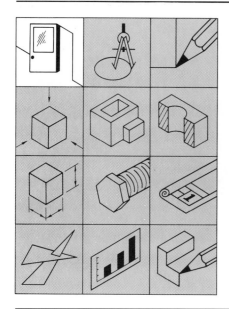

1

GETTING STARTED

1.1 INTRODUCTION

This book is designed for you, the engineering graduate of the
1990s. It will provide you with the instruction in **design** graphics
skills that you will need in your profession. Design is the
essence of engineering, and graphics is the language of design.
Similar to English, or any other language, it is the means we
use to communicate our ideas to others. We have all had the
experience of trying to describe something, such as how to get
to our house, to someone else. We may have ended up drawing
them a map. It has been said that a picture is worth a thousand
words. That saying may seem trite, but it is very true. A single
picture can often illustrate an idea that cannot be
communicated in any other way.

As a graduate engineer, you will enter the engineering
profession at a very exciting time. Historically, new designs
have been created at the drafting table, proven out by
constructing physical models, and tested in the laboratory and
in the field. When the test program was complete, tooling was
designed and fabricated, and production was started. Rapid
technological advances are changing the field to one where the
design is created at a computer-aided design (CAD)
workstation (Figure 1.1). During the design phase, the design
engineer has access to computer-aided engineering (CAE)
design analysis tools that can be used to test the design in

Figure 1.1 Computer-aided design and drafting (CADD) workstations are used by many designers. Courtesy of Ohio State University Engineering Publications

simulation. This means that with **software** that permits **modeling** of geometric solids, a mathematical model of the object can be created on the computer. The mathematical model can then be tested on the computer just as a real physical model might be tested in a laboratory, and design errors can be corrected without having to construct a physical model (Figure 1.2). All of the data that are used in the design are stored in the computer. When the engineer is satisfied that the design is correct, the tool designers can use the same data base to design the needed tooling without ever having seen an actual physical model of the object. When it is time to begin production of the object, the manufacturing personnel can use the same design

Figure 1.2 Editing drawing on CADD workstation. Courtesy of Ohio State University Engineering Publications

Figure 1.3 Computer numerical control milling machine. Courtesy of Ohio State University Engineering Publications

data to program the computer-controlled machines that will be used in its production (Figure 1.3).

Because of the ease with which a design can be created, tested, and modified without ever having to create an actual drawing on paper or a model in the shop, engineers can try out many different designs quickly and economically. They are free to be very creative.

You will need a variety of skills in order to make effective use of the power of the new computer-based tools available for engineering design. The ability to visualize and describe objects will be increasingly important as more use is made of computer-generated images instead of physical models. It will be necessary for you to be able to visualize what an object under design will look like and to be able to create both manually drawn sketches of the object and computer-drawn images.

1.2 *WHAT YOU WILL LEARN FROM THIS TEXT*

This text is written for engineering students. It will teach the graphics skills that engineers must master but will not go into the details normally needed only by drafters. Each chapter will introduce a specific topic area and provide instruction in both manual methods and computer methods, so that you may become proficient in sketching, use of drawing instruments, and use of computer-aided design and drafting (CADD) systems. You will also have learned how to create a drawing or a sketch that communicates the details of your design to others.

1.2.1 Lettering and Geometric Constructions

A drawing is made up of geometric shapes, lettered dimensions, and notes. In Chapter 2 you will learn techniques for lettering, both freehand and with the use of lettering guides and other aids. You will also learn effective methods for performing the most often used geometric constructions. These will include drawing lines parallel or perpendicular to a given line, dividing lines into several parts, bisecting angles, and constructing polygons, circles, and arcs, plus several other constructions.

1.2.2 Sketching Techniques

Engineers and scientists need a quick way to present and study their ideas. They can create a drawing with instruments, but that often takes a lot of time and may require instruments that they do not have with them. Sketching, which is creating freehand drawings, is a powerful tool since it requires only a pencil and some paper. It is a quick way to present an idea. Furthermore, a sketch can be modified quickly and easily. When the idea has been worked out to the designer's satisfaction, the sketch can be used as the basis for detailed drawings. Sketches are also an important part of the records in an engineer's lab book. They document the engineer's work and provide a record for patent applications and other legal purposes. A drafter working from a good sketch can create the drawing without asking many questions.

Engineers increasingly rely on making sketches of their ideas and then completing the design by computer. Because of this, drawing with instruments (triangles, T-squares, compasses) will be done infrequently by engineers. However, the ability to create drawings with instruments will still be important. A very large number of manually created drawings are in existence and must be maintained and modified.

1.2.3 Drawing with Instruments

A well-made sketch can contain as much detail as a final drawing done with instruments. However, a sketch is seldom acceptable for use in manufacturing. Therefore, it is necessary to make a final drawing from the sketch. This drawing might be made on a CADD system or manually with instruments. Basic instrument use will be presented in this chapter. Additional instruction on use of instruments as well as use of a computer drawing system will be presented as needed throughout the text.

1.2.4 Shape Description

Orthographic Projection

A drawing of an object is made so that the graphic image appears to be the same shape as the object that is being represented. Thus, the drawing provides a shape description of the object.

When the drawing is to be used for creation of the object, it will generally provide three different views of the object, most often the front, top, and a side view. The views are placed in a standard arrangement (Figure 1.4). This drawing technique is known as **orthographic projection**. It is covered in Chapter 4.

Figure 1.4 Orthographic projection uses multiple views to describe object.

Pictorial Drawings

Pictorial drawings are generally used to describe an object or assembly, how it goes together, and how it is used (Figure 1.5). There are usually no dimensions or other notes that would tell how to make the object since that information would be on the detail drawings. A person assembling a toaster, for example, does not need to know the exact dimensions of each part, just where each part goes. Techniques for producing the various types of pictorial drawings will be presented in Chapter 5.

1.2.5 Sections and Conventions

Designers often create drawings that show a part or assembly that has had a portion cut away to show interior details (Figure 1.6). Such drawings are known as **section views**. They make it easier to understand how an object is constructed or how an assembly functions. You will learn to make sectional views of single objects and of assemblies.

Sometimes a portion of a drawing would be difficult to understand if it were drawn in a technically correct manner. A number of techniques have been devised to make the drawings more understandable. These special techniques are called conventional practices. They will be covered in detail in Chapter 6.

Figure 1.5 Pictorial drawings are created to resemble picture of object being represented.

1.2.6 Size Description

It is necessary in most instances to indicate the size of the object as well as its shape. Size description is provided in two ways. One is to specify a scale, such as half size, meaning that one unit on the paper represents two units on the actual object. The second

Figure 1.6 Sectional views are drawn so that object appears to have been cut open to display its interior details.

Figure 1.7 Dimensions are added to orthographic drawing to show size of object being represented.

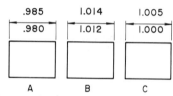

Figure 1.8 Tolerances are added to dimensions to show size of object being represented.

Figure 1.9 Fasteners such as these are used in many assemblies.

method is to mark dimensions on the drawing (Figure 1.7). In most cases, both methods are used. Drawing to a certain scale permits the drafter to lay out the object on the paper using a scale that is graduated in the proper increments. However, determining an object's size by measuring the drawing does not provide accurate dimensions. Therefore, engineering drawings include both dimensions and tolerances. Dimensions show the desired size. Tolerances tell the person making the object how much the size can vary from the dimensions shown (Figure 1.8).

1.2.7 Fastening and Joining

Most of the devices we use are made of two or more parts fastened together into an assembly using fasteners as shown in Figure 1.9. There are many methods of fastening. Bolts, screws, rivets, welds, and adhesives are just a few of the many devices and materials that are used in fastening. In some cases, the fastener does not permit any movement between the objects, while in other cases it is important to permit such movement.

1.2.8 Drawing Notes

Notes are often added to a drawing. Notes are used to indicate the kind of material the object is to be made from, how many units of the object will be needed, and any special requirements (Figure 1.10). For instance, a note on a drawing of a table leg might say that it will be made from oak and that four will be needed for each table.

Figure 1.10 Notes are an important part of a drawing. Courtesy of Jones and Vining, Inc.

1.2.9 Production Drawings

Production drawings, often called working drawings, are used to guide the workers who create the object or assembly (Figure 1.11). They include all the information needed to produce the object: shape description, size description, quantities, notes, and assembly instructions. Mechanical engineers have working drawings of

Figure 1.11 Production (working) drawings provide details needed to make object. Courtesy of Ford Motor Company

mechanical objects such as gears and of assemblies such as transmissions. Civil engineers have working drawings showing how a portion of a highway is to be built. Electrical engineers may have working drawings showing the layout of printed circuit boards or of electrical wiring in a building. Piping diagrams used by chemical engineers are another example of working drawings. Working drawings of airplane wings and other aircraft components are used by aeronautical engineers.

1.2.10 Descriptive Geometry

Many engineering design tasks, such as piping layouts and duct work, often require the techniques of descriptive geometry for their solution. Descriptive geometry is the application of the principles of spatial geometry to the solution of size, shape, and space relationships of objects and parts of objects. For instance, descriptive geometry can be used to determine the necessary size and shape of a flat piece of metal that is to be formed into a transition piece. (Figure 1.12).

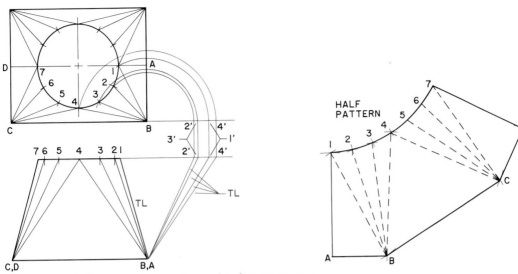

Figure 1.12   Descriptive geometry can be used to lay out pattern.

1.2.11   Creation of Charts and Graphs

Charts and graphs are very valuable tools for presenting information. The information they provide is not intended to take the place of detailed numerical data but rather to supplement it. Charts and graphs are used to show all sorts of relationships. Sales charts show at a glance whether sales are increasing or decreasing. Pie charts are often used to show the expenditures of a budget (Figure 1.13). Relationships between economic data and time are easily shown with charts and graphs (Figure 1.14).

1.2.12   Graphical Mathematics

Many mathematical relationships, particularly in geometry and calculus, can be determined graphically (Figure 1.15). Graphical solutions of such problems may be much faster than performing the calculations by hand and may often give satisifactory accuracy for the task being performed. Graphical analysis is often used to quickly provide or communicate information.

Empirical Equations

Empirical equations are expressions that are written to explain a known experimental relationship. Often it is not possible to write a sound mathematical expression to describe how something be-

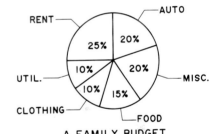

Figure 1.13   Pie charts provide wealth of information in small space.

Figure 1.14   *XY*-coordinate charts show relationship between two variable values.

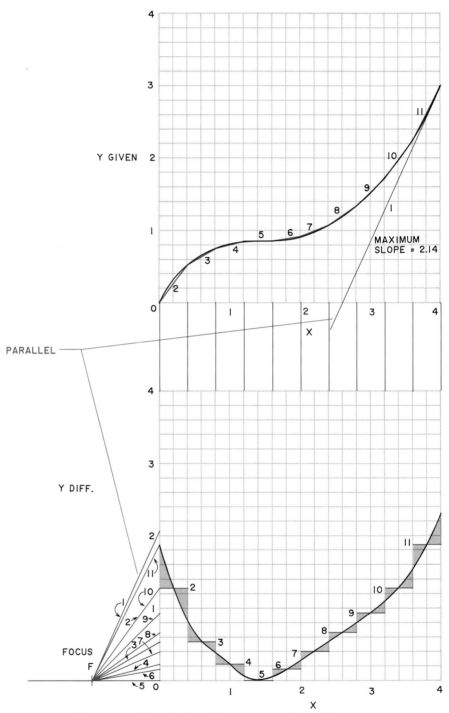

Figure 1.15 Graphical mathematics provides quick answers to many problems.

haves; so it is necessary to investigate how it behaves and write an empirical equation to describe what is observed.

Graphical Calculus
Graphical calculus is a technique for solving both differential calculus and integral calculus problems by graphical means. It can be used when a mathematical expression for the relationship is unknown or does not exist.

Vector Analysis
We are used to solving problems where we are interested only in the values of the variables involved. These variables are called scalars. Often, however, when we are analyzing problems, we find that we must take into account not only the values of the variables but also the directions in which they act. These variables are called vectors (Figure 1.16). Problems involving two or more vectors are solved by a mathematical process called vector analysis. We can also solve these problems graphically. We call the process graphical vector analysis. Chapter 11 deals with this very important problem-solving technique.

1.2.13 Computer Graphics

Computer graphics provides an exciting set of new capabilities that are much more powerful than anything the design engineer has ever had available before. These include both two-dimensional (2D) and three-dimensional (3D) CADD systems (Figure 1.17). Modeling software gives the designer the ability to create a three-dimensional computer model of an object. The model can then be moved or rotated, its size or shape can be changed, or it can be combined with other objects. The designer can examine it from many different angles. Structural analysis software permits the designer to test the design of an object to determine whether it meets the requirements for strength, resistance

Figure 1.16 Vectors show both magnitude and direction.

Figure 1.17   CADD workstations are powerful design tools.
Courtesy of Ohio State University Engineering Publications

to temperature changes, resistance to bending under load, and resistance to shock. These tests can be conducted in simulation, meaning that they are done by the computer on a computer-generated model meeting the designer's specifications.

Computer-aided Design

Many design firms now use computer systems to store their design data and create their drawings. The CAD systems or CADD systems are in use in most large companies. As microcomputers become more powerful, an increasing number of smaller companies are getting these systems. The main difference between the two types of systems is that CAD systems provide more tools to aid in the design process while CADD systems provide more tools to aid in the drafting process.

The CADD systems can produce drawings that are equivalent

to those produced in the "conventional" way with pencil and paper (Figure 1.18). However, with a computer-based design system, the drawing is created on the screen of the computer display terminal, and the information (data) needed to reproduce the drawing is stored in a data file on a computer data disk or on magnetic tape. When the drawing is complete, it can be drawn on paper or other drawing media using a plotter that is connected to the computer (Figure 1.19). If the drawing is small and a "rough" copy is all that is needed, the drawing may be reproduced on a graphics printer.

It is no longer enough for the designer to be able to produce a set of drawings showing how an object is to be made. Historically, that is what has been done. It was then necessary for a tool designer to determine what tools would be needed to produce the object and to design any special tools that were required. Then the drawings of the object and the necessary tools to make it were provided to the manufacturing people. They would have to study the drawings and decide how to make it. Sometimes the tooling would not work. In that case, the manufacturing people and the tool design people would need to work together to modify the tooling or to develop new methods for making the object.

Now, it is becoming very common for the tool design people to use computer files of design data created by the design engineer to design the needed tools. Just as the designer can test the product design in simulation, the tool designer can test the tooling by having the computer simulate the necessary steps to

Figure 1.18 Drawing produced on CADD system. Computer generated image.

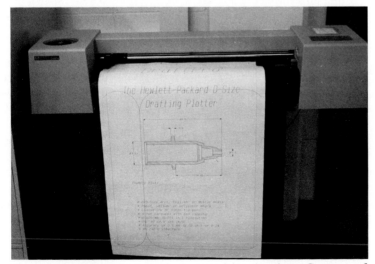

Figure 1.19 Pen plotters producing precision drawings. Courtesy of Ohio State University Engineering Publications

Figure 1.20 Computer numerical control (CNC) machines provide automated manufacturing. Courtesy of Ohio State University Engineering Publications

create the object from the raw material. The best method of manufacturing the object can be determined and the special tools can be designed correctly before the manufacturing process is started.

Computer-aided Manufacturing

Computer-aided manufacturing (CAM) involves using the computer to control the machines that are making the object. For several years, machines have been controlled by creating a tape specifying the required operations and then having the machine follow the instructions on the tape. These machines are known as numerical control (NC) machines. Newer versions of this type of control are computer numerical control (CNC) and direct numerical control (DNC) (Figure 1.20). With CNC or DNC a computer is connected to the NC machine and is programmed to direct its actions.

Computer-integrated Manufacturing

In many of the preceding cases, it is still necessary for a human to study the design drawings and create the necessary tape or computer program to guide the operation of the machines that will produce the object. However, the CAD and CAM systems are now combined in many companies. The result is known as CAD/CAM or computer-integrated manufacturing (CIM). With these systems the same data file is used throughout the design process and the manufacturing process. The file is first created by the designer using the design and analysis capabilities. If new tools are needed, the tool designer uses the file and the same design and analysis software to design the tools. Then the file is used with the manufacturing software to create the instructions for the NC machines to produce the parts. It is possible to complete the entire process without ever producing a drawing on paper. The ability to use the actual design data through all phases of the design-and-manufacturing cycle eliminates the errors that are caused when drawings have to be read and interpreted by humans and the information has to be transferred manually from system to system.

Flexible Manufacturing Systems

Flexible manufacturing systems (FMSs) permit the concepts and techniques of CIM to be used by small companies and others that do not produce enough objects of any one design to make it profitable to set up a complete production line for the object. Flexible manufacturing systems permit rapid and inexpensive changeover from production of one part to production of another, often by simply reprogramming the machine(s) involved (Figure 1.21).

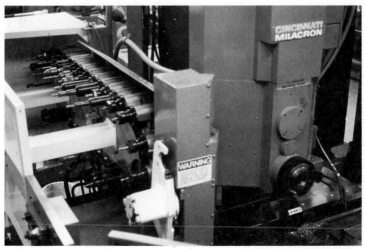

Figure 1.21 Flexible manufacturing center. Courtesy of Ohio State University Engineering Publications

1.3 TOOLS YOU WILL NEED

The tools you will need will depend on whether you are going to produce sketches, final drawings using instruments, or drawings using a computer system. This section will discuss the tools required for each of these processes.

1.3.1 Tools for Sketching

The tools you will need to produce acceptable sketches will include paper, pencils, and an eraser. An erasing shield is also useful. The paper may be ruled with a rectangular or an isometric grid (Figure 1.22). Thus, you can follow the lines and will find it easier to produce a neat sketch. Some people use unruled paper laid over a ruled sheet. This provides both the advantages of a ruled sheet and the better final appearance of the unruled sheet.

Many types of pencils are designed especially for creating

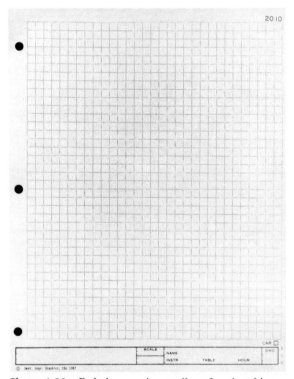

Figure 1.22 Ruled paper is excellent for sketching.

Figure 1.23 Erasers used by drafters.

sketches, drawings, and illustrations. The type of pencil you select will have a strong effect on your ability to produce good work. The thickness and hardness of the pencil lead are also important. Lead hardnesses vary from 6B (very soft) to 9H (very hard). The softer leads (HB to F) are generally used for sketching. The harder leads are better for detail drawing. A more detailed discussion of pencils and leads will be found in Chapter 2.

The type of eraser you use is also very important. Some will erase cleanly, others will not. Some are designed to erase small areas. Others are designed for bigger jobs. The three types of eraser material most often used for drafting are gum, pink rubber, and white plastic (Figure 1.23).

1.3.2 Tools for Instrument Drawing

Instrument drawing requires many more tools than sketching. Figure 1.24 shows a typical set of tools for a beginning student. It is important that you become familiar with the tools and their uses.

Chapter 2 will describe the media, that is, the material, on which drawings are generally made. Paper, vellum, and polyester film are the most commonly used graphics media. Much work in computer graphics is done on the screen of a video display (Figure 1.25).

Figure 1.24 Drawing board and tool kit for student use.

Figure 1.25 CRT displays are used when drawing with CADD systems. Computer generated image.

Whether you are sketching, doing detail drawing, or drawing illustrations, it is important that you be able to measure distances and be able to draw lines perpendicular or parallel to each other. You will need to be able to draw ellipses, irregular curves, and other special shapes. A number of special drafting tools are available to help you. They include scales for measuring distances (Figure 1.26); T-squares; triangles; templates for circles, ellipses, and other shapes; irregular curves (Figure 1.27); drafting machines (Figure 1.28); parallel bars (Figure 1.29); and drawing instruments such as compasses and dividers (Figure 1.30).

Figure 1.26   Designers use variety of scales.

Figure 1.27   Templates and irregular curves make drafting faster and easier.

Figure 1.28 Many professional designers and drafters use drafting machines. Courtesy of Ohio State University Engineering Publications

Figure 1.29 Parallel bars are used by some drafters. Courtesy of Ohio State University Engineering Publications

Figure 1.30 Drafting instruments are available as boxed sets.
Courtesy of Ohio State University Engineering Publications

1.3.3 Tools for Drawing by Computer

Central Processing Unit

The heart of a computer graphics system is the central processor (CPU or system unit). It may be a very large mainframe computer, a medium-sized minicomputer, or even a small microcomputer. The larger computers can usually support more simultaneous users and jobs. However, the power of computers is increasing so rapidly that today's microcomputers are often more powerful than the mainframes of just a few years ago. Workstations designed around powerful microcomputers can support multiple simultaneous tasks and even multiple simultaneous users.

Input Devices

A computer system must have more than just a CPU. In order to communicate with users and other devices, it must have input devices and output devices. It must also have data storage devices.

There are several input devices in common use on CADD systems. Certainly the most common is the keyboard. Nearly all computer programs expect some input from the keyboard, even if it is only the command to run the program.

A joystick can also be used as an input device (Figure 1.31).

Figure 1.31 Joysticks are found on some systems. Courtesy of Ohio State University Engineering Publications

Figure 1.32 Mouse is popular CADD system input device. Courtesy of Ohio State University Engineering Publications

Most people are familiar with the use of joysticks, which are very popular for computer games. They provide very fast response and thus fast action.

Most CADD systems use a mouse as a primary input device. (Figure 1.32). Mouse input is similar in many respects to joystick input but gives the user better control. The mouse is generally preferred to the joystick for CADD systems.

Digitizing tablets are found on many systems (Figure 1.33). They are often used as menu boards for selecting tasks to be done. Another use for digitizing tablets is to copy a drawing from paper to the computer.

Scanners are another type of input device. They are used to automate the process of copying a drawing from paper to the computer. They can often interpret and copy text and symbols as well as graphics.

Still another method of inputting data is imaging. This process uses a video camera and a computer with appropriate software to process and store the data. Laser-based systems can provide three-dimensional (X, Y, Z-coordinate) data at high speeds.

Output Devices

The primary output devices that computer systems use are cathode-ray tube (CRT) displays. The name refers to their television-like display tube. There are two main types of CRTs. An alpha-

Figure 1.33 Digitizing tablets are versatile CADD system input devices. Courtesy of Ohio State University Engineering Publications

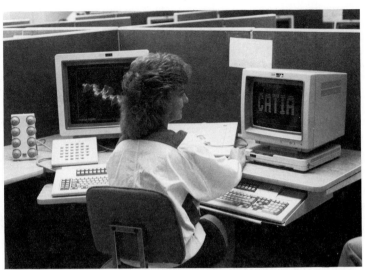

Figure 1.34 Some CADD workstations use two displays. Courtesy of Ohio State University Engineering Publications

Figure 1.35 Pen plotter for small drawings. Courtesy of Ohio State University Engineering Publications

numeric CRT can display letters, numbers, and punctuation marks but cannot display graphics images. A graphics CRT is designed to display lines, circles, curves, and other drawing features, often in color. In most cases, it can also display alphanumeric information. Some systems use only a graphics display. Others use both an alphanumeric display and a graphics display (Figure 1.34).

As was stated earlier, it is often necessary to produce a copy of the drawing on paper or other material. The two types of devices that are most often used for this purpose are plotters (Figure 1.35) and graphics printers (Figure 1.36). Two types of plotters in general use are pen plotters and electrostatic plotters.

Figure 1.36   Dot matrix graphics printer. Courtesy of Ohio State University Engineering Publications

The pen plotters use special drafting pens and create the drawing one line at a time on paper or other drawing media somewhat as you would do with your drawing pencil. Electrostatic plotters produce the image by placing an electrical charge on the media. Plotters can produce very high quality drawings.

Graphics printers are capable of printing a complex page as a pattern (matrix) of dots such as on the screen. Some of them use pins to strike an inked ribbon. Others spray a ''jet'' of ink on the paper. Still other printers use a laser beam. The laser printers can combine graphic and nongraphic (text) material and produce very high quality output. The quality of the drawing a printer produces depends on the capabilities of both the printer and the software, so it will vary from system to system, but it is generally not as good as drawings produced by plotters.

A third type of copy-producing device works somewhat like an office copier. It produces on paper an image of what is on the display screen.

Data Storage Devices
The two types of data storage devices most commonly found on computers are disk drives (Figure 1.37), which store data on disks that have been coated with a magnetic material, and tape drives (Figure 1.38), which store data on tape that has been coated. The magnetic tape is similar to the audio tape used in tape recorders and players.

Disk drives are found on virtually every computer system. It

Figure 1.37 Disk drives come in variety of sizes and types. Courtesy of Ohio State University Engineering Publications

is not possible to keep all the needed programs, data, and other information in the main memory of the computer. Even if it were, the information would all be lost if the system failed due to a power failure or other problem. Disk drives provide permanent storage and fast access to data. They are used both as input devices to load programs and data into the computer and as output devices to store programs and data.

Tape drives are used on many systems. They provide backup information in case something happens to the disk drive. It is standard procedure to copy the files from the disk to the backup tape on a regular basis. Tape drives are not generally used to replace disk drives. They lack the random-access capability of disk drives, so it usually takes much more time to get information from a tape then it does from a disk.

Software

Of course, no computer system can be complete without software. The term *software* refers to the programs that provide instructions to the computer. There are two types of software. The first is system software. It provides the basic instructions to the computer that tells it how to read from the disk drives, how to display information on the screen, how to send information to the printer, and so on. The second type is application software. It includes the programs and data files that provide the specific instructions and information for the job that is to be done. Computers generally have one set of system software. They would

Figure 1.38 Tape drives are used to provide backup for disk drives. Courtesy of Ohio State University Engineering Publications

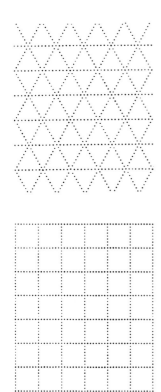

Figure 1.39   Rectangular and isometric grid paper.

have several sets of application software, one set for each type of task to be performed. Word processing would require one set, computer-aided design would need a different set, and so on.

1.4 *USING YOUR DRAWING TOOLS*

1.4.1   Sketching

Sketches are made without the aid of instruments. Grid paper is often used to aid in sketching straight lines (Figure 1.39). An erasing shield will help you to correct mistakes. The erasing shield is a small piece of flat metal containing a number of different-shaped holes (Figure 1.40). To use it, place it on your drawing so that all or part of the marks to be removed show through one of the holes but parts to be saved do not. Then erase the marks through the hole. This prevents you from accidentally erasing a line you want to keep.

The pencil should be relatively soft, such as HB, and should be sharpened so as to produce a clean dark line. Your normal pocket pencil is somewhat harder but is generally acceptable. Correct pencil movements and other techniques to help you to produce good sketches will be covered in Chapter 2. You should start by sketching lightly so that you can erase mistakes easily. When the sketch is the way you want it, go back and darken the lines, making the visible lines the darkest.

Figure 1.40   Erasing shields help when correcting mistakes.

VISIBLE LINE
HIDDEN LINE
SECTION LINE
CENTER LINE
DIMENSION & EXTENSION LINE
CUTTING PLANE LINE
SHORT BREAK LINE
LONG BREAK LINE
PHANTOM LINE

Figure 1.41 Alphabet of lines.

1.4.2 Alphabet of Lines

The weight, or thickness, of a line provides valuable information. Dark, wide lines are used to show the outlines and other visible lines that form a part of the object. Slender, lightweight lines are used if they are not to be part of the final drawing. Dark, thin lines are used to mark the centers of circles and arcs and for other purposes that will be covered in later chapters. Similarly, the style in which the line is drawn is important. Visible lines are drawn solid. Hidden lines, that is, lines that describe edges that are hidden behind other surfaces, are shown dashed. Still other line styles are used to show the locations of circular features, a cutting plane for sectional views, and so on. Drafters use a standard set of styles and widths of lines in their work. Each type of line is designed for a specific purpose. The set of line styles and widths is referred to as the alphabet of lines (Figure 1.41).

1.4.3 Instrument Drawing

Straight Lines

The T-square historically has been a most basic tool in instrument drawing. It is used to draw horizontal lines on the paper and is also used to provide a reference for lines at other angles. The T-square is placed on the drawing board with the blade going from left to right and the top of the T pressing firmly on one side of the board (Figure 1.42). The paper is attached to the board so that its edge or reference line is parallel to the blade of the T-square. The T-square may be slid up or down on the board as needed. It may also be used as a straight edge to draw a line between any two points (Figure 1.43). It should not be used as shown in Figure 1.44.

Figure 1.42 Using T-square.

Figure 1.43 T-squares can be used at an angle.

INCORRECT

Figure 1.44 Incorrect use of T-square.

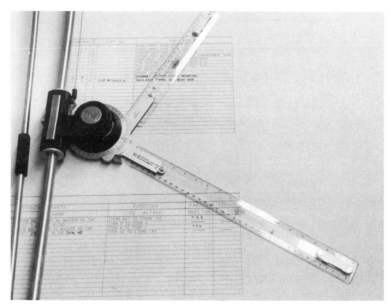

Figure 1.45 Drafting machine blades are designed to aid drafter. Courtesy of Ohio State University Engineering Publications

The T-square is gradually falling into disuse. Many student exercises are designed so they are not needed. Professional drafters use drafting machines or parallel bars (Figure 1.45). And, of course, CADD systems do not use instruments at all.

Your triangles will permit you to accurately draw lines at 15° increments from any reference line. The 30°–60° triangle can be used with the T-square, parallel bar, or another triangle to draw lines at 30° increments (Figure 1.46). Similarly, the 45° triangle can be used to draw lines at 45° increments (Figure 1.47). The two triangles can be used together to provide the 15°-increment capability (Figure 1.48). You can also use your triangles to draw lines parallel to another line. To do that, place two triangles in position as shown in Figure 1.49, with one edge of one triangle along the line. Then hold the other triangle to keep it from moving and slide the first one along it. The edge of the triangle will move away from the line but will remain parallel to it. When the edge of the triangle is the desired distance from the line, draw the new line.

The two triangles can also be used to draw lines that are perpendicular to other lines, as illustrated in Figure 1.50. Two of the edges of each triangle are perpendicular to each other, so when one of these edges is along or parallel to a line, the other edge is perpendicular to that line.

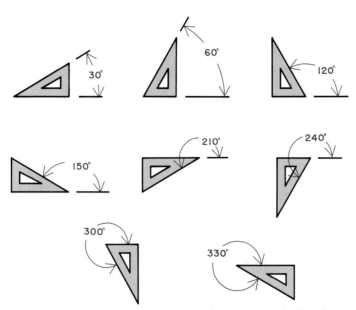

Figure 1.46 Constructing angles at 30° increments with triangle.

Figure 1.47 Constructing angles at 45° increments with triangle.

The protractor is shaped similar to a half circle or a circle with a flat bar through it (Figure 1.51). The flat bar contains a straight edge with a mark in the center to show the location of the center of the circle or arc. To use the protractor to measure the angle between two lines, place the straight edge along one of the lines forming the angle, with the mark on the intersection of the two lines. Read the angle on the circular portion of the protractor where the other line crosses it.

The drafting machine combines the capabilities of a T-square, triangles, and a protractor. The head of the machine has two blades fixed at right angles to each other. They can be rotated to

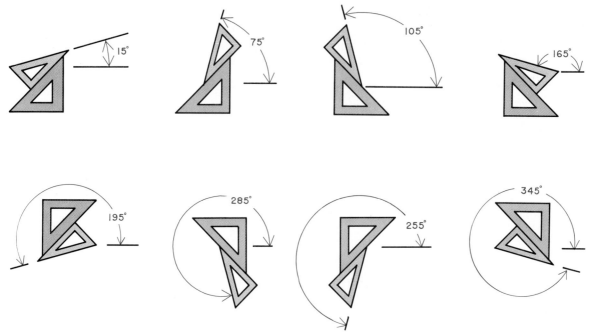

Figure 1.48 Using two triangles to construct angles at 15° increments.

Figure 1.49 Two triangles can be used to draw parallel lines.

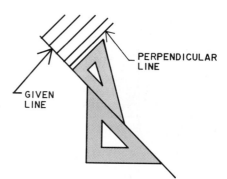

Figure 1.50 Two triangles can be used to draw lines perpendicular to other lines.

fixed settings selected by markings on the head (Figure 1.52). The arms of the drafting machine are constructed so that the angle of the blades does not change as the head is moved from position to position.

Figure 1.51 Using protractor to measure angles. Courtesy of Ohio State University Engineering Publications

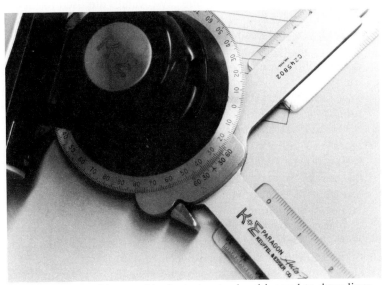

Figure 1.52 Drafting machine's protractor head is used to draw lines at set angles. Courtesy of Ohio State University Engineering Publications

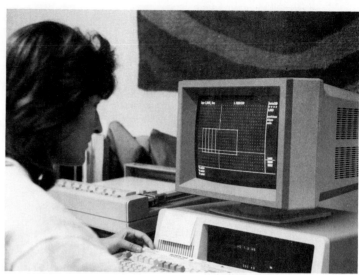

Figure 1.53 Drawing straight lines on CADD system. Courtesy of Ohio State University Engineering Publications

Drawing Straight Lines with Your CADD System

In order for your CADD system to draw a line, it must know the location of each end of the line. Two values are used to locate each end of the line. The first value is the X coordinate, which is the horizontal position of the point on the drawing, and the second value is the Y coordinate, which is the vertical position of the point. If you were drawing with a three-dimensional drawing system, you would also have a Z coordinate (Figure 1.53). Your system will provide more than one way for you to specify the location of the points.

One way to specify the location of a point is to determine what the X and Y coordinates are and then enter the values from the keyboard. That is usually the most accurate way to do it but not always the easiest. A more commonly used method is to use the graphics cursor, which is a cross-hair-like pair of lines that can be moved around the screen with the keyboard arrow keys, a mouse, or other input device. When the intersection of the cursor lines is on the desired spot on the screen, the space bar or mouse button is pressed to select the point. This is less accurate since the cursor does not move smoothly. The screen surface is made up of phosphor dots called picture elements, or pixels. The cursor jumps from pixel to pixel on the screen. The number of pixels, and thus the distance between them, will vary from system to system. To overcome this problem, CADD systems provide dot grids and the ability to snap the cursor to the grid or to another known location such as the end of an existing line.

To draw a line on your system at a specified angle, you can use the graphics cursor to locate one end of the line and then move the cursor until the angle and line length displays on the screen are showing the desired values. Then press the space bar or mouse button to select the point. A line will be drawn between the two points. On some systems, you will need to type in the angle and line length from the keyboard.

Circles and Arcs

Your instrument set should contain at least one and perhaps two compasses (Figure 1.54). They are called bow compasses. Look at your bow compass. One leg has a metal point. The other leg has a sharpened pencil lead. To draw a circle, the metal point is pressed into the paper at the center of the desired circle and the compass is swung about it with the lead pressing on the paper (Figure 1.55). If an arc of a circle is desired, the compass is swung only through the range of the arc (Figure 1.56).

Circle templates (Figure 1.57) are often used to draw circles and arcs of common sizes. They are somewhat easier to use than compasses but provide only a limited range of common circle diameters. They do not have a mark for the center of the circle but have four marks at 90° increments around the edge of the circle. The drafter first draws horizontal and vertical reference lines through the center point and then lines up the marks on the template with these lines to locate where the circle is to go.

Figure 1.54 Compasses are available in variety of styles and sizes. Courtesy of Ohio State University Engineering Publications

Figure 1.55 Using compass to draw circles. Courtesy of Ohio State University Engineering Publications

Figure 1.56 Using compass to draw arcs. Courtesy of Ohio State University Engineering Publications

Figure 1.57 Drawing with circle template. Courtesy of Ohio State University Engineering Publications

Drawing Circles and Arcs on Your CADD System

Similar to lines, circles are described by two points. They are the center and one point on the circle. Your system may offer you several alternate ways of selecting a circle, such as three points on the circle, two points on opposite sides of the circle, or its center and radius. If you select one of the alternate ways, the software will use the information you provide to calculate the locations of the points it needs.

To draw a circle, select one of the methods from the menu and follow the prompts. Each system is slightly different, but most give you the option of selecting each point with the cursor or typing its coordinates from the keyboard. If you are using the center and radius, you may be required to type in the value of the radius.

Arcs are drawn in the same way as circles except that the system needs to know three points, the center and the two end-points of the arc. Again, your system may offer several alternatives for entering the information and will calculate the needed values if they are not provided directly.

Drawing Ellipses

Ellipses are much more difficult to draw than circles. They are typically drawn in one of three ways. The first is to construct the ellipse by plotting several points on its surface and then connecting them with a smooth curve. This is a slow process. Reasonably good ellipses can be made by drawing four arcs of circles (four-center method) as described in Chapter 2.

Many ellipse templates are available and are widely used by drafters. Due to the wide range of sizes and shapes of ellipses, it may be impractical to have a full set of ellipse templates. If so, either the construction method or the four-center approximation just mentioned can be used.

Drawing Ellipses on Your CADD System

Not all CADD systems provide ellipse drawing capability. If yours does not, you may need to approximate the ellipse using the same techniques described for manual methods. With some three-dimensional CADD systems you can draw the ellipse in an orientation where it appears as a circle and then rotate it into the desired position.

Some CADD systems allow you to create ellipses in the following way: Draw a square by specifying one corner (vertex) and then the opposite one. A circle will be drawn in the square. Now, move the cursor (which is "attached" to the second vertex) around the screen. The square will be distorted into a rhombus and the circle into an ellipse. When the ellipse appears as you want it, press a key to fix it in that position.

Drawing Irregular Curves

Irregular curves, often called splines, are smooth curves that are made up of various shapes and thus cannot be drawn as combinations of arcs and straight lines. Irregular curves are drawn by plotting several points on the curve and then fitting a smoothly curved line to the points. A thin, flexible strip of material, called a spline, is sometimes bent to fit the points. It is then held in place with weights while the curve is being drawn. More often, an irregular curve, often called a French curve (Figure 1.58), is used. There are many different shapes and sizes of irregular curves. However, none is likely to fit the entire set of points, so the curve is drawn in sections by fitting the curve to as many points as possible, drawing a portion of the curve, fitting the curve to more points, drawing another portion of the curve, and so on, until the curve is complete (Figure 1.59).

Drawing Irregular Curves with Your CADD System

Not all CADD systems offer spline drawing capability. For those that do, the input techniques vary somewhat. A typical system

Figure 1.58 Drafters use wide variety of irregular curves. Courtesy of Ohio State University Engineering Publications

Figure 1.59 Many shapes can be created by using irregular curve.

will ask you to locate each point for your spline either by typing its coordinates from the keyboard or by locating it with the graphics cursor. When you have entered all of the points, the software will perform calculations to select locations for many additional points between the ones you entered. It will then connect all of the points with straight lines, but the lines will be so short that it will appear to be a smooth curve. Other systems permit you to "sketch" or draw a line freehand by moving the graphics cursor about the screen. In this case, the quality of the curve is dependent on your ability to move the cursor in a smooth manner.

Lettering Text

Most drawings and sketches contain letters and numbers in addition to lines and circles. The title block will contain the drawing name, number, and other information. Dimensions and notes are used to show the information needed to construct the object. Many times, this text is lettered by hand. You will learn in Chapter 2 how to create clear, legible text.

Lettering Text on Your CADD System

The CADD systems provide lettering by allowing you to type the text from the keyboard. The lettering style can be chosen from the various fonts (styles) that the system provides. The height and width of the letters can also be specified on most systems. When you have selected the size and style of text, you can specify where the text is to go by selecting a starting point with the graphics cursor and then typing the text from the keyboard. The system will automatically put it at the specified location. Most systems also allow you to move, rotate, or change the size of the text after you have created it.

Measuring

Most drawings are created to a scale, that is, a certain size. If the drawing is the same size as the object, the scale is full size. If each line on the drawing is half as long as the corresponding edge on the object, the scale is half size. Architects often use a scale of 1 in. on the drawing equals 8 ft on the object. Engineers work in fractional or decimal inches. They may also work in the metric system. Civil engineers and others working on large projects may work in feet, miles, or kilometers. Figure 1.26 shows a variety of scales that are used by engineers and architects.

Drawings should not be scaled, that is, you should not place a scale on the drawing to determine the size of some feature. The size should be shown as a dimension. However, it is important to draw the object accurately to a selected scale. Failure to do so will result in the shape of the drawing being different than the shape of the object being represented. This can cause confusion for anyone using the drawing.

Correct use of the scale is important if you are to maintain accuracy in your drawing. Some drafters insist that dimensions be "picked off" the scale with the dividers and transferred to the drawing (Figure 1.60). This method is very accurate due to the sharp points on the dividers but can damage the scale if you are not careful. Many drafters feel that sufficient accuracy can be obtained by placing the scale on the drawing and carefully marking

Figure 1.60 Dividers can be used to transfer measurements from scale to drawing. Courtesy of Ohio State University Engineering Publications

Figure 1.61 Scale is placed on drawing and distance required is carefully marked by points. Courtesy of Ohio State University Engineering Publications

points at the markings on the scale (Figure 1.61). If this latter method is used, it is important to use a newly sharpened pencil and to hold it absolutely vertical so as to get the point precisely at the desired location.

Measuring on Your CADD System

The CADD systems allow you to measure distances by several means. In a typical case, the units in which you are working are selected from a listing. Then the system establishes an "origin" whose X, Y- or X, Y, Z-coordinate values are all zero (0). You can elect to have the system print out the cursor coordinates as it is moved about the screen. On some systems, once you enter the first point for a line, circle, or other entity, the distance and angle of the current cursor location from that starting point can be displayed. Often, you can key in the location of a point in terms of either (1) its coordinates relative to the origin or (2) its distance and angle from another point.

Inking Drawings

Drawings that are going to be placed in a permanent file as master drawings are generally traced on vellum or polyester film from the original drawing. Copies of the drawing are then made on a reproduction machine, and the original is placed in the master file. Technical ink pens are used for this purpose (Figure 1.62).

Figure 1.62 Technical ink pen set. Courtesy of Ohio State University Engineering Publications

Drawings made on a CADD system are stored on disk or tape. It is not necessary to maintain a permanent file of tracings. In fact, it is undesirable since the copy on disk or tape is really the master. However, if desired, high-quality plots can be made on vellum or polyester film.

1.5 SUMMARY

As a student who will be practicing in your chosen field well into the twenty-first century, you will be in the work force at a very exciting time. Engineering design graphics will become increasingly important as automation continues to mechanize the production floor. Machines controlled by computer rely very heavily on geometric descriptions of the objects to be produced. More and more frequently, these descriptions are created during the design process by a designer using a CADD workstation. While the designer may be drawing by moving a mouse rather than triangles and a pencil, the graphics principles are the same. The designer must be able to visualize and graphically describe three-dimensional objects. Sketching will also continue to be very important since designers continue to create designs on paper even if they modify them on a CADD system.

Throughout this book, we shall present graphics techniques using both manual and computer-based methods. We will try to keep the description of computer-based techniques general enough so that they will be applicable to whatever CADD system you are using.

PROBLEMS

Problem 1.1
What is a data base? How is it used?

Problem 1.2
Why do engineers make sketches? What tools and materials are needed to make a sketch?

Problem 1.3
What is meant by shape description?

Problem 1.4
What is meant by size description?

Problem 1.5
What does tolerance mean?

Problem 1.6
Why is a drawing drawn to scale?

Problem 1.7
If a 6-in. long object is drawn to half scale, how long will it be on paper?

Problem 1.8
For what are section views used?

Problem 1.9
What are conventional practices?

Problem 1.10
Why are notes added to a drawing? Name three uses for drawing notes.

Problem 1.11
What is a working drawing? For what are working drawings used?

Problem 1.12
What are illustrations? How do illustrations differ from working drawings?

Problem 1.13
For what are illustrations used?

Problem 1.14
What is meant by fastening and joining? Name four types of devices, materials, or processes that are used for fastening or joining.

Problem 1.15
What is descriptive geometry? How is it used by the designer?

Problem 1.16
What is a pie chart? Name a use for pie charts.

Problem 1.17
What are empirical equations? How are they developed? For what are they used?

Problem 1.18
What are scalars? What are vectors? What is the main difference between them?

Problem 1.19
What is graphical calculus? For what is it used?

Problem 1.20
What is vector analysis? Can vector analysis be performed (1) mathematically, (2) graphically, or (3) both?

Problem 1.21
What is meant by modeling? For what are physical models used? For what are computer-generated models used?

Problem 1.22
You are asked to sketch three orthographic views of an object. What equipment and materials will you need for the job?

Problem 1.23
You are creating a drawing with instruments and must add several arcs and circles to it. What tool(s) could you use to draw the arcs and circles?

Problem 1.24
What is simulation? How is it used in computer-aided design? Is it used in manual design?

Problem 1.25
What is meant by numerical control? How does computer numerical control differ from numerical control?

Problem 1.26
What is computer-integrated manufacturing?

Problem 1.27
What are some similarities and differences between computer-integrated manufacturing and flexible manufacturing?

Problem 1.28
How is a T-square used? Can you use it to draw horizontal lines? Vertical lines? Lines at various angles?

Problem 1.29
You have just purchased a drawing instrument kit. It contains two triangles. What size angles can you draw with one triangle? What size angles can you draw with the other?

Problem 1.30
Using just your T-square, the two triangles, and a pencil, how many different angle lines can you draw? What angles are they?

Problem 1.31
You are to draw a line at 75° to the horizontal using your T-square and your two triangles. Sketch a rectangle representing the drawing board and show the positions of the T-square and each triangle.

Problem 1.32
Your drawing instrument kit contains a pair of dividers. For what are they used?

Problem 1.33
You must lay a distance of 3.6 in. on your drawing. Name the tools from your kit that you will use for this task and how you will use them.

Problem 1.34
Name two types of erasers commonly used when creating drawings.

Problem 1.35
Your drawing contains two lines that are very close together. What tool do you have in your kit that will permit you to erase one line without erasing the other?

Problem 1.36
Name three types of drawing media (material) on which drawings are commonly made.

Problem 1.37
You have plotted several points on your drawing paper and must connect them with a smooth curve. The curve will be of irregular shape. What tools do you have in your drawing kit for this purpose?

Problem 1.38
Drawing pencils are made with several different hardnesses. Name three hardnesses that are available.

Problem 1.39
You are going to create a sketch. What lead hardness will you choose? Why?

Problem 1.40
Name the principal components of a computer-aided drawing system.

Problem 1.41
Name three input devices that are commonly used with a computer.

Problem 1.42
Name three output devices that are commonly used with a computer.

Problem 1.43
You have created a drawing on your computer. Name two types of output devices that can be used to make a copy of it.

Problem 1.44
Name two important differences between disk drives and tape drives.

Problem 1.45
What are tape drives generally used for on modern computer systems?

Problem 1.46
Name two types of computer software that might be used by a designer.

Problem 1.47
How is a digitizing tablet used?

Problem 1.48
How is a mouse used?

Problem 1.49
Name a task that could be performed with either a digitizing tablet or a mouse.

Problem 1.50
Name a task that could be performed with a digitizing tablet but not with a mouse.

Problems 1.51–1.70
Draw the figures shown using either manual instruments or a CADD system as directed by your instructor. Use the dimensions shown or other dimensions as directed.

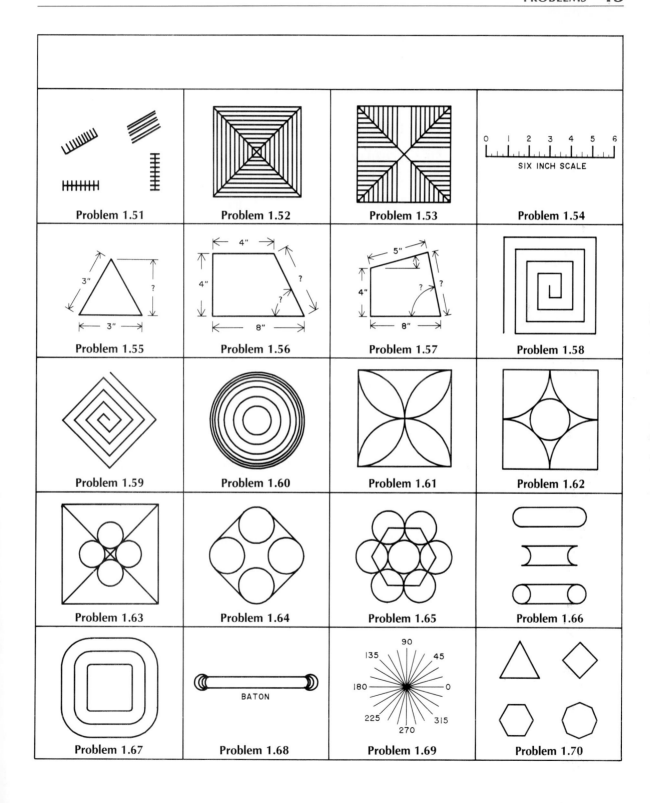

Problem 1.51

Problem 1.52

Problem 1.53

Problem 1.54

SIX INCH SCALE

Problem 1.55

Problem 1.56

Problem 1.57

Problem 1.58

Problem 1.59

Problem 1.60

Problem 1.61

Problem 1.62

Problem 1.63

Problem 1.64

Problem 1.65

Problem 1.66

Problem 1.67

Problem 1.68

BATON

Problem 1.69

Problem 1.70

TERMS YOU WILL SEE IN THIS CHAPTER

LETTERING TERMS

UPPERCASE LETTERS   Capital letters.

CAPS

LOWERCASE LETTERS   Style of letters that is smaller and of different shape from capital letters.

Lower

LETTER SPACING   Distance between letters in a word.

DIMENSION

WORD SPACING   Distance between words.

TWO WORDS

GEOMETRIC CONSTRUCTION TERMS

PLANE   Two-dimensional surface; flat, not curved.

SOLID   Three-dimensional object.

POLYGON   Closed plane figure with straight sides.

PARALLEL   Same distance apart.

PERPENDICULAR   Meeting at right angles.

2

LETTERING AND GEOMETRIC CONSTRUCTIONS

2.1 INTRODUCTION

Before you can make a complete engineering drawing, you need to learn about two elements that are found on most drawings. The first is lettering, which you may have called printing before you studied graphics. This is the way words and numbers are put on a drawing. The second is geometry. Many of the constructions we need in graphics and for building or producing useful devices are the same ones learned in plane geometry.

You are already familiar with some parts of the graphics language. You have been making letters of the alphabet for years. In graphics we try to make very readable letters in standard styles. The lettering and numbering on the face of the drawing should be dark and clear so that anyone reading the drawing will read exactly the same message.

You will use some of the basic geometric constructions you (hopefully) learned in plane geometry. You can build on this knowledge and learn new uses for geometry. Several geometric figures and shapes you may know by name are shown for reference in Figure 2.46. If you do not know these figures and shapes and their names, you should learn them.

When you finish the work in this chapter, you will know the basic lettering styles used in engineering graphics presentation

Figure 2.1   Samples of lettering.

(Figure 2.1). You will also review and learn about some simple geometric shapes and constructions that are useful for solving graphics problems.

Computer-aided drawing programs can easily provide uniform lettering for finished design drawings. Intermediate work, sketches, and rough drawings generally are drawn by hand. This chapter will cover the basics of lettering using manual freehand and manual mechanical methods. The principles of good lettering apply to CAD as well as manual methods.

2.2   LETTERING INSTRUMENTS

Common lettering instruments for engineering drawings are pencils and pens. There are three types of pencils: the wooden pencil, the lead holder with a clutch, and the fine-line pencil (Figure 2.2). Wooden pencils have mostly been replaced by the mechanical pencils—the lead holder and fine-line pencil.

The lead holder is a type of mechanical pencil that holds long pieces of lead of varying degrees of hardness. The leads range from the very hard 9H to the very soft 6B (Figure 2.3). The engineer will use leads that fall in the center range of hardness (HB to 4H). It is common for the engineer to use three pencils. One pencil has hard lead for very light layout lines, another pencil has medium-hardness lead for thin and medium-width black lines, and the third has soft lead for wide black lines and lettering (Figure 2.4).

The leads are inserted into the lead holder from the working end of the mechanism. Care must be taken *never* to push the lead into the holder with the thumb or finger on the end of the lead. It is possible for the blunt end of the lead to get caught in the tube of the pencil, and the pointed end of the lead could then puncture the finger or thumb. To avoid this problem, place the thumb and index finger on the sides of the lead. If the blunt end of the lead stops for any reason, the lead will then slip between the thumb and index finger causing no injury (Figure 2.5). The working end of the lead must be pointed using a pencil pointer.

Figure 2.2 Types of pencils used for lettering on engineering drawings.

Figure 2.3 Leads for lead holder.

	•	•	•	•	•	•
HARD	9H	8H	7H	6H	5H	4H
	HARDEST					SOFTEST
	•	•	•	•	•	•
MEDIUM	3H	2H	H	F	HB	B
	HARDEST					SOFTEST
	•	●	●	●	●	●
SOFT	2B	3B	4B	5B	6B	7B
	HARDEST					SOFTEST

Figure 2.4 Lead hardness chart.

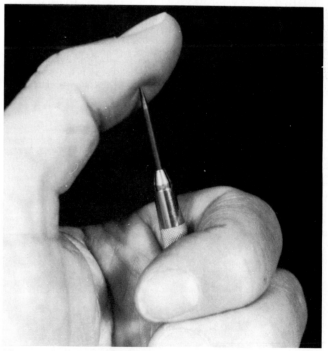

Figure 2.5 Inserting lead safely into lead holder.

Figure 2.6 Pointers used to prepare working end of lead held in lead holder.

There are many styles of pointers on the commercial market for a reasonable price. Some pointers are designed only for desk use, and others are designed to be portable. Desk-top models hold the graphite (lead) shavings, but if the pointer is turned over, the shavings will spill out. Pointers designed to be portable physically retain the graphite shavings regardless of how the pointer is turned or moved (Figure 2.6). The most obvious problem with the graphite shavings escaping is that they smear on everything that they contact, and this can create a mess.

The pointer puts a very sharp conical point on the end of the lead. The point should be slightly dulled before use. If the sharp point is not dulled, it may snap off during the initial strokes made with the lead. This lead will not only make a rough line, but the point could snap and cause injury (Figure 2.7).

While lettering with this style of pencil, you should rotate the pencil in your hand after making several letters or numbers. Rotate the pencil between each word. Some people find that they must rotate the pencil between letters in a single word. This rotation process helps keep the point on the pencil uniform and prevents the lead from developing a flat spot. If the lead point becomes dull, the width of the strokes will be too wide; then the lettering will not be uniform and may not be clear (Figure 2.8).

The newest style of mechanical pencil is the fine-line pencil. The fine-line pencil is available in four metric sizes: 0.3, 0.5, 0.7,

Figure 2.7   Point should be slightly dulled.

UNIFORM LETTERING BY ROTATING PENCIL
CORRECT

SKINNY LETTERS
TO WIDE STROKES
INCORRECT

Figure 2.8   Rotating pencil produces equal line widths.

and 0.9 mm (Figure 2.9). These fine-line pencils are not available in inch sizes. Note that the lead for each of these pencils is a different diameter. A 0.5 mm pencil should be used for lettering. The other widths are used for drawing lines of different widths.

A softer lead (such as HB lead) is used in fine-line pencils in order to produce black lines. The lines will be more uniform because the lines become as wide as the diameter of the lead (Figure 2.10). Another advantage of using fine-line pencils is the elimination of the need for a pointer; this saves time and reduces graphite dust. For students as well as others who carry their drafting instruments from location to location, this is an advantage

Figure 2.9   Fine-line pencils are manufactured to hold leads of different diameters.

Figure 2.10 Fine-line pencils produce lines of different widths.

Figure 2.11 Erasers for erasing pencil lines from vellum or film.

because the graphite shavings from the previously required pointer will not spill and smear all over the other equipment.

Many of the commercial leads available for fine-line pencils have a plastic base instead of a graphite base. This lead is designed for use on drafting film, a clear medium, as well as vellum, a type of drawing paper that is translucent. The soft plastic erasers are very effective when erasing lines from either the film or vellum and will not scratch the surface of the film such as some other erasers (Figure 2.11).

Figure 2.12 Technical pens for drawing ink lines.

.005 IN.	.007 IN.	.010 IN.	.012 IN.	.014 IN.	.020 IN.	.024 IN.	.028 IN.	.031 IN.	.039 IN.	.047 IN.	.055 IN.	.079 IN.
6x0 .13 mm	4x0 .18 mm	3x0 .25 mm	00 .30 mm	0 .35 mm	1 .50 mm	2 .60 mm	2.5 .70 mm	3 .80 mm	3.5 1.00 mm	4 1.20 mm	6 1.40 mm	7 2.00 mm

Figure 2.13 Technical pens are used to draw ink lines of different widths.

Technical pens have replaced the ruling pen for drawing or tracing ink lines and ink lettering (Figure 2.12). Technical pens have drawing tips of different diameters that produce lines of different widths (Figure 2.13). The technical pens need not be cleaned after every use. If you will not be using your pens for a period of time, take the time to draw a few lines with each pen every few weeks or so to keep the ink from drying in the point. If this is done, you can go for several years without having to clean the pens and can just add more ink occasionally.

Erasing ink lines from vellum requires a little more work and an abrasive eraser. It is better to pass the eraser over the ink line to be removed 100 times lightly than 10 times pressing so hard that the surface of the vellum is damaged (Figure 2.14). Erasing

Figure 2.14 Erasing ink lines from vellum.

Figure 2.15 Erasing ink lines from film.

ink lines from film is easy and clean when using one of the plastic chemically treated erasers that has been manufactured and treated for removing a particular manufacturer's ink (Figure 2.15).

2.3 *DRAWING SURFACES*

There are three types of drawing surfaces upon which the engineer will draw lettering: drafting film, vellum, and opaque paper.

Vellum is a special type of paper that is semitransparent. Both lead and ink produce dark opaque lines on this type of surface and are easily erased.

Drafting film is more expensive but much more durable and dimensionally stable; that is, it does not expand or shrink when temperature or humidity changes. It is the best surface to use if a great number of drawings must be made from the same master. It will also last much longer than vellum. A master is a final drawing that is the result of design work and revisions. This drawing is used as the master for reproduction purposes. The master drawing usually is stored in a safe place such as a fireproof file in order to protect it. The master drawing is never used in the shop; it is protected so clean copies can be made for shop use.

Opaque paper is used for smaller applications, such as common $8\frac{1}{2} \times 11$-in. (216 × 280-mm) standard paper. This paper is commercially available in a variety of grades and is more absorbent than vellum. Erasures on paper are not as clean as on either film or vellum.

2.4 *LETTERING GUIDELINES*

Lettering guidelines are drawn for all numbering and all lettering on the face of all drawings. Lettering guidelines are thin, very

LETTERING IS MADE BETWEEN GUIDE LINES

Figure 2.16 Lettering guidelines should be drawn before lettering freehand.

CORRECT INCORRECT CORRECT

Figure 2.17 Lettering is never placed adjacent to border line.

light lines that only serve as a guide to the person doing the lettering. These lines should not be dark enough to interfere with the visibility of the lettering or even dark enough to be seen when copies are made (Figure 2.16).

Lettering guidelines are usually $\frac{1}{8}$ in. (3 mm) apart. Lettering is never placed in the space adjoining a line such as a border line (Figure 2.17). A space of $\frac{1}{8}$ in. (3 mm) is usually left between lines of lettering. Lettering is never placed in adjacent $\frac{1}{8}$-in. (3-mm) spaces (Figure 2.18).

Time-saving devices are commercially available to help draw

NEVER LETTER IN ADJACENT SPACES BECAUSE NO ONE CAN READ IT

INCORRECT

Figure 2.18 Do not letter in adjacent line spaces.

Figure 2.19 Lettering guidelines are made more uniformly with lettering devices.

Figure 2.20 Draw guidelines by sliding device against straightedge.

these lettering guidelines. The Ames lettering guide and the Braddock–Rowe triangle are examples (Figure 2.19). To use these devices, position the row of holes and simply draw the guidelines by sliding the device along a straight edge (Figure 2.20).

2.5 LETTERING STYLES

Nearly all pencil lettering drawn by the engineer is drawn free-hand. The style of letters used is single-stroke Gothic style (Figure

ABCDEFGHIJKLMNOPQRS
TUVWXYZ
1234567890 $\frac{1}{2}$ $\frac{1}{3}$ $\frac{1}{4}$ $\frac{1}{8}$ $\frac{1}{16}$ $\frac{1}{32}$ $\frac{1}{64}$

Figure 2.21 Standard single-stroke Gothic lettering.

Figure 2.23 Capital letters and numbers are generally used when lettering engineering drawings.

Figure 2.22 Common errors are often made when forming letters and numbers.

Figure 2.24 Letters and numbers larger than $\frac{1}{8}$ in. are used for titles and drawing identifers.

2.21). Note the common errors made while lettering; they are illustrated in color (Figure. 2.22).

With rare exception, lettering is drawn using uppercase letters (capital) or numbers (Figure 2.23). Most lettering is $\frac{1}{8}$ in. (3 mm) high, although some titles and the drawing number identifier in the title block may be made using $\frac{3}{16}$ in. (5 mm) or even $\frac{1}{4}$ in. (6 mm) characters (Figure 2.24).

Upper- and **lowercase letters** are used in civil engineering when making plot plans. These are usually made in ink with a mechanical lettering device of some sort (Figure 2.25).

The lettering may be drawn with either vertical or inclined strokes (Figure 2.26). With the vertical characters, the stems of the letters (a vertical line extending from guideline to guideline) are vertical. With inclined characters, these stems are drawn at a uniform incline. The standard incline is 68 degrees. One edge

Figure 2.25 Civil engineering drawings are usually lettered in upper- and lowercase letters. Courtesy of Timothy C. Whiting

A B C D E F G H I J K L | *A B C D E F G H I J K L*
M N O P Q R S T U V W X | *M N O P Q R S T U V W X*
Y Z 1 2 3 4 5 6 7 8 9 0 | *Y Z 1 2 3 4 5 6 7 8 9 0*

Figure 2.26 Either vertical or inclined letters or numbers are acceptable.

of the Ames lettering guide is manufactured at this angle so that you may use it as a guide (Figure 2.27). Whether you elect to use vertical or inclined letters and numbers makes little difference, but you must be consistent and use only one style on any one drawing.

Figure 2.27 One edge of Ames lettering device is manufactured at 68° angle to be used as guide for inclined letters.

$1 \ 1\frac{1}{8} \ 1\frac{13}{16} \ 2\frac{3}{4} \ 6\frac{11}{16}$

$2.500 \ \ 3.125$

Figure 2.28 Fraction numbers extend above and below $\frac{1}{8}$-in. lettering guidelines.

Whole numbers are drawn $\frac{1}{8}$ in. (3 mm) high as are the capital letters. However, fractions must be treated differently. The fraction numbers are nearly as large as the whole numbers but extend above and below the lettering guidelines. Note that the fraction line is a horizontal line (Figure 2.28). Also, the height of the fraction is twice the height of the whole numbers.

2.6 *CHARACTER UNIFORMITY*

Good lettering is characterized by uniformity. The appearance of a well-lettered drawing will be directly related to how uniform it appears. The uniformity of line widths has already been addressed.

Uniform blackness is very important. All of the characters must be lettered *black*, not gray. This is very important because characters not dark enough may not copy well and may cause errors in reading the drawing. The characters must be lettered uniformly and black enough to copy under any conditions.

A uniform slant to the vertical letters will enhance the appearance of the drawing. Uniformity when drawing the curved portion of the characters having curves is also important (Figure 2.29).

Stability in the appearance of lettering is shown with example characters such as B, P, and the number 8. If these characters are drawn with the middle stroke drawn at the horizontal center,

UNIFORMITY INCLUDES LETTER BLACKNESS, PARALLEL STROKES, AND GOOD LETTER SPACING

Figure 2.29 Uniformity is very important consideration.

they will appear below center. For this reason, the top part of these letters is drawn slightly smaller to provide space for drawing the middle stroke above the horizontal center of the letter. If there is any question about this, turn the lettering upside down and observe the illusion (Figure 2.30).

2.7 *SPACING*

Both **letter spacing** and **word spacing** must be addressed. Letter spacing is the spacing between individual letters. This space should be very small if the words are to appear as words and not just as random letters (Figure 2.31).

Figure 2.30 ''Rule of stability'' requires lower part of some letters and numbers to be larger than top.

SMALL SPACES SHOULD BE USED FOR GOOD LETTER SPACING

CORRECT

POOR LETTER SPACING RESULTS FROM SPACES BEING TOO BIG

INCORRECT

Figure 2.31 Letter spacing should be very small to make letters visually appear as words, not letters.

Figure 2.32 Letters with adjoining stems require more letter spacing.

Figure 2.33 Certain letter combinations require overlapping of adjacent letters.

LETTERS AND
NUMBERS ARE
USED WITH PICTURES
ON A DRAWING

Figure 2.34 Word spacing requires more space than letter spacing.

A few letter combinations must be treated with special attention. When two letter stems are adjacent, more space must be left between the characters (Figure 2.32). When an L is followed by a T, the top of the T actually overlaps the L (Figure 2.33). When letters follow the letter A or L, they should follow very closely because of the open space appearing between the characters (Figure 2.33).

Word spacing should be wider than letter spacing. Letter spacing should be so small that the characters nearly touch. However, word spacing should be approximately the width of a medium-wide character such as the letter E (Figure 2.34).

2.8 SUGGESTIONS

There are several techniques that you can use to improve your lettering. When making an engineering drawing, all the lines are drawn dark first, and then the lettering is drawn last. This prevents smearing the lettering since the most lead is put on the paper where lettering and numbering appears.

When lettering on the drawing, place a clean sheet of paper under the hand holding the pencil. This prevents your hand from ever touching the drawing and prevents sweat, hand oil, and other foreign materials from being transferred to the surface of the drawing (Figure 2.35). The use of eraser chips from a dry cleaning

Figure 2.35 Protective sheet under lettering hand helps keep surface of sheet receptive to ink or pencil.

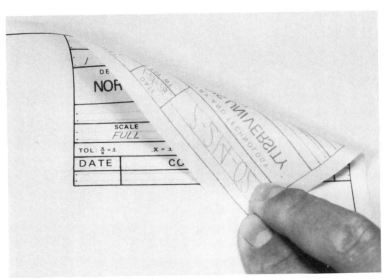

Figure 2.36 When much erasing is anticipated, draw border and guidelines on one side of sheet and lettering on other side.

pad can aid significantly in keeping the surface of the drawing clean.

If you are working on a translucent medium, such as vellum, and a significant amount of lettering will appear on the drawing surface, such as on a specification sheet, it may be a good idea to draw the border lines and lettering guidelines on the *back* of the drawing sheet. This allows for many erasures or changes to be made on the front side of the sheet without damaging the guidelines. Remember, the copy machine will not know on which side of the vellum or film you are drawing (Figure 2.36).

After the drawing is complete, the tape holding it to the board may be removed. You may then find it much more comfortable to rotate the sheet slightly on the board before starting to letter (Figure 2.37).

You will find that your lettering will be be much more uniform if your entire forearm is resting on the surface of the drawing table. This simply offers more support from your arm and drawing hand (Figure 2.38).

2.9 MECHANICAL LETTERING

There are many mechanical devices and aids available on the commercial market for use while making letters by mechanical

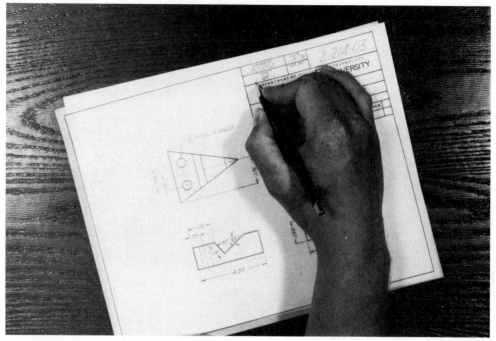

Figure 2.37 Rotate paper to most comfortable position for lettering.

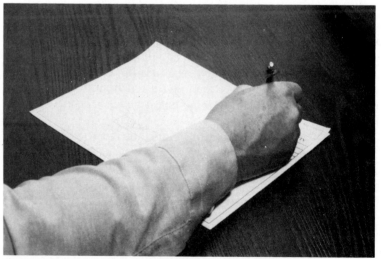

Figure 2.38 Place paper on table in such a position as to allow your entire forearm to rest on table.

means. This mechanically produced lettering is almost exclusively used when drawing with ink for presentation purposes. The Leroy type of lettering equipment is one type used most often (Figure 2.39). Technical pens are inserted into a scribe having three points extending from its lower side (Figure 2.40). The point at the end of the handle (1) slides in the wide groove at the bottom of the scale. The thin point (2) (called the stylus) is then inserted in the groove representing the desired letter in the scale. The third point is near the pen point (3) and acts as a support for the pen point (Figure 2.41).

The letter is formed by moving the stylus within the groove of the desired letter with the pen point resting on the drawing surface. The scribe is then tilted back, and the stylus is placed in the next letter. Then slide the scale until it is in the correct po-

Figure 2.39 Hand lettering equipment is used to produce mechanical lettering.

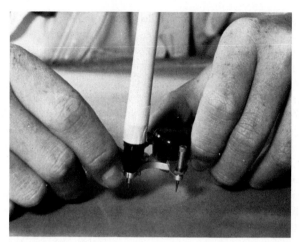

Figure 2.40 Pen is secured in arm of lettering device.

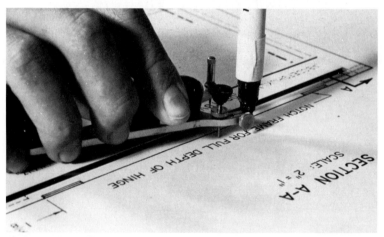

Figure 2.41 Adjusting screw keeps pen point at correct distance from paper.

sition. Trace that letter and repeat this step as many times as necessary. Once the origin point for the new letter has been established, keep your eye on the stylus and engraved template while drawing the letter or number (Figure 2.42).

The skill that the person making the letters must develop is the spacing of the letters and words. For letter spacing, first letter the first character. Then place the stylus in the left extreme of the new letter on the template. Then slide the scale until you line up by eye with the right extreme of the old letter. Remember, align the left extreme of the new letter with the right extreme of the previously drawn letter (Figure 2.43).

Those of you who must draw a lot of mechanical lettering and can afford the expense of an automatic lettering machine will benefit greatly from the features available on such an automated device. Options such as proportional spacing, degrees of slant and memory, as well as predrawn symbols for many disciplines are available (Figure 2.44).

Civil and mechanical engineers comprise the primary group that uses ink letters and numbers that are mechanically generated. This group also uses upper- and lowercase letters more than any of the other engineering disciplines (Figure 2.45). Ink drawings

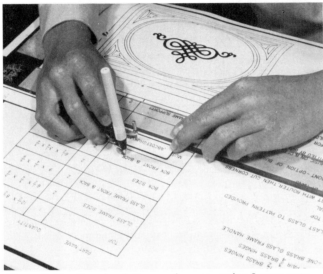

Figure 2.42 Forming letters and spaces using Leroy-type lettering device.

Figure 2.43 Align left extreme of new letter with right extreme of previously drawn letter.

Figure 2.44 Automatic lettering device is used by professionals. Courtesy of MUTOH America, Inc.

UP

BULKHEAD

Center line – Girt steel I beam pref. Objective – No center supporting
columns. If one is required, post must be in center of span,
and be placed on appropriate footing. Substitute for this girt
or additional columns – subject to owner approval. Appropriate
pockets required in foundation for this beam.

BASEMENT AREA
⟨20⟩

Furnace flue.

Basement fireplace
flue.

4 Anderson basement window units

Footings 2'-0" square x
8" deep. 8" Sono tubes
to level of foundation
wall.

8'-6"

8'-0"

8'-6"

14'-4"

10'-8"

5'-8"

6"

14'-0"

10'-0"

1'-0"

8'-0"

10'-0"

4' Frost wall to support slab at upper level
of foundation wall.

Figure 2.45 Civil engineers often use both upper- and lowercase
letters and also make presentation drawings using mechanical
lettering. Courtesy of Timothy C. Whiting

can be produced easily using CAD systems, and today many
industries are using CAD for documentation.

2.10 COMPUTER-GENERATED LETTERING

Lettering may be generated from a computer by typing the desired
letters or numbers on a keyboard. Many variations of lettering
styles are available depending on the printer device or software.

2.11 BASIC CONSTRUCTIONS

Now that you know about making readable words and numbers
on drawings, let us consider some of the basic constructions that
are useful for describing common objects. You may have already
seen many of these constructions; others may be new to you.
There are many other construction methods that may be of interest; here we present only those needed to understand the theory
and solve the problems presented later in this book. The **plane**
figures and **solid** shapes you will need to know are shown in
Figures 2.46 and 2.47.

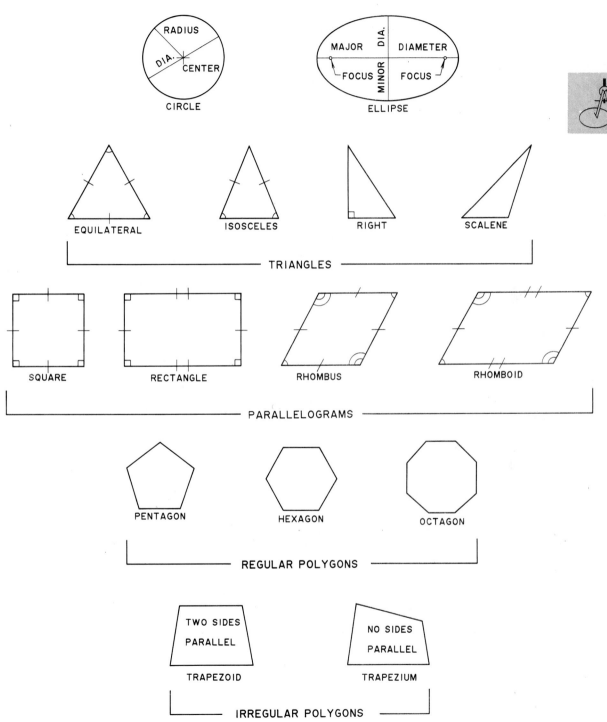

Figure 2.46 Common plane figures.

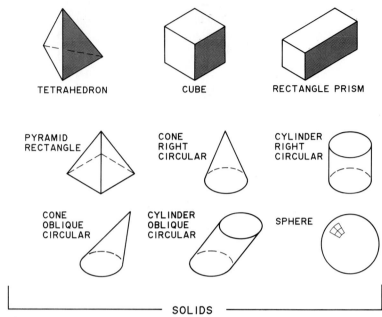

TETRAHEDRON   CUBE   RECTANGLE PRISM

PYRAMID RECTANGLE   CONE RIGHT CIRCULAR   CYLINDER RIGHT CIRCULAR

CONE OBLIQUE CIRCULAR   CYLINDER OBLIQUE CIRCULAR   SPHERE

SOLIDS

Figure 2.47   Some common solids.

Figure 2.48   Door with parallel sides.
Photo by Howard Keith Williams

2.12 *TO DRAW A LINE PARALLEL TO A GIVEN LINE*

A line parallel to a given line is a construction you will use very often. Many objects around us include **parallel** lines. Consider a door—the sides are usually parallel. If they are not, the door may not close or it may have large cracks that create drafts (Figure 2.48).

A parallel bar or a T-square makes it easy to draw parallel lines. A T-square provides parallel horizontal lines as you slide it up and down on the drawing board, keeping the head of the T-square firmly against the edge of the board (Figure 2.49). When you put a triangle on the T-square and slide the triangle, it will generate a series of parallel lines. These parallel lines may be vertical or at the angle of the triangle (Figure 2.50). If you are not using a T-square, you can make parallel lines by holding one triangle solidly in place and sliding another triangle along the edge of the first one (Figure 2.51). Figure 2.52 demonstrates parallel lines with computer graphics.

Figure 2.49 To draw horizontal parallel lines using T-square, (1) align T-square with given horizontal line on paper, (2) tape paper on top two corners with T-square in place, and (3) move T-square to desired position and draw new line.

Figure 2.50 To draw vertical parallel lines using T-square and triangles (tape paper down before starting), (1) slide triangle along upper edge of T-square with 90° angle on T-square until vertical edge of triangle is aligned with given line and (2) slide triangle, still on T-square, to desired position and draw new line.

Figure 2.51 To draw parallel lines using only triangles, (1) place two triangles on drawing surface hypotenuse to hypotenuse; (2) align one edge of upper triangle with given line; and (3) hold lower triangle firmly in place, slide upper triangle to desired position, and draw line.

Figure 2.52 To draw parallel line using computer graphics, (1) create line on screen using LINE task and (2) move cursor to desired position for one end of new line and use DUPLICATE, COPY, or PARALLEL function to create parallel line. Alternate method: Note angle of line on screen and create new line at same angle. Computer generated image.

2.13 *TO DRAW A LINE PERPENDICULAR TO A GIVEN LINE*

Perpendicular lines are important features in drawings for making buildings, machine parts, and most other man-made objects. We are used to walls being plumb (perpendicular to floors) and are uncomfortable if they appear to lean. Carpenters, masons, and ironworkers all learn to plan and build structures that have square corners (Figure 2.53). As engineers and designers, we need to be

Figure 2.53 Masons must learn to make square corners (perpendicular) when building walls. Photo by Howard Keith Williams

Figure 2.54 To draw line perpendicular to given line at given point using compass, (1) set compass to any convenient radius, (2) place compass point on given point P and draw two arcs cutting given line on each side of given point, (3) increase radius of your compass and draw two arcs from cut points on line (Q and R) that intersect on same side of line S, and (4) draw line from given point P to intersection S. This line (PS) is perpendicular.

Figure 2.55 To draw line perpendicular to given line at given point using triangles, (1) place two triangles on drawing surface hypotenuse to hypotenuse, (2) slide pair of triangles until side of one triangle is parallel to given line, (3) hold second triangle firmly in place and slide first triangle along hypotenuse until side perpendicular to given line lies on given point, and (4) draw perpendicular along edge of triangle.

able to draw perpendicular lines quickly and accurately (Figures 2.54–2.56). Later in this book you will learn to make orthographic drawings and use descriptive geometry. Both of these subjects require accurate perpendiculars.

2.14 *TO DIVIDE A LINE INTO A NUMBER OF PARTS*

We often need to divide a line into a number of parts and very often the number of parts is not easily divisible into the length of the line. Dividing a 4-in.-long line into two parts is easy; dividing

Figure 2.56 To draw line perpendicular to given line using computer graphics, (1) create inclined line 3–4 in. long in upper part of screen; (2) pick point on line where perpendicular is to intersect and create perpendicular line at this point (program may allow this by specifying 90° angle); (3) if more than one perpendicular line is needed, the first one is created and then additional lines parallel to first one; and (4) perpendiculars may also be created using ARC or CIRCLE function to duplicate compass arcs described in manual method. This is much slower on screen than on paper. Computer generated image.

a line $4\frac{3}{16}$ in. long into exactly five parts does not appear to be as easy. If you were planning a graph, you might have only so much space on the paper and need to divide the axes into a number of parts that are not whole numbers or simple fractions (Figure 2.57). Highways usually have a stripe down the center. To be sure that the stripe is in the middle of the road, the highway crew has to bisect the highway, divide it into two equal parts (Figure 2.58).

The method for dividing a line into any number of parts is based on the principle of similar triangles, which you may remember from plane geometry (Figures 2.59–2.61). The method for bi-

Figure 2.57 To design graph, axes must be divided evenly.

Figure 2.58 Center stripe on highway should bisect road. Photo by Howard Keith Williams

Figure 2.59 To divide line into any number of parts: (1) Draw line of known length (XZ) from one end of given line (XY) at any convenient angle. Line of known length (XZ) chosen should contain easy multiple of number of parts needed. (If four parts are needed, line could be 2 in. long, each part $\frac{1}{2}$ in. long.) (2) Connect open end of given line to open end of line of known length (Y to Z). (3) Draw lines parallel to connector line from each of divisions on known line. Where these lines cross given line, you have divided it into the required number of parts.

Figure 2.60 To divide line into any number of parts by successive approximations also works for dividing a circular arc: (1) Estimate how long one part would be. Set dividers to this length. (2) Step off number of parts on line by "walking" dividers along line. Distance will probably be too long or too short. (3) Change setting of dividers by estimate of total error divided by number of parts (error/*n*). If error is $\frac{1}{2}$ in. short and four parts are needed, lengthen dividers by about $\frac{1}{8}$ in. (4) Step off parts again on line. If still not exactly right, change divider setting by (new error)/*n*. Three tries will usually give good fit.

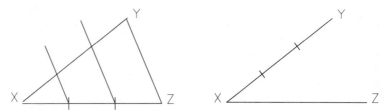

Figure 2.61 To divide line into any number of parts using computer graphics: (1) Create line on screen about 3–4 in. long using line task. It should be near middle of screen and inclined. (2) Name line *XY*. (3) Create horizontal line *XZ* exactly 6 in. long. (Line at any angle can be used; however, most graphics programs give distances more readily if line is horizontal or vertical.) (4) Name endpoint of line *Z*. (5) Create line from *Z* to *Y* (start at *Z*). (6) Move cursor to point on *XY* exactly 2 in. from *Z* (use distance counter) and duplicate line *ZY* or create parallel to it, creating line parallel to *ZY*. (7) Move cursor to point on *XY* exactly 4 in. from *Z* and repeat step 6. Parallel lines cut *XY* into three equal pieces. Computer generated image.

secting a line into exactly two equal parts is one you may also remember (Figure 2.62).

2.15 *TO BISECT AN ANGLE*

Whenever there are two lines or two pieces of material intersecting in space, there is an angle between them. For neatness and strength we often make a mitered joint in which we split the angle and have the parts meet on a line that cuts the angle exactly in two. Steel framing for buildings and other structures can be cut this way (Figure 2.63). From plane geometry, you may remember the construction for the bisector of an angle (Figures 2.64–2.66). You may also remember that the bisector of an angle is the locus of all points equidistant from the sides of the angle. We shall use this fact when we solve some descriptive geometry problems.

AE = BE

Figure 2.62 To divide line into two parts (bisect line), (1) set compass to radius greater than one-half length of line *AB*, (2) swing intersecting arcs with this radius from ends of the line, and (3) join intersections of arcs with straight line (*CD*) (this line is perpendicular bisector of given line; midpoint is at *E*).

Figure 2.63 Mitered joint in roof truss.

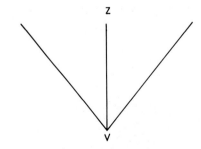

Figure 2.64 To bisect angle using compass, (1) set compass to any convenient radius less than length of sides of angle, (2) place compass point on vertex *V* of angle and swing arcs cutting each side of angle, (3) place compass on each of cut points (*W* and *X*) and swing arcs intersecting with angle (*Z*), and (4) draw line from vertex *V* to intersection *Z* (this is bisector).

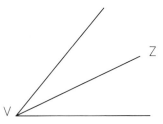

Figure 2.65 To bisect angle using computer graphics: (1) Create two intersecting lines 3–4 in. long near middle of screen. (2) Name intersection vertex *V* using TEXT. (3) Create arc with center at *V* and radius a little shorter than shortest line; make arc cross both lines. (4) Name cut points on lines *W* and *X*. (5) Create two more arcs of same radius with centers at *W* and *X*; make them long enough to intersect inside angle. (6) Name intersection point *Z*. (7) Create line *ZV*. This is bisector. Computer generated image.

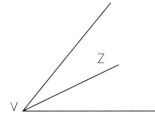

Figure 2.66 Alternate method to bisect angle using computer graphics: Follow steps 1–4 of Figure 2.65. Then create line *W–X*, find midpoint of line *WX* using midpoint or center function and name point *Z*, and create line *ZV*, which is bisector. Computer generated image.

Figure 2.67 Sailboat with triangular sail. Photo by Howard Keith Williams

2.16 *TO CONSTRUCT A TRIANGLE GIVEN TWO SIDES AND THE INCLUDED ANGLE*

Sail makers and other technical people apply graphic trigonometry in their work (Figure 2.67). A triangle has three sides and three angles. If we know the three sides, two sides and the included angle, or two angles and the included side, we can draw the triangle. We may find the information in different forms—two sides and the included angle is a fairly common way of describing a triangle. Our sail maker may know the length of the mast, the length of the boom, and the angle between them. The sail maker would then lay out the triangular pattern for the sail as described here (Figures 2.68 and 2.69).

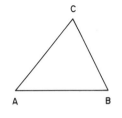

Figure 2.68 To construct triangle given two sides and included angle using manual graphics: (1) Draw line equal in length to one of given lines (*AB*). (2) Set compass to radius shorter than sides of angle (and shorter than line *AB*) and swing arc from vertex *T*, which cuts both sides *E* and *F*. (3) Do *not* change compass setting; now swing same-size arc from *A* on one side of line *AB*, cutting *AB* at *H*. (4) Place compass point on *E* and *F* and set compass to distance *EF*. (Dividers may be used for this operation.) (5) Do *not* change compass setting; now swing arc (or transfer divider distance) along arc from point *H*, cutting arc at *D*. (6) Draw line from point *A* through point *D*. (7) Set compass to length of second given line and swing arc from *B*, which cuts line *AD* at some point *C*. Triangle *ABC* is desired triangle containing two given lines and given angle.

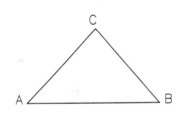

Figure 2.69 To construct triangle using computer graphics: (1) Create one line horizontally on lower part of screen and use counter to measure length. (2) Name line *AB*. (3) Create arc with center at *A* and length equal to second line and use counter to measure arc length. Make angle of arc greater than given angle. (4) Create line starting at *A* that crosses arc and makes given angle with line *AB*. Use angle counter to measure angle. (5) Name intersection point *C*. (6) Create line *CB*. Triangle *ABC* is desired triangle containing two lines of given length and given angle. Computer generated image.

2.17 *TO CONSTRUCT A REGULAR POLYGON*

Regular **polygons** are closed plane figures having sides of equal length and equal internal angles. You will see polygons in nature and in man-made objects (Figure 2.70). Consider the honeycomb built by bees; it is a collection of hexagons (six-sided figures) attached to each other to provide maximum efficiency in the use of space. Each cell in the honeycomb provides food storage for a new insect. A large polygon built by man is the Pentagon outside Washington, DC, which houses the Department of Defense. Construction methods are illustrated in Figures 2.71–2.73.

Figure 2.70 Cells in honeycomb are series of hexagons; many bolt heads are hexagons also.

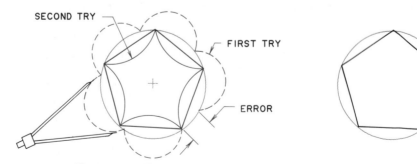

Figure 2.71 To construct polygon of any number of sides using manual graphics, (1) draw circle that contains polygon, (2) divide circumference of circle into as many parts as there are sides in polygon using method of successive approximations, and (3) connect points on circumference.

Figure 2.72 To construct polygon of six sides using manual graphics (special construction for just six sides), (1) draw circle that contains (circumscribes) polygon, (2) step off radius of circle along circumference (it will fit exactly six times, think about angles and number of sides in terms of plane geometry), and (3) connect points on circumference.

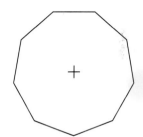

Figure 2.73 To construct polygon of any number of sides using computer graphics, use a POLYGON task (available with several programs) and create polygon with any number of sides by following prompts on screen. Typically, you would be asked for the number of sides, the location of the center, and the location of one vertex. Computer generated image.

2.18 TO CONSTRUCT A CIRCLE THROUGH THREE GIVEN POINTS NOT IN A STRAIGHT LINE

The circle is common in nature and in man-made devices. When we cut through a tree trunk or a flower stem, we see a circle. Every time a bicycle or an automobile passes, we see circles in the wheels. The circle rolls, and it is the most efficient way of enclosing an area. (The ratio of the circumference of a circle to the area inside the circle is a minimum; no other plane figure can enclose the same area with a smaller perimeter.)

The construction shown here will help you build a circle that passes through three given points or to find the center of a circle if three points on the circumference are known (Figures 2.74 and 2.75). This is useful when we need to repair or replace a broken wheel or gear (Figure 2.76).

Figure 2.74 Center and diameter of broken wheel can be found by using three points on circumference.

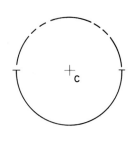

Figure 2.75 To construct circle through three points using computer graphics: (1) Create points on screen and connect them to create two straight lines. (2) Construct perpendicular bisectors of the lines (Figure 2.56). (Intersection of bisectors is center of circle.) (3) Use CIRCLE task to create required circle. *Note:* Some programs have routine that creates circle directly from three points. Computer generated image.

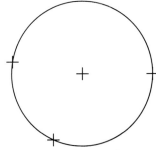

Figure 2.76 To construct circle using manual graphics: (1) Connect two points on circle to make straight line; repeat to form another straight line. (2) Construct perpendicular bisector to each line. (Intersection of bisectors is center of circle.) (3) Set compass to length from center (intersection) to end of any of lines and swing circle.

Figure 2.77 Bicycle chain is tangent to sprockets. Photo by Howard Keith Williams

2.19 *TO CONSTRUCT A LINE TANGENT TO A CIRCLE THROUGH A GIVEN POINT*

A line tangent to a circle is visible when you look under the hood of a car and see the fan belt going around a pulley. You also see a tangent when you look at the chain on a bicycle going around the sprockets (Figure 2.77). The most useful devices have contours that are a combination of straight sides and curves that are tangent to the straight-line contours.

The word *tangent* means to touch, so the point of tangency is where the circle just touches the straight line. A radius drawn from the center of the circle to this point will always be perpendicular to the tangent line. You will use this fact in your constructions (Figures 2.78–2.80).

Figure 2.78 To construct tangent line using manual graphics (two-triangle method): (1) Place triangles on drawing surface hypotenuse to hypotenuse (as for drawing parallel or perpendicular lines). (2) Slide triangles so that one edge of upper triangle passes through given point and touches circle. (3) Hold lower triangle firmly in place and slide upper triangle until other edge passes through center of circle. (4) Mark point where this edge cuts circle arc using short dash. This is point of tangency. (5) Draw line from point of tangency to given point. (Slide upper triangle back to its original position; now you know where to stop tangent line.)

Figure 2.79 To construct tangent through point on circle: (1) Draw line from given point to center of circle. (2) Bisect line (Figure 2.61). (3) Set compass to radius one-half length of line and swing half circle with line as diameter and that cuts given circle. Point of intersection of two arcs is point of tangency. (4) Draw line from given point to point of tangency. (5) Draw line from center of circle to point of tangency. This line makes right angle with tangent line and confirms point of tangency. *Note:* Theorems of plane geometry teach that any triangle inscribed in semicricle is right triangle. This construction gives perpendicular at point of tangency.

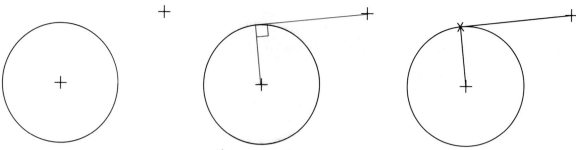

Figure 2.80 To construct tangent using computer graphics: (1) Create circle on screen using CIRCLE task (record coordinates of center). (2) Create point on screen at least 2 in. away from circle. (3) Through point create line that touches circle. Make construction line longer than just touching circle. Use zoom function to check that line just touches circle. This is tangent line. (4) Create line through center of circle perpendicular to tangent line. Intersection with circle (and tangent line) is point of tangency. Computer generated image.

2.20 *TO CONSTRUCT A LINE TANGENT TO TWO CIRCULAR ARCS*

If you examine the fan belt in a car, you will find that it is not only tangent to one circle (pulley) but is also tangent to one or more other pulleys (Figure 2.81). The next time you are looking at machine tools—lathe, milling machine, or big drill press—note that the straight-sided contours are rounded wherever they join another surface. Here you have a straight line tangent to two circle arcs (Figures 2.82 and 2.83).

Figure 2.81 Fan belt on automobile engine is tangent to all pulleys it touches. Courtesy of Ohio State University Engineering Publications

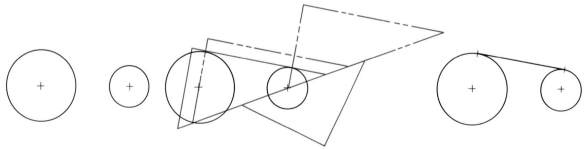

Figure 2.82 To construct tangent line using manual graphics: (1) Place triangles on drawing surface hypotenuse to hypotenuse. (2) Align one edge of upper triangle so that it touches both arcs. Draw light line connecting two arcs. (3) Hold lower triangle firmly in place and slide upper triangle so that other edge passes through center of circle; mark tangent point with short dash. (4) Slide upper triangle so that it passes through center of other circle and mark this tangent point. (5) Connect two points.

Figure 2.83 To construct tangent line using computer graphics: (1) Create two circles on screen, preferably of different diameters and not touching each other. (2) Create line that touches both circles. Check accuracy with ZOOM task. Delete line and recreate it if it does not just touch both circles. (3) Create line through center of each circle perpendicular to tangent line. Points of tangency occur where these lines touch tangent line. (Some programs have a TANGENT task that locates tangent.) Computer generated image.

2.21 *TO CONSTRUCT AN ARC TANGENT TO A LINE (OR LINES)*

Nature does not produce plants or animals with sharp corners. Sharp corners are easily broken. When you study the mechanics of materials, you will learn that destructive forces increase at internal sharp corners (stress concentration). When we design mechanisms that will be heavily loaded, we avoid sharp corners, just as nature does. We do this by rounding outside corners and filling in interior corners with a curve (a fillet) (Figures 2.84–2.88).

Figure 2.84 Exterior and interior corners on machinery are rounded with arcs tangent to straight surfaces. Courtesy of Ohio State University Engineering Publications

Figure 2.85 To construct arc tangent to one line using manual graphics, (1) draw perpendicular to given line that is longer than radius of arc, (2) set compass to desired radius, (3) swing arc from intersection of two lines that cuts perpendicular line, and (4) set compass point on cut point and draw tangent arc.

Figure 2.86 To construct arc tangent to one line using computer graphics, if line is horizontal or vertical, read coordinates on screen and move cursor distance of radius away from line to locate center for arc. Alternate method: If line is not horizontal or vertical, use manual graphics method to locate center; then create arc. Computer generated image.

Figure 2.87 To construct arc tangent to two intersecting lines using manual graphics: (1) Follow steps 1–3 in Figure 2.85 to find point that is radius distance away from each line. (2) Draw line parallel to each given line through cut points. These lines will intersect at point equidistant from both given lines. (3) Draw perpendicular from this point to each given line. (These are tangent points.) (4) Swing arc with its center at intersection that touches each given line where perpendicular cuts it.

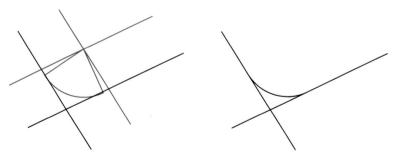

Figure 2.88 To construct arc tangent to two intersecting lines using computer graphics, if lines are horizontal and vertical, read coordinates on screen to locate center of arc at radius distance from each line. Create arc. (Some programs have FILLET function that locates arc.) Alternate method: If lines are not horizontal and vertical, use manual graphics method to locate center and then create arc. Computer generated image.

2.22   *TO CONSTRUCT AN ARC TANGENT TO A LINE AND AN ARC*

Not all roads are straight. Some roads have curves, and some have curves leading into other curves. We need to know how to lay out such patterns. If you ever need to design cams (such as those that move the valves in an automobile engine), you will also need to know how to fit an arc tangent to a line and another arc (Figures 2.89–2.91).

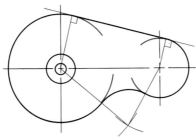

Figure 2.89   Cam development may require arcs tangent to straight lines and other arcs.

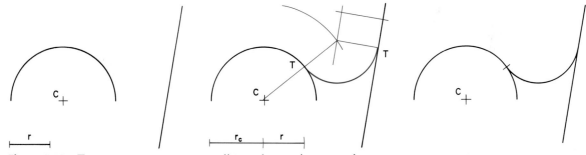

Figure 2.90   To construct arc tangent to line and arc using manual graphics: (1) Set compass to radius of existing arc. (2) Increase (or decrease) radius by length of radius of arc to be drawn. (3) Swing new arc with increased (decreased) radius from original center; be sure it is in area of line. (4) Draw line parallel to existing line and exactly radius distance away from existing line; this line should cut new arc. Cut point is center for desired arc. (5) Align triangle with center of desired arc and center of existing arc and mark where this line crosses existing arc. This is point of tangency. (6) Draw perpendicular from center of desired arc to given line. Cut point (intersection) is point of tangency. (7) Set compass to radius of desired arc and swing it from center, drawing from tangent point to tangent point.

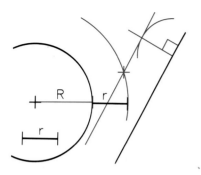

Figure 2.91 To construct arc tangent to line and arc using computer graphics: Follow the method using manual graphics in Figure 2.90 and (a) record coordinates of circle center and ends of lines and (b) use ZOOM function to increase accuracy when creating desired arc. You may be able to use MOVE, COPY, PARALLEL, or PERPENDICULAR functions available in some programs. Computer generated image.

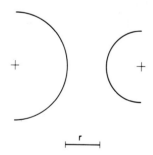

Figure 2.92 Two idler gears driven by third gear.

2.23 TO CONSTRUCT AN ARC TANGENT TO TWO ARCS

The design of some gears requires that we locate the pitch circle of one tangent to the pitch circles of two other gears (Figure 2.92). (The pitch circles of circular gears are the lines along which the gears transmit motion.) We may also use this construction to avoid the sharp corners that would occur if we had a straight-line contour connecting two circular shapes (Figures 2.93 and 2.94).

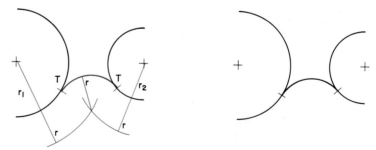

Figure 2.93 To construct arc tangent to two given arcs using manual graphics: (1) Increase or decrease radius of each given arc by length equal to radius of desired arc. (2) Swing concentric arcs around each of given arcs with new radii. Intersection of these arcs locates center of desired arc. (3) Align triangle with new center and each of given centers. Make short dash where these lines of centers cross given arcs. These are tangent points. (4) Set compass to radius of desired arc and swing arc from tangent point to tangent point.

Figure 2.94 To construct arc tangent to two given arcs using computer graphics, follow the method using manual graphics in Figure 2.93. Computer generated image.

2.24 TO CONSTRUCT AN ELLIPSE

An ellipse is a closed plane curve. Every point on its curve is located so that the sum of the distances from the point on the curve to two points (the *foci*) on the long axis (the major diameter) is constant (Figure 2.95). The ellipse is also a conic section, the curve produced when you cut a right-circular cone with a plane not perpendicular to the axis but making a greater angle with the axis than the elements of the cone (Figure 2.96).

We often use ellipses in making drawings because every view of a circle except the one looking straight at it (normal) is an ellipse (Figure 2.97) or an edge view. Ellipses are an attractive

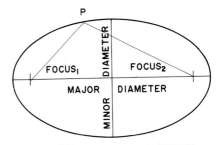

$$\overline{F_1P} + \overline{F_2P} = K = \text{MAJOR DIAMETER}$$

Figure 2.95 Ellipse showing its foci and major and minor diameters.

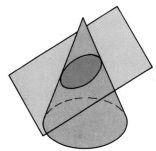

Figure 2.96 Ellipse as conic section.

Figure 2.97 Coin is round when viewed normally; from any other angle it is seen as ellipse or edge.

shape and make a pleasing design element. Some gears are also made with an elliptical shape. Several constructions for drawing ellipses are given in Figures 2.98–2.102. Commercial design rooms use ellipse templates for drawing ellipses. A common template is one used for drawing circles in isometric pictorial drawings (Figure 2.103).

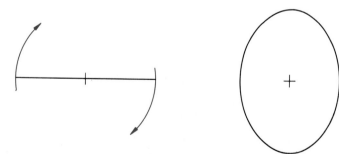

Figure 2.98 To construct ellipse using foci method: (1) Set compass to one-half length of major diameter ($\frac{1}{2}$MD). (2) Swing arcs from ends of minor diameter (points m and d) intersecting major diameter at points f_1 and f_2, the foci. (3) On one end of major diameter mark points between focus and intersection of diameters. (Three points are shown here; more points provide greater accuracy.) (4) With foci as centers, set compass to lengths f_1A and f_2A and swing arcs intersecting in each quadrant at points APV_1PVPY, A_2, A_3, and A_4. Repeat for points B and C so that point is plotted in each quadrant for each selected point. (5) Connect intersection points with irregular curve.

Figure 2.99 To construct ellipse using computer graphics, use foci method given in Figure 2.98 and connect points using curve-plotting task. (This method is very time consuming; if alternate is available, try it.) Alternate method: Some graphics programs create round and edge views of circle and then view edge from any angle to create true ellipse by projection. This method is shown in Chapter 10, "Descriptive Geometry." Some programs also have an Ellipse command. Computer generated image.

TRUE ISOMETRIC ELLIPSE

Figure 2.100 To construct approximate ellipse using four-center method (there is general method for making this construction that is of little interest; method shown here will be needed when making isometric pictorial drawings, discussed in Chapters 3 and 5): (1) Lightly draw isometric square that contains circle of given diameter; each side of square is same length as diameter and interior angles are 60° and 120°, as shown in square *ABCD*. (2) From each of large-angle vertices (*A* and *D*) draw lines to midpoints of each side of square. These lines are perpendicular bisectors of sides. (3) Set compass to length of bisector lines and swing arcs from vertices *A* and *D* to feet of perpendiculars (points of tangency). (4) From points where bisectors intersect each other (*X* and *Y*) as centers, swing small arcs to points of tangency. Approximate four-center ellipse results. Note how well it matches true ellipse; it is good approximation and faster for this application than any other method, except template.

Figure 2.101 To construct ellipse using concentric circle method: (1) Set compass to one-half major diameter ($\frac{1}{2}$MD) and draw circle with center at 0, intersection of diameters. (2) Set compass to one-half minor diameter ($\frac{1}{2}$md) and draw circle with center at 0. (3) Use triangles to divide circle into equal parts, drawing radial lines through center intersecting both large and small circles (30° increments give 12 equal parts; 15° increments, using triangles in combination, give 24 equal parts.) (4) From intersection of each radial line with large circle draw line perpendicular to major diameter. (5) From intersection of each radial line with small circle draw line perpendicular to minor diameter. Where two perpendiculars for any pair of points cross is point on ellipse. (You may want to construct one pair of perpendiculars at time and find point where they cross in order to reduce confusion.) (6) Connect points using irregular curves.

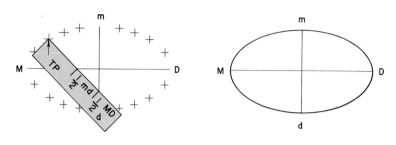

Figure 2.102 To construct ellipse using trammel method: (1) Select small card or piece of paper with straightedge longer than one-half major diameter for building tool (trammel is used for transferring measurements). (2) Mark one end of trammel as tracing point with arrow. This point will locate points on ellipse. (3) From tracing point measure one-half major diameter and mark it with dash and MD. (4) From tracing point measure one-half minor diameter and mark it with dash and md. (5) Lay trammel across diameters so that MD point is on the minor diameter and md point is on major diameter; tracing point now gives point on curve. Mark it with a short dash. (6) Move trammel around axes keeping MD point on minor diameter and md point on major diameter and mark points on ellipse; mark three or more in each quadrant. (7) Connect points with irregular curves.

Figure 2.103 Ellipse template for drawing circles as ellipses in isometric pictorial drawings.

2.25 TO INSCRIBE A CIRCLE IN A TRIANGLE

A classic construction from Euclidean geometry is the circle inscribed in a triangle. Inscribed means that the circle is tangent to each of the sides of the triangle.

You may need to inscribe a circle in a triangle if you are designing tools to hold circular parts in a shop (Figure 2.104). A common shipping container for rolled drawings is a triangular carton. How large a roll of drawings could you ship in a triangular carton shaped similar to an equilateral triangle with 2-in. sides? The constructions for this problem are given in Figures 2.105 and 2.106.

Figure 2.104   Triangular set of surfaces encloses cylindrical pin on this test fixture.

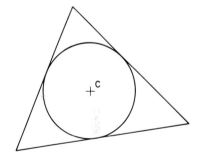

Figure 2.105   To inscribe circle in triangle using manual graphics: (1) Construct bisector of each angle of triangle (Figure 2.59). (2) Swing circle from intersection of bisectors tangent to sides of triangle. *Note:* Theorems of plane geometry teach that bisector of angle is locus of all points equidistant from sides of angle. Point where bisectors intersect is then equidistant from all sides of angles. This is proper location for center of circle with radius distance that is the same to all sides.

2.26   TO CIRCUMSCRIBE A CIRCLE ABOUT A TRIANGLE

Another construction from Euclidean geometry is fitting a circle around a triangle so that the vertices of the triangle all lie on the circle. This construction is also useful for designing tools and packaging (Figures 2.107–2.109).

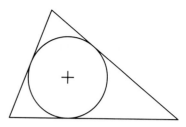

Figure 2.106 To inscribe circle in triangle using computer graphics: (1) Bisect angles of triangle. (2) Create circle with center at intersection of bisectors and that just touches the sides. You can locate points of tangency by creating perpendiculars from center to sides of triangle. Use ZOOM function to improve accuracy. Computer generated image.

Figure 2.107 Round shipping container protects triangular part.

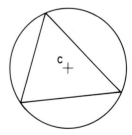

Figure 2.108 To circumscribe circle around triangle using manual graphics: (1) Construct perpendicular bisectors of sides of triangle (Figure 2.61). (2) Set compass to distance from intersection of perpendiculars to one vertex of triangle (distance should be same to all vertices). (3) Construct circle passing through all three vertices. *Note:* Theorems of plane geometry teach that perpendicular bisector of line segment is locus of all points equidistant from ends of line segment. Point where bisectors intersect is equidistant from ends of all sides. This is proper location for center of circle with radius distance the same from all vertices.

Figure 2.109 To circumscribe circle around triangle using computer graphics: (1) Find midpoints of sides. (2) Create perpendiculars at these midpoints. (3) Create circle with point of intersection of perpendiculars as center and radius that just touches sides. Use ZOOM function to improve accuracy. Computer generated image.

Figure 2.110 Involute of circle is path traced by string unwound from it.

2.27 TO CONSTRUCT AN INVOLUTE OF A CIRCLE

The involute of a circle is the path traced by unwinding a string wrapped around the circle (Figure 2.110). This shape is used for gear teeth to assure true rolling motion between circular gears. You can develop involutes for lines, triangles, and other shapes, but the involute of the circle has the greatest application for engineers. Methods for constructing involutes are shown in Figures 2.111 and 2.112.

Figure 2.111 To construct involute of circle by manual graphics: (1) Divide circle into number of parts using radial lines or using method of successive approximations on circumference (Figure 2.60). (2) Construct tangents (perpendiculars) at each division point (Figure 2.78). (3) Set dividers to length of arc of one part of circle, and beginning at any point on circle measure this distance on tangent line. Mark distance with short dash. (4) Go to next tangent line and measure two parts of length of arc. Mark it with short dash. (5) Continue around circle adding one more part to number measured on tangent lines, completing as much of the circle as directed by instructor. (6) Connect points with irregular curves (curve plotted may get quite large; more than one irregular curve may be needed to get a smooth fit among the points).

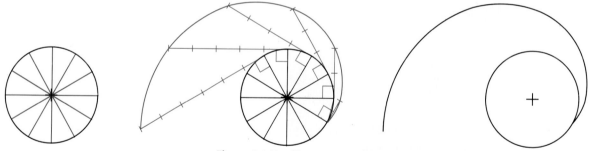

Figure 2.112 To construct involute of circle by computer graphics, follow the method for manual graphics in Figure 2.111 and connect plotted points with curve-fitting task. Computer generated image.

2.28 *SUMMARY*

After practicing the material presented in this chapter, you will be able to letter text and numbers on drawings so they are readable. You will also have an understanding of lettering spacing and proportions that will be useful when you prepare text using a computer graphics program. The constructions will be useful for making your first complete drawing. They will be used often when you apply descriptive geometry to solving problems. Some of them may have application to solving problems later in your professional career when you may be designing gears, tooling for manufacturing, machine parts, or highways. You may even decide to build an elliptical garden plot using the foci method.

PROBLEMS

Problem 2.1

Draw a series of $\frac{1}{8}$-in. spaces on your paper using very light horizontal lettering guidelines. Using capital single-stroke Gothic lettering (inclined or vertical is your choice), letter the following paragraph (remember to only letter in every other space):

> Dimensions are added to detailed engineering drawings to specify the size or location of all features. Notes may also be included to explain additional details about the object. Dimensions and notes become the communication link between the engineer and the craftsman.

Problem 2.2

Draw the following sets of numbers and labels. Use $\frac{1}{8}$-in.-high whole numbers, but let the fractions extend a little above and a little below the guidelines:

Whole numbers	1,234,567,890
Fractions	$\frac{1}{4}, \frac{3}{8}, \frac{5}{16}, \frac{7}{32}, \frac{9}{64}$
Mixed numbers	$6\frac{5}{16}, 4\frac{1}{2}, 22\frac{7}{32}, 9\frac{7}{8}$

Problem 2.3

Letter the following in every other space of the $\frac{1}{8}$-in. lettering guidelines.

> When dimensioning, I will form the letters, numbers, and mixed numbers correctly. All letters and numbers will be black and crisp so they will copy well. Clear uniform lettering is essential for accurate communication.

Problem 2.4

Letter the following notes within the $\frac{1}{8}$-in. lettering guidelines. Letter in every other space.

> Note: all unspecified radii $R\frac{1}{8}$
>
> 0 4.13, 5 holes, equally spaced
>
> $\frac{1}{2}$-20UNF-2B
>
> Chamfer 3 × 45 both ends
>
> 13.5 drill, spotface 024 × 1.5 deep, 2 holes
>
> Top security, do not copy
>
> For in-house price quote only, not for production

Problem 2.5

Your instructor will provide you with a drawing sheet that has a printed lettering block that will be used in your class. The instructor will also tell you what information is to be lettered in that lettering block. Remember to draw the lettering guidelines before lettering.

Problem 2.6

(Solve these problems on $8\frac{1}{2} \times 11$ drawing paper with the long dimension placed vertically.)

    (a) Draw a line 4.2 in. long in the upper third of your paper; divide it into five equal parts using the method of similar triangles.

    (b) Draw a line $2\frac{15}{16}$ in. long in the middle third of your paper; bisect it using your compass. Erect a perpendicular bisector $1\frac{9}{16}$ in. long on one side of the original line and extend it $\frac{7}{8}$ in. on the other side.

    (c) Draw a line 100 mm long in the lower third of your paper; divide it into seven equal parts by the method of successive approximations.

Problem 2.7

    (a) Use your triangles to draw a 45° and a 30° angle on your paper. Make the legs about 50 mm long and bisect each angle with your compass.

    (b) Draw a 60° angle with your triangles. Make the legs about $1\frac{3}{4}$ in. long and bisect it. Check the results with the 30° angle on your triangle.

Problem 2.8

    (a) Draw a triangle with legs 45 and 72 mm long with an included angle of 45°.

    (b) Think of a new way to build a triangle; draw a triangle with *angles* of 45° and 60° and an included *side* of 64 mm.

    (c) Here is another one to try. Build a triangle with three sides each 64 mm long (an equilateral triangle). What is the size of each angle?

Problem 2.9

    (a) Construct a hexagon with sides 1.30 in. long.

    (b) Construct a square that just fits (inscribed) in a circle 3.00 in. in diameter.

    (c) Construct a square with sides 1.40 in. long. (All the sides are equal; and all the internal angles are 90°. What a great way to practice perpendiculars!)

Problem 2.10

Draw two circles that have 4 in. between their centers. One circle should have a $1\frac{1}{2}$-in. radius; the other should have a 1-in. radius. Construct four tangent lines touching both circles; one pair crosses between the circles, the other pair does not. Mark all tangent points with red dashes.

Problem 2.11

Draw a reduced-size copy of your 30°–60°–90° triangle; the longest leg making the right angle should be 5.00 in. long. Round the 30° angle with a 0.50-in. radius; round the 60° angle with a 0.40-in. radius. Mark the tangent points with red dashes. Bisect the 90° angle.

Problem 2.12
Construct an ellipse with a major diameter of 165 mm and a minor diameter of 95 mm. Use the trammel method. Draw a copy of your trammel, in one position, on your ellipse. Find the foci of this ellipse and mark them with green crosses.

Problem 2.13
    (a) Draw a reduced size copy of your 45° triangle; each of the legs making the 90° should be 4.50 in. long. Inscribe a circle in the triangle. Mark the tangent points where the circle touches the sides of the triangle with blue dashes; draw a small (0.13 diameter) blue circle around the center of the inscribed circle.

    (b) Now circumscribe a circle around your triangle. Draw a small red circle around the center of the circumscribed circle.

Problem 2.14
    (a) Locate the center for a circle 4 in. down from the top of your paper and 3 in. from the left side. Draw a circle of $1\frac{1}{2}$ in. diameter using this center. Starting at the bottom of this circle, draw the involute formed when you unwind the circumference; do not go lower on the sheet than the bottom of the circle. Use your irregular curves to connect the points.

    (b) Draw an equilateral triangle with 1-in. sides that is $2\frac{1}{2}$ in. from the left side of the paper and 2 in. above the title block or border. Start at the lower right corner and unwind the involute.

Problem 2.15
Prepare a graph that converts inches to millimeters (1 in. = 2.54 mm). Make the x axis 7 in. long; divide it into 10 parts. Make the y axis 9 in. long; divide it into 254 parts, but mark only every tenth part.

Problem 2.16
(Turn your drawing paper with the long dimension horizontal for problems 2.16 and 2.17.)

Draw a triangular roof truss as shown. (Scale: 1 in. = 2 ft.) Bisect the angles for miter cuts of the truss members.

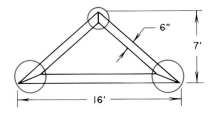

Problem 2.17

Lay out the pattern for the side plate of a link of roller chain. Draw it double size. Mark tangent points with a red dash.

Solve problems 2.18–2.25 on your computer screen using a graphics program.

Problem 2.18

Create a horizontal line 4.50 in. long near the top of your computer screen. Erect a line 2.75 in. long perpendicular to the first line at its midpoint.

Problem 2.19

Create a line 5.20 in. long near the bottom of your screen that is neither horizontal nor vertical. Divide this line into six equal parts.

Problem 2.20

Create a circle 4 in. in diameter near the middle of your screen. Inscribe a hexagon in the circle; connect alternate vertices of the hexagon to form another closed plane figure. What is this figure?

Problem 2.21

Construct a triangle with sides $1\frac{1}{2}$, 2, and $2\frac{1}{2}$ in. long. Circumscribe a circle around this triangle. What is the diameter of the circle? Is there a right angle in the triangle? If your answer is yes, prove it!

Problem 2.22

Construct a pentagon of any convenient size. Join alternate vertices to form a star. Delete lines in the central portion of the star so that the center is open.

Problem 2.23
Two small pulleys are 76.1 mm apart on their line of centers. The smaller is 25.4 mm in diameter; the larger is 50.7 mm. Show the pulleys and the path of the belt that goes around them.

Problem 2.24
Create a block letter T on your screen 4.32 in. high and 4.32 in. wide. The horizontal and vertical members should be 1.00 in. Round all exterior and interior sharp corners with a radius of 0.25 in. (Rounded exterior corners on machine parts are called *rounds*; rounded interior corners are called *fillets*.)

Problem 2.25
Construct two circles with 1.50 in. diameters 2.50 in. apart on a horizontal line of centers near the top of your screen. Install a circle with a diameter of 3.00 in. tangent to both of the original circles. Does this remind you of a famous amusement park?

Solve problems 2.26–2.32 using pencil and paper or on a computer screen. If you do not have a color monitor or colored pens in your printer, change any color instructions to black.

Problem 2.26
Lay out the center lines for a portion of a highway and fit the needed curves to the drawing. Mark the tangent points with a red dash (Scale: 1 in. = 100 ft.)

Problem 2.27
Draw a view of the trammel base. Round all external corners with a 6-mm radius and all internal corners with an 8-mm radius. Show tangent points with a red dash.

Problem 2.28

Make an outline drawing for a cam latch. Show tangent points in red. (Scale: full size.)

Problem 2.29

What is the diameter of the largest circular pipe that can be put through area A of the truss? (Scale: 1 in. = 10 ft.) For more practice check areas B and C.

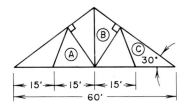

Problem 2.30

A special triangular beam is to be shipped in a round protective tube. What is the inside diameter of the needed tube? (Scale: full size.)

Problem 2.31
Your Aunt Eleanor wants to have the largest possible elliptical flower bed in her front yard. The space available is shown in the accompanying figure. Design a trammel you can use to determine points on the elliptical curve. Also make a scale drawing of the bed using the concentric circle method (edge view method on a computer). (Scale: 1 in. = 10 ft.)

Problem 2.32
Your little sister sends you a broken wheel from her red wagon and asks you to tell her how big a wheel she needs to replace it. You measure two chords on the wheel as shown in the accompanying figure. What is the diameter?

TERMS YOU WILL SEE IN THIS CHAPTER

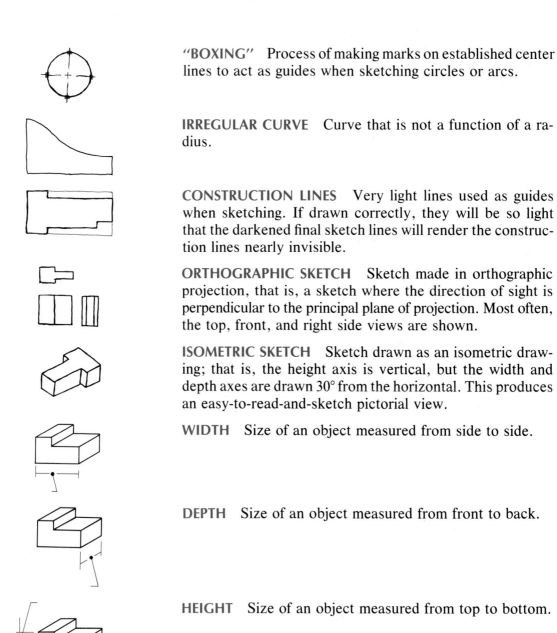

"BOXING" Process of making marks on established center lines to act as guides when sketching circles or arcs.

IRREGULAR CURVE Curve that is not a function of a radius.

CONSTRUCTION LINES Very light lines used as guides when sketching. If drawn correctly, they will be so light that the darkened final sketch lines will render the construction lines nearly invisible.

ORTHOGRAPHIC SKETCH Sketch made in orthographic projection, that is, a sketch where the direction of sight is perpendicular to the principal plane of projection. Most often, the top, front, and right side views are shown.

ISOMETRIC SKETCH Sketch drawn as an isometric drawing; that is, the height axis is vertical, but the width and depth axes are drawn 30° from the horizontal. This produces an easy-to-read-and-sketch pictorial view.

WIDTH Size of an object measured from side to side.

DEPTH Size of an object measured from front to back.

HEIGHT Size of an object measured from top to bottom.

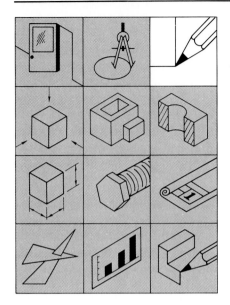

3

ENGINEERING SKETCHING

3.1 INTRODUCTION

The graphic language of the engineering world is used to communicate ideas and concepts through a series of pictures and symbols. After these ideas, concepts, and details for a project have been finalized, precise engineering drawings are produced so that parts may be manufactured. These drawings are usually made from engineering sketches generated during the project-planning stage.

The objective of making an engineering sketch is to make a quick, graphic representation of an idea that can be easily communicated to a co-worker, a drafter, or a client. This freehand sketch is generally made without the aid or use of technical drawing instruments. The primary need for making a sketch is to represent what you see rather than to give technical information. Sketches are an integral part of analytical calculations where they are used to define the physical objects being analyzed. For this reason, sketches tend to show an object as it appears, with little emphasis on hidden or internal surfaces or features, although they may be included if necessary. The sketch also serves as an *immediate* graphic communication, often drawn under varying conditions such as on the job site or in a business meeting (Figure 3.1).

Figure 3.1 Sample of four steps for making sketch.

Sketching, similar to writing, is a skill that can be developed through practice. This chapter will show you how to sketch effectively, and it will address materials and show engineering applications. Techniques and suggestions will be given so your sketches will look more realistic. You can practice sketching anywhere and at any time because the only items needed to make a sketch are a writing instrument, a drawing surface, and possibly an eraser.

Sketching is a process that generally precedes the action of creating a drawing on a computer or on a drawing board. This chapter offers you sketching techniques with only brief explanations concerning drawing theory. Detailed theory of projections will be given in Chapter 4.

Figure 3.2 Sketching parallel horizontal lines.

3.2 SKETCHING HORIZONTAL LINES

Lines commonly used for sketches include straight lines and curved lines. Straight lines can be divided into horizontal, vertical, or inclined lines. The normal way to draw parallel horizontal lines is from left to right for right-handed people or right to left for left-handed people. You should be able to develop enough skill to draw many evenly spaced horizontal lines while leaving uniform spaces between the lines (Figure 3.2).

Care should be used to avoid drawing arcs instead of horizontal lines. This occurs when the elbow becomes stationary. You must move the entire arm while drawing horizontal lines. This is particularly important when sketching long horizontal lines. Long lines may be sketched more accurately by sketching a series of short horizontal lines end to end. An effective technique for drawing parallel lines is to draw the lines while guiding your finger along the edge of the drawing board. Another important tip is to put the pencil on the starting point and look at and draw to the finishing point (Figure 3.3).

Figure 3.3 Using side of board as guide to sketch parallel lines.

3.3 *SKETCHING VERTICAL LINES*

Vertical lines are usually drawn from top to bottom. You should be able to develop enough skill to draw uniformly spaced parallel vertical lines. The edge of the drawing surface can be used for this as well.

3.4 *SKETCHING ANGULAR LINES*

Angular lines are drawn from lower left to upper right. This is a comfortable position for the right-handed person. The most uncomfortable and difficult lines to sketch are angular lines extending from upper left to lower right. A helpful hint would be to rotate the paper to a position where it would be more comfortable to sketch these lines (Figure 3.4).

3.5 *SKETCHING CURVED LINES*

Curved lines are identified as either circles, arcs, or irregular curves. Circles are sketched most accurately by first sketching center lines. The radius is then marked on these center lines as guides. These marks form a box within which the circle can easily be sketched. This process is often called **"boxing"** a figure. Lastly, the circle is sketched in two motions. First, the upper left side of the circle is sketched in a counterclockwise direction, and then the lower right side of the circle is sketched in a clockwise direction (Figure 3.5).

Figure 3.4 Sketching vertical and angular lines.

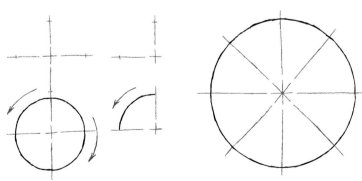

Figure 3.5 Guide marks for circles and arcs.

Large circles may require a second set of center lines rotated at a 45° angle in order to increase the number of guides from four to eight. Also, four arcs may be required. If so, the paper should be rotated for completion of the last two quadrants. Arcs are sketched similar to circles. The orientation of the arc and the size of the arc will govern how the person sketches the arc. For medium and large arcs, it is recommended that center lines and guides be drawn.

3.6 SKETCHING IRREGULAR CURVES

Irregular curves (those without a given radius) should be sketched using pencil strokes that are most comfortable for you. Very light lines should be sketched first to check for the correct shape. Draw dark lines over the light lines to complete the curve.

3.7 DRAWING MEDIA

After becoming familiar with the most comfortable sketching strokes, identify an appropriate media for the sketch. Drawing media are the materials upon which you will make your sketch. We all know that sketches are made on everything from match book covers to expensive papers or films. In Chapter 2 we learned about films and vellums. However, the interest here is to identify those paper or film materials that are commonly used in the engineering field.

Technical sketches are most often made on $8\frac{1}{2} \times 11$-in. opaque paper. For copying purposes, the original sketch is referred to as the master. Sketches made on this opaque paper may be copied on a normal copy machine. Normal copy machines generally accept $8\frac{1}{2} \times 11$-in. masters; however, some office copy machines will copy sheets $8\frac{1}{2} \times 14$ or even 11×17 in.

Some paper may be printed with a **grid** of thin, lightly colored lines, which are a significant aid to the person doing the sketching. Common grid patterns are square or isometric (Figure 3.6). The printed grids may be printed on opaque paper or on a translucent paper or film. Again, sketches made on opaque paper may be copied on a standard office copy machine, and sketches made on translucent stock can be copied on copy machines or on Diazo equipment.

Figure 3.6 Sample grid patterns that are useful for making refined sketches.

3.8 GRIDS FOR SKETCHING

A special printed grid is the "dropout" grid. This consists of a light blue grid printed on translucent paper or film. The grid serves as a guide to the person sketching, but when the copies are made, the grid will not copy and the sketch stands by itself. This is a great aid to making better appearing sketches.

Translucent paper or film may be placed over a printed grid as an overlay during the sketching process. Then any printed grid may be used. Special large printed grids are available for this purpose, such as perspective grids for interior or exterior views at various angles. This allows you to have the advantage of using a selected grid system without the grid appearing on the master or on the copies (Figure 3.7).

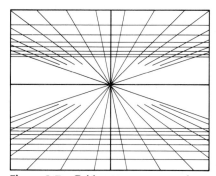

Figure 3.7 Grid patterns are used to make more accurate perspective sketches.

EM 273 STATICS | WJK | 10-15-87 | SMITH, R. | ¼

PROB 4.12

GIVEN: A BLOCK LOCATED ON A
PLANE AS SHOWN.
COEF. OF FRICTION, μ = 0.2

REQ'D: DETERMINE THE FORCE
P TO MOVE THE BLOCK
UP THE PLANE AT
CONSTANT VELOCITY

ANALYSIS: DRAW A FREE-BODY
DIAGRAM SHOWING
ALL FORCES APPLIED
TO THE BLOCK

F – FRICTIONAL FORCE
N – NORMAL FORCE

$F = \mu N = 0.2N$

N   R = REACTION
γ = ARCTAN 0.2
= 11.3°

RESOLVE W INTO COMPONENTS
PARALLEL TO AND PERPENDICULAR
TO THE SLOPE

$W_{||} = W \sin 30° = 0.5W$
$W_{\perp} = W \cos 30° = 0.866W$

W COS 30°   W SIN 30°

$\Sigma F_{||} = 0, \; P = W_{||} + F = 0.5W + 0.2N$
$\Sigma F_{\perp} = 0, \; N = W_{\perp} = 0.866W$

SUBSTITUTE N INTO FIRST
EQUATION

KNOWN
VECTOR, W

DIRECTION
OF P

$P = 0.5W + 0.2(0.866)W$
$= 50 + 17.32$
$\underline{P = 67.32\# \longleftarrow \text{ANSWER}}$

CHECK: WITH VECTOR DIAGRAM

$\underline{P \approx 65\#}$

DIRECTION OF
REACTION FROM
COEF. OF FRICTION

Figure 3.8 Example of typical engineering problem.

Figure 3.9 Use both sides of drawing media to protect guidelines if many erasures are anticipated.

Sometimes a special grid may be printed or drawn on the back side of the paper. Engineering problem paper is a typical example. In this case, the grid is printed on the back of the sheet, and you sketch on the front of the sheet. One such example might be when a number of lettering guidelines are used on a large specification sheet. If many erasures or changes must be made, the guidelines will not be disturbed when the erasures are made (Figures 3.8 and 3.9).

3.9 *LINE WEIGHTS*

Line weight refers to the line blackness or width of a line. The very light lines used to sketch for size, proportion, and position are called layout lines, or **construction lines**. They are so light that they would not appear if copied using a standard copy process. Construction lines are the first lines drawn. After the shape of the object has been sketched, the lines are refined. Then the visible lines are darkened with black solid lines. Hidden lines represent nonvisible edges or surfaces and are drawn with black dashed lines. If the original construction lines are drawn light enough, none should have to be erased after the sketch has been darkened and refined (Figure 3.10).

3.10 SKETCHING INSTRUMENTS

Sketches are generally made with black lines from a lead pencil. The common wood pencil has a number 2 lead, which may be used to draw the layout lines simply by allowing the weight of the pencil to pass over the paper. Then during the darkening process, the pencil must be pushed harder against the paper to gain the proper blackness of the final sketch.

Sketches made using a lead holder are commonly drawn with 2H lead. Pencils with a harder lead may be used for the layout lines, and pencils with a softer lead may be used for shading or illustrating after the sketch has been outlined. These, similar to wooden pencils, must be sharpened with a pointer. The more contemporary lead pencils used in technical graphics are the fine-line mechanical pencils. The 0.3-mm fine-line pencil may be used for the construction lines, and a 0.5- or 0.7-mm pencil will likely be used to make the dark refined lines. Fine-line pencils do not require a pointer because the lines take the width of the diameter of the lead (Figures 3.11 and 3.12).

Colored pencils are not commonly used for sketching in the engineering field. One reason for not using colored pencils is that the purpose of the sketch is to quickly convey information, not to make a finished illustration. Also, some copy machines are

Figure 3.10 Line weight conventions used for finished drawings are also used for sketching.

Figure 3.11 Wood pencil and lead holders both require special pointers.

Figure 3.12 Fine-line pencils are manufactured in different sizes to produce given line widths and require no pointers.

"color blind" to certain colors such as blue or red. However, if color is important for communication, such as for comments or corrections, the use of color should be considered on analytical sketches created by the engineer. As the lines are darkened, remember that all the final lines should be dark so they are clear and copy well. The width of the lines on detail sketches should follow the accepted conventions for line widths. This means the object lines will be the widest, hidden lines will be of medium width, and lines such as center lines, dimension lines, extension lines, and so on, will be thin lines (Figure 3.10).

3.11 ENLARGING AND REDUCING USING A GRID

Several methods may be used to enlarge or reduce the size of a sketched image. One of the easiest methods of increasing or decreasing the size of the image, especially if the original was made on a grid, is to construct a larger or smaller grid and simply plot enough points to reconstruct the original image (Figure 3.13).

Figure 3.13 Given image presented on grid may be enlarged or reduced by simply altering size of grid.

3.12 *ENLARGING AND REDUCING WITH A RECTANGULAR DIAGONAL*

The **rectangular diagonal** is another useful aid for sketching. If an extended diagonal is drawn across opposite corners of a given rectangle, a proportionally larger or smaller rectangle can easily be constructed (Figure 3.14).

3.13 *ENLARGING AND REDUCING BY CHANGING THE FOCAL DISTANCE FROM THE OBJECT TO THE PLANE*

If you are sketching large buildings from a distance, many proportions may be transferred by holding the pencil at arms distance from the eye. This will help estimate the proportion of heights of buildings, bridges, or other similar large objects (Figure 3.15). The greater is the focal distance between the object and the picture plane, the shorter the height will appear on the picture plane (Figure 3.16).

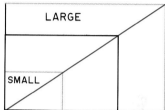

Figure 3.14 Rectangle may be enlarged or reduced by changing size of rectangle by using diagonal that is drawn through opposite corners of given rectangle.

Figure 3.15 Proportion of large object may be proportionately reduced by holding pencil at arms length and sketching shortened heights accordingly.

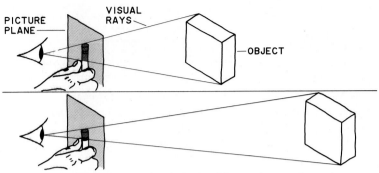

Figure 3.16 The further object is located from picture plane, the shorter height will appear on picture plane.

3.14 *SKETCHING SMALL OBJECTS WITH A HAND FOR PROPORTION*

A very common sketching technique employed when showing small objects is to show them in relation to a hand. For instance, an object pictured by itself may not be representative of its actual size. However, the association of size may easily be illustrated by showing a hand holding the object. Other familiar objects may be used in place of the hand such as a pencil or a scale. For large objects, for example, tractors or buildings, it may be helpful to sketch a human figure next to the object to indicate proportion (Figures 3.17 and 3.18).

Figure 3.17 Relative proportion of small objects may be represented by showing them with a hand.

Figure 3.18 Proportion of large objects may be represented by showing them with a person.

3.15   *SKETCHING SOLIDS*

You must learn to sketch two- and three-dimensional shapes accurately. One approach to sketching a manufactured product is to assume that all the components making up the shape of the assembly are simply a collection of outlines of various basic shapes. These shapes are comprised of cubes, rectangular solids, cylinders, cones, and pyramids.

The major concern when drawing cubes or rectangular solids is that the sides must appear parallel in the sketch (vertical lines are always vertical even though the angular lines may vary) (Figure 3.19). The cylinder also requires lines to be drawn parallel to each other. This is most easily accomplished by sketching layout lines that define the two ends of the cylinder and then connecting the end surfaces (Figure 3.20). For the cone and pyramid, the center lines of the base should be sketched first, then the altitude, and lastly the sides (Figure 3.21). Using these sketching layout procedures, it is easy to sketch such objects as a sander and drill press (Figures 3.22 and 3.23).

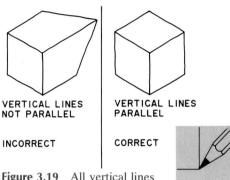

VERTICAL LINES
NOT PARALLEL

VERTICAL LINES
PARALLEL

INCORRECT

CORRECT

Figure 3.19   All vertical lines must be parallel or distorted view will result.

Figure 3.20   Define ends of cylinder, and connect ellipses with lines to complete cylinder.

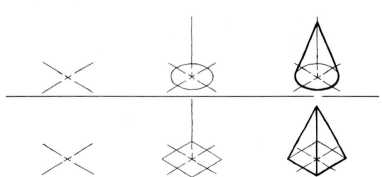

Figure 3.21   When sketching cone or pyramid, first construct center lines, then base and altitude, and finally sides of cone or pyramid.

Figure 3.22   Simple sketch of electric sander.

Figure 3.23 Making sketch: (1) block out outline, (2) block out major features, (3) refine shapes, and (4) add details and darken.

A good exercise to improve sketching techniques is to sketch basic shapes over and over until you are comfortable with the steps involved in the process. You will also develop speed through this practice.

Figure 3.24 Sketch of drill press with shading.

3.16 *FOUR STEPS TO SKETCHING AN OBJECT*

There are four major steps to follow when making a sketch of most objects or manufactured products. The first step is to *block out the outline* of the object or assembly to be sketched with light construction lines. This includes orienting the view. For instance, you may outline the object in isometric or perspective before drawing a detailed sketch. Proper orientation is very important in order to show the most descriptive view of the object in the front view.

The second step when sketching a product is to *block out the major features* of the product. This is generally accomplished by sketching light construction lines to outline the major shapes in the form of solids (cubes, rectangular solids, cylinders, cones, or pyramids) or variations of these and other solid shapes.

The third step is to *refine the shapes* already sketched. This step takes into account particular details such as rounded edges, sharp edges, finished and cast surfaces, and so on.

The fourth and last step is to *darken* only the layout lines you want to appear in the final sketch. This should be done with black lines, in accordance with the accepted line widths (Figures 3.23 and 3.24).

3.17 *WHERE DO IDEAS COME FROM THAT REQUIRE SKETCHING?*

There are many reasons for making engineering sketches. The major reason is to produce a quick graphic picture of an idea or object to convey to another person or group. One might ask: From where does the original idea or concept come? There are many sources from which you should be able to make a sketch of an object or product. The normal classroom procedure of sketching one or more views from a view given in a textbook is totally unrealistic when compared to the real world. After a sketch has been made using some of the proportional techniques discussed, additional information in the form of dimensions or notes may be added (Figure 3.25).

One source for the original image is from your mind. Another source is from a picture or technical drawing. This happens frequently in the classroom. Product modification is a good example of a real-world experience requiring a need for sketching. A third source is from a photograph of the object. Recall that translucent paper or film can be placed over the surface of the photo in order to make a refined sketch. Alterations can easily be made to the sketch of a photograph. Another source is a model or prototype of the object. Now you are asked to sketch from a three-dimensional object, which means you must visualize and be able to draw what you see. Making a sketch while looking at the object is essential. Sketches needed to make product modifications or alterations to products are very important. You should be able to make a better graphic image if the actual object is within sight during the sketching process.

Another method of receiving information from which a sketch is to be made is from a written description. This means you must visualize from a written description without the aid of any visuals such as the product, photos, or drawings.

Often the information required for a sketch is given by a verbal description. In the real world a person receives information over the telephone daily and often has to make a sketch representing this information. Unfortunately, this skill is rarely developed in the classroom. Some helpful tips include first orienting either top, bottom, left, or right (or north, south, east, or west). Also, have the person first briefly describe the entire object to be sketched before you even try to draw a line. This enables you to visualize the general concept of the object. Then have the person describe the parts by major components before refining the details. After you have sketched the item, take the time to describe your sketch back to the originator in order to check that you correctly sketched the features required. It would also be appropriate to send a copy of the sketch in the mail so that you and your correspondent are working from the same picture (Figure 3.26).

There are several descriptions provided at the end of the chapter for practice. These exercises require you to make sketches using many of the methods described in this chapter.

NOTES:

NO. REQ'D: 15,000
MATERIAL: PLASTIC
COLOR: RED

Figure 3.25   Dimensions and notes may be added to sketch.

Figure 3.26   Sketches are often made from verbal information given over phone.

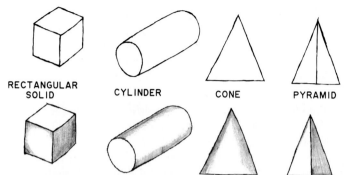

Figure 3.27 Shading techniques applied to basic solids.

3.18 BASIC SHADING OF SOLIDS

Shading is the process of indicating surface texture, shape description, or depth by adding certain sketching patterns. Seldom will the engineer take the time to shade various parts being sketched. However, if such basic shading is necessary, the rectangular shapes should be shaded showing a light, medium, and dark surface. A cylinder will be darker on the upper and lower edges of the length of the axis view. A similar treatment is applied to the cones, except the shading gets narrower toward the point. On the pyramid, the two adjacent surfaces should be shown by the contrast between medium and dark shading. A suggestion for making the shading look more uniform is to dull the point of the pencil before trying to shade large surfaces. The development of shading technique, as all parts of the sketch, requires practice. Daily observation of objects in the sunlight or in light areas can aid in this process (Figure 3.27).

There are several construction methods for determining shades and shadows on a drawing. However, these techniques are normally considered technical illustration rather than engineering sketching and will not be addressed.

3.19 QUICK ORTHOGRAPHIC PROJECTION

Sketches made in **orthographic** projection are generally the top, front, and right side views of an object. However, the back, bottom, and left side views may have to be sketched occasionally. The front view is the most critical. It should be the most characteristic view of the object. The front view indicates the **height** and **width** dimensions. The right side view is projected directly to the right of the front view and indicates the height and **depth** of the object. The top view is projected directly above the front view and indicates the dimensions for width and depth (Figure 3.28).

Before sketching an object's front, top, and right side views, first determine the best direction to view the front view. Then sketch the front view (Figure 3.29).

This is a sketch, so follow all four steps of the sketching process already described. First, outline the extremes of the box, which will include the entire front view, with light layout lines. Then sketch a layout of the major features. Next sketch a layout of the minor details. Lastly, darken the lines that should remain on the final sketch remembering to employ the correct line widths. You must develop these lines for all the required views. Both visible and hidden surfaces and edges are represented when sketching orthographic views.

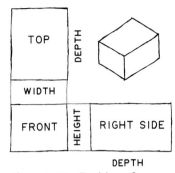

Figure 3.28 Position of commonly shown views in orthographic projection.

3.20 *QUICK ISOMETRIC PICTORIAL SKETCH*

The most common pictorial view to sketch is an **isometric** view. This is a sketch having a vertical axis and two axes at 30° from the horizontal. The axes shown are located at the near bottom corner of the object. When sketching, height dimensions are sketched along or parallel to the vertical axis, and the width and depth dimensions are sketched on or parallel to each of the two angular axes. Remember, the four steps still apply to making the sketch.

There are a few differences and similarities when drawing isometric views. Note that the parallel lines of rectangular surfaces are still parallel, but they are not 90° to each other. However, they are parallel to one of the three original axes lines (Figure 3.30).

The axes for circles drawn in isometric are parallel to one of the three isometric axes. This means the circles will appear as ellipses. Note the orientation of the long and short distances across

Figure 3.29 It is important to select best position for front view.

Figure 3.30 Isometric axes are vertical and 30° above horizontal (to left and right).

Figure 3.31 When sketching ellipses (circles shown on isometric planes), it is important to have ellipses oriented correctly. This is done by orienting center lines for each ellipse parallel to axes of that isometric surface.

the ellipses when they are drawn on the three isometric planes (Figure 3.31).

The four steps for making a sketch are still in effect. Sketch the outline box first using the extreme height, width, and depth. Then the major solids are sketched using layout lines. The details are then added. Lastly, darken (to black) the lines you want to appear in the final sketch. Only the visible lines are darkened when making an isometric pictorial sketch.

3.21 *QUICK OBLIQUE PICTORIAL SKETCH*

An oblique pictorial view is quickly drawn and shows one of the principal views (front, top, or side view) in the plane of the paper. This means two of the principal dimensions (height and width if the front view is drawn in the plane of the paper) will be sketched on the horizontal and vertical axes. The depth dimension will be sketched on an angular axis slanting away from the object at a selected angle (often 45°). When most of the detail is on a single surface, an oblique sketch can be effective, especially if most of the arcs or circles fall on this plane. In this case, the front surface is sketched as a front view; then the depth axis is added, at 45°. Examples of objects that could be appropriately sketched in oblique would be a set of kitchen cabinets, a desk, a speedometer, or a gauge. When sketching in oblique, the same four steps are applicable. It is common practice to darken only visible lines when sketching oblique pictorials. The oblique sketch may be easier when only rectangular grid paper is available. The normal surface can be sketched accurately as can be any selected receding axis (Figures 3.32–3.34).

Figure 3.32 Depending on desired representation, receding axis of oblique sketch may vary from 15° to 75°. Most common receding angle for oblique sketching is 45°.

Figure 3.33 Oblique sketch of desk.

Figure 3.34 Oblique sketches of some gauges.

3.22 SPECIFIC ENGINEERING APPLICATIONS

A typical engineering problem follows. The problem is done on engineering problem paper, and the information on top of the page gives the class, the student's name, the date, and the problem number. The problem itself contains the given information, including a sketch, a description of what is required, the analysis with equations, some sketches, and the solution (Figure 3.35).

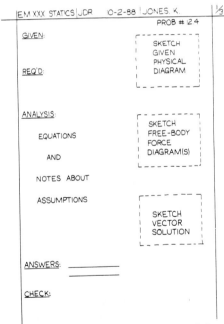

Figure 3.35 Common layout used for stating and solving engineering problems.

Some examples of analytical sketches relating directly to specific engineering disciplines are illustrated. These represent the type of sketches you might be asked to produce in industry or on the job site (Figures 3.36–3.43).

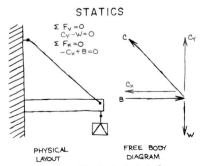

Figure 3.36 Typical sketch for general engineering problem.

THREE–DIMENSIONAL VIEW OF PROBLEM

TWO–DIMENSIONAL VIEWS OF PROBLEM

FREE BODY DIAGRAMS

VECTOR SOLUTIONS

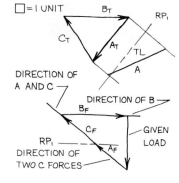

Figure 3.37 Three-dimensional statics problem.

STRENGTH OF MATERIALS

MACHINE DESIGN

Figure 3.39 Sketch of typical mechanical engineering application.

FLUID MECHANICS

Figure 3.38 Sketch of typical engineering application.

Figure 3.40 Mechanical, chemical, or civil engineers might use this type of sketch.

TYPICAL MICROSTRUCTURE

Figure 3.41 Metallurgical engineers sketch microstructures of materials.

CIRCUTS

DYNAMICS

D – DRAG
L – LIFT
T – THRUST
W – WEIGHT

Figure 3.42 Electrical, welding, and industrial engineers often sketch circuits.

Figure 3.43 Aeronautical engineers typically sketch effects of forces acting on aircraft.

3.23 *SUMMARY*

Sketched images generally show what the viewer visualizes rather than technical details. Often the engineer does not have time to make detailed engineering drawings so the sketches are given to a drafter who will make the technical drawings. Therefore, the engineer must be able to sketch clearly and accurately. Also, the engineer may be required to sketch in the field, on the job, or in a meeting in order to better communicate ideas or concepts to others. Engineering sketching is an extremely useful skill when used to make quick graphic images to help enhance communication.

PROBLEMS

Problems 3.1–3.20
These problems are drawn in isometric on an isometric grid. You are to sketch these objects in orthographic (draw the top, front, and right side views) using ¼-in. square grid paper. Each space on the isometric grid equals one square on your grid paper. Leave two spaces between the views. *Note:* Problems 3.1 through 3.5 have only normal surfaces, problems 3.6 through 3.10 also have inclined surfaces, problems 3.11 through 3.15 have oblique surfaces, and problems 3.16 through 3.20 have curved surfaces.

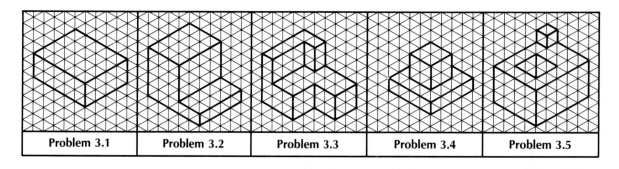

| Problem 3.1 | Problem 3.2 | Problem 3.3 | Problem 3.4 | Problem 3.5 |

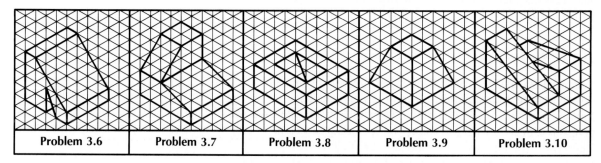

| Problem 3.6 | Problem 3.7 | Problem 3.8 | Problem 3.9 | Problem 3.10 |

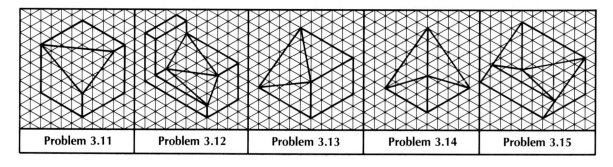

| Problem 3.11 | Problem 3.12 | Problem 3.13 | Problem 3.14 | Problem 3.15 |

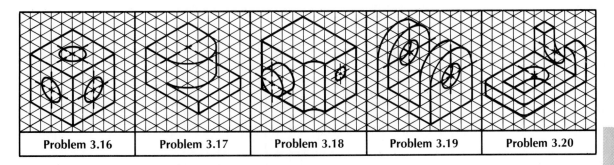

| Problem 3.16 | Problem 3.17 | Problem 3.18 | Problem 3.19 | Problem 3.20 |

Problems 3.21–3.40

These problems are drawn in orthographic on square grids. You are to sketch these objects in isometric on isometric grid paper. *Note:* Problems 3.21 through 3.25 have only normal surfaces, problems 3.26 through 3.30 also have inclined surfaces, problems 3.31 through 3.35 have oblique surfaces, and problems 3.36 through 3.40 have curved surfaces.

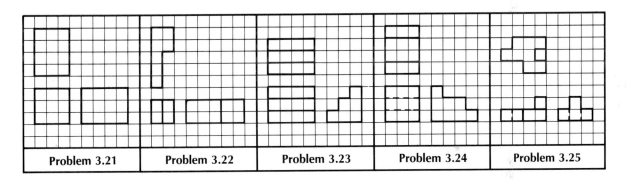

| Problem 3.21 | Problem 3.22 | Problem 3.23 | Problem 3.24 | Problem 3.25 |

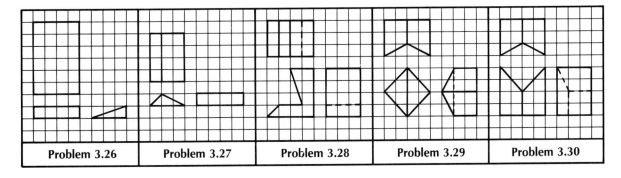

| Problem 3.26 | Problem 3.27 | Problem 3.28 | Problem 3.29 | Problem 3.30 |

Problems 3.21–3.40 (*continued*)

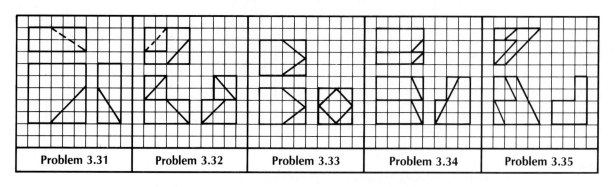

| Problem 3.31 | Problem 3.32 | Problem 3.33 | Problem 3.34 | Problem 3.35 |

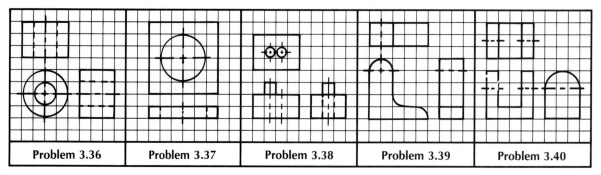

| Problem 3.36 | Problem 3.37 | Problem 3.38 | Problem 3.39 | Problem 3.40 |

Problems 3.41–3.45
Sketch these photographed objects in isometric and add simple
shading to the appropriate surfaces of each object.

| Problem 3.41 | Problem 3.42 | Problem 3.43 | Problem 3.44 | Problem 3.45 |

Problems 3.46–3.50
Sketch these photographed objects in isometric and add a sketch of
your hand holding each of these objects.

| Problem 3.46 | Problem 3.47 | Problem 3.48 | Problem 3.49 | Problem 3.50 |

Problems 3.51–3.55
Sketch these photographed objects in isometric.

| Problem 3.51 | Problem 3.52 | Problem 3.53 | Problem 3.54 | Problem 3.55 |

Problem 3.56
Sketch in isometric a cube (length of sides X) that has a cylinder
extending from the center of each of the six faces of the cube. The
axis of each of the cylinders is $2X$, and the diameter is $\frac{1}{2}X$.

Problem 3.57

Sketch in isometric a hand-held remote control channel selector for a television or VCR. This is to be a rectangular solid with a width of $4X$, depth of $2X$, and height of X. A rectangular window appears at the top, and three rows of three keys each are slightly raised from the top surface of the rectangular solid.

Problem 3.58

Sketch a cylindrical salt and pepper shaker in isometric. They are both the same diameter and have glass on the lower two-thirds. A shiny metal cap covers the upper third of each shaker. Sketch some holes in the top of each shaker. Both shakers are one-third full of salt or pepper. Sketch the salt shaker upright and the pepper shaker on its side.

TERMS YOU WILL SEE IN
THIS CHAPTER

PRINCIPAL PLANE   Plane in space that captures a view or is used for reference; the three principal ones are frontal, horizontal, and profile.

LIMITING ELEMENT   Boundary line of a curved surface shown in the noncircular view.

INCLINED SURFACE   Plane surface not parallel to any of the principal planes but perpendicular to one of them, where it appears as an edge.

OBLIQUE SURFACE   Plane surface not parallel or perpendicular to any of the principal planes.

CONVENTION   Accepted practice that violates the strict rules of graphics in order to improve clarity.

FILLET   Interior rounded corner on an object.

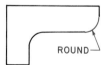

ROUND   Exterior rounded corner on an object.

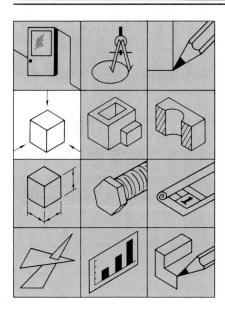

4

ORTHOGRAPHIC PROJECTION

4.1 INTRODUCTION

Engineering graphics is the "language" of the engineer. It is through graphics that engineers communicate ideas to other engineers and technical people. Just as English or Spanish has rules of grammar that must be followed, the graphics language also has rules. English and Spanish have exceptions to some rules to avoid confusion. This is also true of the graphics language.

4.1.1 Communicating in Three Dimensions

The world we live in is a three-dimensional world. All the objects we see every day have width, depth, and height. The world and all the objects in it are three-dimensional. The engineer must know how to work and think in three-dimensional space.

How can we represent the width, height, and depth of objects on a two-dimensional medium such a sheet of paper or the computer screen? Over the centuries, engineers have developed a way in which objects can be represented on paper and now on the computer screen. This method is called orthographic projection. It is the the international language engineers use to communicate their thoughts and ideas in the design of real-world objects to enhance all of our lives. Orthographic projection is

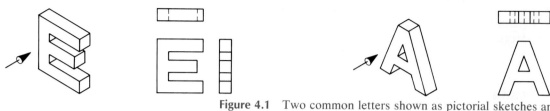

Figure 4.1 Two common letters shown as pictorial sketches and in orthographic projection.

Figure 4.2 Orthographic projectors, straight and parallel.

truly international in that an engineer in the United States can understand the shape description provided with orthographic projection methods by an Italian engineer. They may not understand each other in normal conversation, but the drawing "speaks" a common language understood by all who have been trained in the techniques and procedures of orthographic projection (Figure 4.1).

The term *orthographic* comes from the Greek words *ortho* meaning straight and *graphos* meaning written. Orthographic projection means that the object is described using straight parallel projectors (Figure 4.2).

4.2 THEORY OF ORTHOGRAPHIC PROJECTION

There are several methods of projection. They all require an object, an observer, a picture plane, and projectors from the object to the observer.

Figure 4.3 shows an object in space with an observer at a known distance from the object. There is also a picture plane between the observer and the object. Projection lines extend from the object through the picture plane to the observer. A projected image is formed on the picture plane. Since the observer is at a known or finite distance from the object, the projection lines meet at the eye of the observer and the projected image on the picture plane is smaller than the actual object. This projected image is

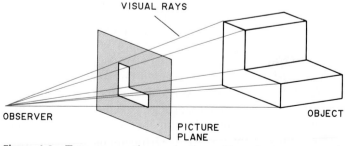

Figure 4.3 True perspective projection with projectors converging at observer's eye.

called a true perspective projection. The image on the picture plane is similar to a photograph of the object. The exact size and shape of the image are a function of

the distance between the object and the observer,

the angle between the object and the observer, and

the location of the picture plane.

A change in any one of the these factors will change the image. Forty people in a classroom looking at an object in the front of the room will each see it differently, for none of them will have exactly the same point of view.

Figure 4.4 shows the same object, picture plane, and observer with the observer located at an infinite distance from the object. The projection lines from the object through the picture plane going to the observer are now parallel. The image shown on the picture plane is not a perspective projection, but it is an orthographic projection of the object. The projectors, or "visual rays," as they are sometimes called, intersect the picture plane at right angles (perpendicular). The image on the picture plane is a one-to-one mapping of the points projected from the object. Every observer viewing this object from the front will be considered to be an infinite distance away. The image is now the same for all observers:

All are at the same distance from the object,

the angle of viewing is the same for all, and

the location of the picture plane does not affect the image.

Every observer viewing the object from the front will now have exactly the same image.

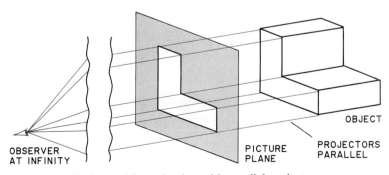

Figure 4.4   Orthographic projection with parallel projectors and observer at infinity.

4.2.1 Third-Angle and First-Angle Projection

Orthographic projection uses a horizontal picture plane and a vertical picture plane that are perpendicular to each other to capture images. Figure 4.5 shows these planes in their proper orientation. Think of the planes as being infinitely large and intersecting each other so as to divide space into four areas. Each of these four areas is referred to as a quadrant. When the object is placed in the first quadrant, it is between the observer and the picture plane. Projectors from the observer pass to the object and then to the picture plane to create an image. This projection method is referred to as first-angle orthographic projection. It is used in Europe.

When the object is placed in the third quadrant, the picture plane is between the observer and the object. Projectors from the object pass through the picture plane to the observer and show an image on the picture plane. This projection method is referred to as third-angle orthographic projection. Third-angle orthographic projection is the standard in the United States and Canada (Figure 4.6).

Third-angle orthographic projection will be used throughout this book. The top view will always be above the front view, right side views will be to the right of the front view, and left side views will be to the left of the front view. A first-angle drawing will have the top view *below* the front view, the right side to the *left* of the front view, and the left side view to the *right* of the front view. If you encounter a drawing that appears to be first-angle projection, check for the international symbol for projection (near the title block). Study Figure 4.7 and note the little symbol at the bottom of the drawing. When the small end of the cone frustum is to the right, it indicates third-angle projection; when the small end is to the left, it indicates first-angle projection.

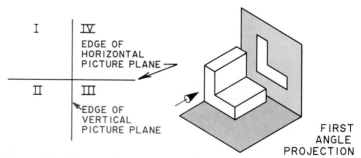

Figure 4.5 First-angle projection with object between observer and picture plane.

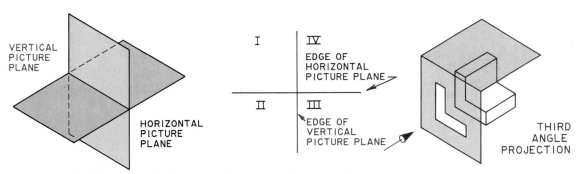

Figure 4.6 Third-angle projection with picture plane between object and observer.

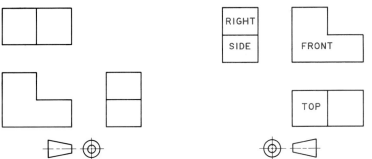

Figure 4.7 International symbol that identifies method of projection.

4.3 PRINCIPAL VIEWS AND PLANES

When using third-angle orthographic projection, you should imagine that the picture plane is always between you and the object and the projectors pass through the picture plane from the object to you. The projectors "map" the points of the object on the picture plane.

4.3.1 Front View and the Frontal Plane

Figure 4.8 shows a picture plane in a position that gives an image of the object showing its characteristic shape. This view of the object is called the front view, and the picture plane on which the view is projected is called the frontal plane.

4.3.2 Top View and the Horizontal Plane

Figure 4.9 shows a picture plane in a position that gives an image showing a "bird's eye" view of the object. This view of the object

Figure 4.8 Frontal plane with parallel projectors capturing front view.

Figure 4.9 Horizontal plane with parallel projectors capturing top view.

Figure 4.10 Profile plane with parallel projectors capturing side view.

Figure 4.11 Another position for horizontal plane; underneath object to capture bottom view.

Figure 4.12 Rear view with picture plane and observer behind object.

is called the horizontal view, and the picture plane is called the horizontal plane. Other names for this view include the top view and the plan view. Regardless of the name, the view is always looking down on the object; that is, the observer is imagined to be overhead with the horizontal plane between the object and the observer.

4.3.3 Side View and the Profile Plane

Figure 4.10 shows the picture plane in a position that gives an image of the object showing its depth. This view is called the side view, and the picture plane is called the profile plane. A side view may be drawn of the right side of the object or the left side of the object. Both views are considered to be side views.

4.3.4 Other Principal Views and Planes

Figures 4.11 and 4.12 show the picture plane in a position that gives an image of the object as viewed from below and from the rear, respectively. These views are called the bottom view, and the rear view and the picture planes are called the bottom plane and the rear plane. Notice that the bottom view and the rear view of the object are nearly the same as the top view and front view, respectively. The bottom and rear views are used only when there is interesting information on the bottom or rear of the object, which would be difficult to understand without these views.

The six views discussed in the preceding are called principal views. An object has six principal views: front, top, right side, left side, bottom, and rear. Only the views that best describe the object are used to make an orthographic drawing or sketch of the object.

4.3.5 Relation of the Principal Views

When the picture planes of the principal views are assembled in their proper position around the object, a projection box, or "glass box," is created. Each of the principal views is projected onto each of the **principal planes** (Figure 4.13). When the glass box is unfolded and laid flat on the paper, the orientation of the orthographic views can be seen (Figure 4.14). *This orientation is the only acceptable way to describe shapes using orthographic projection. Placing views of an object anywhere on the paper and*

Figure 4.13 Collection of planes that forms projection box, or glass box.

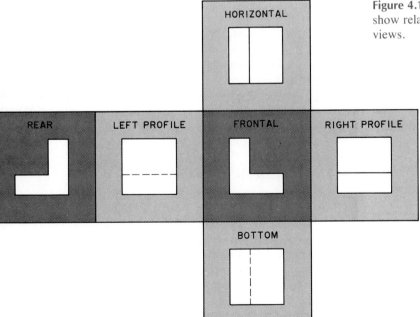

Figure 4.14 Projection box opened to show relationship of orthographic views.

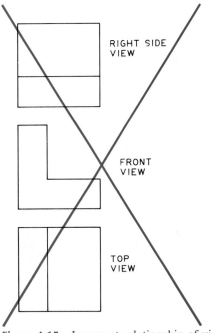

Figure 4.15 Incorrect relationship of views.

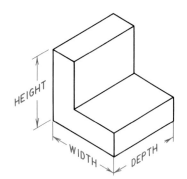

Figure 4.16 Principal dimensions that describe overall maximum size.

Figure 4.17 Principal dimensions on views where they appear.

labeling them is unacceptable, as shown in Figure 4.15. The relationship of the views and the orientation of the views in proper orthographic position is as important to the person "reading" the drawing as the actual shape of each view.

Later in this chapter you will be introduced to views that are not principal views. They are called auxiliary views; they also follow all the principles of orthographic projection.

4.4 PRINCIPAL DIMENSIONS

The principal dimensions of an object define the overall maximum size of the object. In other words, if you imagine that the object you are drawing just fits inside a box, the principal dimensions of the object would be the dimensions of the box (Figure 4.16).

The principal dimensions of an object are the overall width, height, and depth of the object. The width of the object is shown in the frontal and horizontal principal views. The height of the object is shown in the frontal and profile views. The depth is shown in the horizontal and profile views. Each principal view shows two of the principal dimensions. All of the principal views together describe the three-dimensional shape of the object (Figure 4.17).

4.5 NONLINEAR FEATURES

Complex objects often have nonlinear features described in their shape. Circles and arcs are nonlinear features that need special attention when drawing them orthographically. Figure 4.18 shows the orthographic views of a cylinder. The top view shows the cylinder as a circle, and the front view shows the cylinder as a rectangle. Simply looking at the front view does not describe the object. Both views together define the object as a cylinder.

4.5.1 Center Lines

A significant feature shown in the orthographic views of a cylinder or any circular object is the center line. Figure 4.18 shows the correct way to show a center line in the circular view and in the longitudinal view (the view where the cylinder resembles a rectangle). The center line is a thin black line and is broken as shown in Figure 4.18. The view in which the cylinder appears as a circle has the center lines intersecting to form a cross. This cross is the point where the end of a compass would be positioned in order to draw the circle. The center line shown in the longitudinal view of the cylinder is simply the axis of the cylinder; it may be called the axial center line. A cross is not shown in this view.

Parts of cylinders or arcs are shown in a similar way. Center lines are required if half a circle or more is shown. A quarter of a circle does not require a center line. The center lines used with arcs are the same as the center lines used with cylinders (Figure 4.19).

4.6 VISIBILITY

Visibility refers to what you can see when you look at the object. You will be able to see some parts of the object; other parts will be hidden from your eyes. When you change your location, some parts of the object that were hidden will become visible and some parts that were visible will be hidden. All features that can be seen in a view are shown with solid lines. The outline or boundaries of the view will always be shown with solid lines. Other features inside the outline that are visible in that view will also

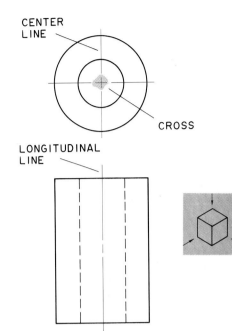

Figure 4.18 Center lines for cylindrical objects.

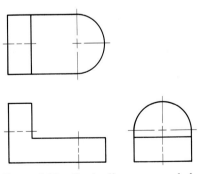

Figure 4.19 Center lines are needed for half a cylinder or more.

Figure 4.20 Pictorial view of concrete block.

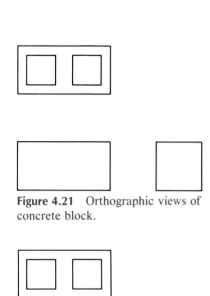

Figure 4.21 Orthographic views of concrete block.

Figure 4.22 Orthographic views of concrete block showing hidden details.

Figure 4.23 Pictorial view of hollow cylinder.

be shown with solid lines. Look at a concrete block (Figure 4.20). It has two rectangular holes in it. You can see the shape of these holes in the top view. You cannot see the holes in the front or side views (Figure 4.21). The holes are still there; this set of views does not tell you everything about this concrete block.

4.6.1 Describing Hidden Features

The next set of views gives you more information (Figure 4.22). Here you see dashed lines in the front and side views, which represent the holes. Dashed lines are used to show features of an object that the observer should know about but are not visible in a given view. In this case, they give you more information about the block. You now know that the holes go all the way through the block; that information was not available from the top view. This example demonstrates an old saying in graphics: In orthographic projection, when you have one view of something you have nothing. We need two or more views to describe a three-dimensional object completely in orthographic projection. Always remember that you are using a two-dimensional sheet of paper or computer screen to describe a three-dimensional object.

Now look at Figure 4.23. There is a hole shown in the top of the cylinder in the pictorial view. You do not know from this view if the hole goes all the way through or only part way through. (If it only goes part way through, you cannot see through the block. A hole that does not go all the way through is called a blind hole. A hole that does go all the way through is called a through hole.) The orthographic views without hidden lines (Figure 4.24) do not tell whether you have a blind hole or a through hole. The next figure (Figure 4.25) shows clearly that the hole goes only part way through the cylinder. The dashed lines here complete the description of the object.

4.7 *THE ALPHABET OF LINES*

All languages have alphabets. The English language has 26 letters in its alphabet. Russian, Japanese, and Arabic have much different alphabets. The graphic language of engineers also has its own alphabet, to which you were introduced in Chapter 1. At this time you will need to use only four parts of this alphabet. These lines are shown in Figure 4.26. Another kind of line will be described when you learn about section views.

4.7.1 Visible Lines

Visible lines are the most important lines in our alphabet. They are drawn as solid black lines. Drawing standards specify that the lines be 0.032 in., or 0.7 mm, wide. Mechanical drawing pencils and pens are specified in metric sizes and are available in the 0.7-mm width.

4.7.2 Dashed Lines for Hidden Features

Dashed lines are used to represent hidden features. They are drawn as $\frac{1}{8}$-in.- (3-mm-) long dashes separated by $\frac{1}{16}$-in.- (1.5-mm-) long spaces. They may be shortened a little on short lines and lengthened a trifle on long lines. They should be solid black lines 0.022 in., or 0.5 mm, wide. Standard mechanical pencils and pens are available in a 0.5-mm size.

4.7.3 Junctions of Lines

There are established practices for the junctions of solid lines and dashed lines. Solid lines should always meet completely with no gaps. Dashed lines should meet other dashed lines at the dashes; the junction should not be at an open space. Dashed lines should cross solid lines at a space if possible. When a dashed line is a continuation of a solid line, a gap should be left to show that these are different lines (Figure 4.27).

Figure 4.24 Orthographic views of cylinder.

Figure 4.25 Orthographic views of cylinder showing hidden details.

Figure 4.26 Partial alphabet of lines.

Figure 4.27 Line junctions for visible and dashed lines.

4.7.4 Center Lines

Center lines are used to show the center of circles and the axes of cylindrical objects. They are composed of $\frac{3}{4}$-in. (19-mm) dashes and $\frac{1}{8}$-in. (3-mm) dashes with a $\frac{1}{32}$-in. (1-mm) space between them. Center lines should be solid black lines 0.016 in., or 0.35 mm, wide.[1] Standard pens are available that are 0.35 mm wide; standard pencils are available with leads measuring 0.3 mm. The pencils require care when you use them; it is easy to break the fine leads. The symbol for center line is also shown on the alphabet of lines. Sometimes it is placed on a center line to call attention to it. The symbol may also be used as a shorthand way of writing center line in a document (₵).

4.7.5 Construction Lines

Construction lines are the lines you will use when you start a drawing and are planning the views graphically. They are made very light gray (just dark enough to be seen by the person making the drawing). They should be thin. You might use a 0.5- or a 0.3-mm mechanical pencil to draw them or a 6H wooden pencil or lead holder with a fine point. The original engineering documents never leave the engineering office, only copies are sent to a shop or construction site. If your construction lines are properly drawn, they will never show on a copy of the drawing. There is no need to erase construction lines on an engineering drawing unless it is a presentation drawing that would be shown directly to a client to explain a project.

4.7.6 Precedence of Lines

The graphics alphabet has a feature we call precedence. Some lines are considered more important than others. If they occur in the same place, one type of line can cover up another. Visible lines are the most important and take precedence over, or cover up, hidden lines or center lines. Hidden lines take precedence over center lines but not over visible lines. The views of the "inverted V" block in Figure 4.28 contain a line *PQ* that is the

---

[1]This text recommends the 0.022-in. width for hidden lines. The current American National Standards Institute (ANSI) standards specify only two widths of lines: thick, 0.032 in. (0.7 mm), and thin, 0.016 in. (0.35 mm). Note that these widths are not direct conversions between measurement systems but are specified separately in the standards. Your authors believe that pens available for computer drawing and the pens and pencils available for manual graphics make it relatively easy to provide more line widths. The larger alphabet of lines makes it easier to understand the drawings.

intersection of the two sloping sides inside the block. This line appears in the side view but not in the top view. It is coincident with line *NL* in the top view, and line *NL*, being visible, covers up line *PQ*.

The views in Figure 4.29 show some more examples of precedence. The horizontal center line of the circular hole is at the same elevation as line *EF*, so in the side views line *EF* is visible and covers up the center line of the circle. Note that it is desirable to show the ends of the center line extending beyond the boundaries of the side view in line with the side view of *EF*. Line *GH* is the intersection of two surfaces in the front view; it appears as a solid line in the left side view. The boundary of the circular

Figure 4.28 Inverted V-block; in top view visible line *NL* covers hidden line *PQ*.

Figure 4.29 Visible lines cover hidden lines and center lines; hidden lines also cover center lines.

hole is in line with *GH* and would appear as a dashed line in the side view. The solid line *GH* takes precedence, and you see the solid line *GH* but not the hidden boundary line of the circular hole. Line *CD* is coincident with the upper boundary line of the circular hole. Both would appear as dashed lines in the left side view, so one dashed line is shown in that view. In the right side view line *CD* would be visible, and the upper boundary line would be covered by the solid line $C_R D_R$. Line *EF* is dashed in the right side view to represent the hidden line; it takes precedence over the center line in this view.

Construction lines are always covered by visible, dashed, or center lines for they are on the drawing only for your convenience in building the views of the object. Hopefully, there will be no conflict on your drawings, and the construction lines will all disappear when the drawings are copied.

4.8 USING SUBSCRIPTS

The last time line *CD* was described, it was shown with subscripts ($C_R D_R$). This is standard practice when describing points on an engineering drawing. Points in the front view are labeled F, in the top view T, in the right side view R, and in the left side view L. When you label points, add the subscripts for each view in which the point is shown.

Figure 4.30 Line in space connecting two points.

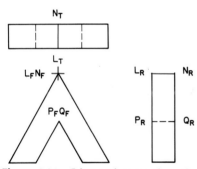

Figure 4.31 Line as edge view of plane surface.

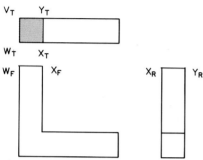

Figure 4.32 Line as intersection of plane surfaces.

Figure 4.33 Line as limiting element of curved surface.

4.9 *WHAT A LINE REPRESENTS*

As you look at a drawing, you will learn to "read" it just as you learned to read written language. A line on a drawing can represent several different features in our three-dimensional world. Look at Figure 4.30, and you will see points *A* and *B* connected by a line. This line would be called line *AB*. A line is the shortest distance between two points. You will use lines to connect points when you learn about descriptive geometry later in this book.

4.9.1 A Line as the Edge View of a Surface

A line can represent the edge view of a plane surface. Look at Figure 4.31, and you will see the flat, horizontal surface on the top of the L-shaped block. The corners are named *V*, *W*, *X*, and *Y*. In the front view there is a visible line $W_F X_F$. This line describes the flat, horizontal surface in the front view. The visible line $X_R R_R$ describes the flat, horizontal surface in the right side view.

4.9.2 A Line as an Intersection

A line can also represent the intersection of two surfaces. Look at Figure 4.32, and you will see the two **inclined surfaces** on the outside of the block meeting at the top. They meet in line *LN*. This line is the intersection of the two plane surfaces and appears as line $L_T N_T$ in the top view and as line $L_R N_R$ in the right side view. Sometimes a line can represent more than one feature. Look at Figure 4.31 again; the line $X_R Y_R$ represents the edge view of the plane *VWXY* in the right side view, and it also represents the intersection of plane *VWXY* with the vertical plane surface on the rising leg of the L-block.

4.9.3 A Line as a Limiting Element

A line can also represent the **limiting element** of a curved surface. Look at the cylinder in Figure 4.33. Trace the outline of the circular top view with your finger or a pencil point. As you go from the front of the cylinder to the rear, you go farther to the right (or left) until you reach a maximum distance from the center and then your pointer starts going back in the other direction. The point of change is the outermost point on the boundary of the circle. In the three-dimensional cylinder the outermost vertical line on the surface is called the limiting element. It represents the outermost vertical line, or element, that can be seen in the non-circular view. The colored vertical lines are the limiting elements in the front view. They are the outermost boundaries of this view of the cylinder.

4.9.4 What Can a Line Represent?

You have seen that a line on a drawing can represent several different ideas. It can be a line (a two-dimensional object). It can be the edge view of a plane surface. It can be the intersection of two surfaces, or it can be the limiting element of a curved surface. These possibilities are summarized in Figure 4.34. When you see a line on an orthographic drawing, it has to mean one or more of the following:

1. a line in space,
2. the edge view of a plane surface,
3. the intersection of two surfaces, and/or
4. the limiting element of a curved surface.

These are the only things a line can represent!

E - EDGE VIEW OF A SURFACE

N - INTERSECTION OF TWO SURFACES

L - LIMITING ELEMENT OF A CURVED SURFACE

Figure 4.34 What can a line represent?

Figure 4.35 Point in space shown by small cross.

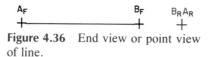

Figure 4.36 End view or point view of line.

4.10 *WHAT A POINT REPRESENTS*

A point on a drawing can represent a single point in space (Figure 4.35). It can also represent the end view of a line. Look at Figure 4.36. The right side view of line AB is $B_R A_R$, a point that is the end view of the line. An end view of a line is sometimes called the "point view." You can see this on a three-dimensional object in Figure 4.37. The lines LN and PQ, which appear as lines in the top and side views, are points in the front view. Notice that line PQ is not visible in the top view. It is covered by line $L_T N_T$, which is visible in this view, and by the rules of precedence, the visible line covers the hidden line. Look at $P_R Q_R$ in the side view. It is shown as a dashed line for it is hidden. The dashed line can be seen for there is no visible line to cover it.

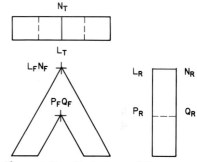

Figure 4.37 End views of lines on object.

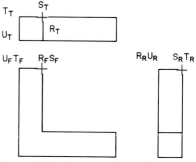

Figure 4.38 Intersection points on object.

4.10.1 A Point as the Intersection of Lines

A point can also represent the intersection of lines. Look at Figure 4.38. Lines RS and ST intersect at point S. This point appears in all three views. If you look carefully, you will see that a vertical line on the object also passes through point *S*. So a point can represent the intersection of two or more lines.

4.10.2   What Can a Point Represent?

A point can represent three things: a point alone, the end view of a line, or the intersection of two or more lines. Every point on an orthographic drawing represents one or more of these three things.

4.11 *CREATING A MULTIVIEW ORTHOGRAPHIC DRAWING*

Engineering drawings should have enough views to provide all the information needed for fabrication or construction. For three-dimensional objects this means that more than one view is required. For example, when making a bolt, the designer must specify the length of the bolt and the shape of the bolt head (Figure 4.39). In a construction project there are normally a plan view and elevation views of a building to be built (Figure 4.40). When-

Figure 4.39   Two views are needed to describe bolt.

ever you look at an object, you must think about which view of the object will provide the most information. An automobile is normally viewed from the side. This is the view that gives the most information about the intended use of the vehicle. It is easy to see the number of doors and windows and determine whether it is a two-door sedan, four-door sedan, station wagon, or convertible. If all of these vehicles were viewed from the front, they would all appear to be the same. A view from the rear or the top of the car would help distinguish between the convertible, the station wagon, and the sedan but would not identify them as completely as the view from the side does (Figure 4.41).

PLAN VIEW

FRONT
ELEVATION

SIDE
ELEVATION

Figure 4.40 Three views describe outside of building. Courtesy of Timothy C. Whiting

Figure 4.41 Three views describe automobile.

4.11.1 Choice of Views

You should choose the most descriptive or characteristic view and use it as the front view. It is also good practice to select the front view so that the longest dimension of the object appears as the *width* in the front view. The other views required should then be chosen. These will usually be right or left side views and top or bottom views. A simple cylindrical object such as a can requires two views: one to show the circular shape and the other to show the length (Figure 4.42).

Figure 4.42 Cylinder can be described with two views.

Figure 4.43 Planning drawing takes quick sketch of views and some simple calculations.

4.11.2 Planning a Drawing

Before you start a drawing, you should plan how it is going to look and how it will fit on the paper. This is best done by sketching the views to be drawn and deciding how they will fit on the paper (Figure 4.43). This sketch may be made on scrap paper or the back of an old drawing. The object in Figure 4.43 is best described in a set of views that are wider than they are high, so the paper would be turned with the long dimension horizontal. If the dimensions of the views were as shown, the views alone would require 14 in. in width and 9 in. in height, and space is required between the views. What size paper do you have on which to draw? If it is 11 × 17 in., you can probably fit the drawing on the paper. What do you do if your paper is only $8\frac{1}{2}$ × 11 in.? The drawing will not fit the paper so it must be drawn less than full size (drawn to scale). Often-used scales are half size and quarter size—which is the best choice? Drawn half size, the width becomes 5 in. on the paper, the height becomes $2\frac{1}{2}$ in., and the depth would be 2 in. (note the numbers in color). Now will it fit on $8\frac{1}{2}$ × 11-in. paper? It appears that it will, and now we need to plan for the space between the views. An $8\frac{1}{2}$ × 11-in. sheet will have about 8 × 10 in. of usable space. Subtracting the space required for views from the total available space leaves 3 in. of width and $3\frac{1}{2}$ in. of height. You could distribute this space as shown by the circled numbers on the sketch. *Only now are you ready to make a drawing; the planning or layout step is important.*

4.11.3 Developing the Drawing

Tape your paper to the drawing surface and set up your straight edges for horizontal and vertical lines. If you plan to use a border line, draw it lightly first. Now use your circled dimensions to locate the principal dimensions of the object. Draw the box that will contain each view with light construction lines and put in the center lines for circular features (Figure 4.44a). Add light con-

Figure 4.44 (a) Block in views and center lines as first step in making drawing. (b) Construct circles and arcs before adding straight lines. (c) Add visible and hidden lines. (d) Finish drawing with lettering and notes.

struction lines for major features on each view; some features cannot be drawn directly on a view but must be projected from another view (e.g., tangent points in noncircular views must be projected from the tangent found in the circular view).

Using your compass or circle template, draw the circles and arcs that appear in each view (Figure 4.44*b*). Then draw the lines tangent to the circular arcs. Find the tangent points by constructing a perpendicular from the center of the arc to the tangent line. (It is much easier to fit a line to an arc than to try to adjust the center of an arc to fit a line.) Now check projections for any features not shown with construction lines. Add them. Start down the paper making thick horizontal lines for visible lines; then make thinner dashed lines for hidden features. Work across the paper making thick vertical lines and thin dashed lines. Draw any lines that are not horizontal or vertical. Check your center lines and draw them as thin lines with long and short dashes (Figure 4.44*c*). If you have a border line, now is the time to make it a wide, black line. The last task is to complete your title block and notes with neat lettering (Figure 4.44*d*).

4.11.4 Checking Your Work

Your drawing should be complete. Check it carefully by locating each feature in every view and making sure the features are in projection. In a work environment you would make a copy of the drawing to be checked by a more experienced drafter; this is usually called a "check print." The experienced checker examines the copy of your drawing and notes in red any needed corrections. When those changes are made, the drawing is rechecked before any copies leave the design room.

4.12 CREATING A DRAWING ON THE COMPUTER SCREEN

You will need a computer terminal or personal computer that has graphics capability and a software package, known as a computer-aided design and drafting (CADD) program designed to produce engineering graphics. There are many software programs available for engineering graphics. Some are intended for student use; these programs are fairly simple and will lead you through the steps in making a drawing much as you would draw with pencil and paper. Other programs were designed for professional drafters; these programs have many options and great flexibility. They take longer to learn to use but offer advantages in speed and features when you take the time to master them. In order to print your work, you will need a printer capable of graphics or a plotter primarily intended for graphics but that can also plot words or numbers.

4.12.1 Selecting a Grid

When using a CADD program to create a three view drawing, first determine the grid size to be used. If the dimensions are to be in inches, normally a 0.125-, 0.25-, or 0.50-in. grid is used. If you are working in millimeters, you would usually select a 10-mm grid. Then select one of the grid dots on the screen as the starting point, or origin (0.0.0.0). This is usually the first visible dot in the lower left corner of the screen. When you have chosen the dot and reset the origin, it is a good idea to mark that dot so you can easily find the origin again. This can be done with two crossed lines or a small polygon.

4.12.2 Selecting the Layers

Once the grid has been selected, it is helpful to decide how many layers will be needed for the drawing. A CADD program allows you to draw on many different layers. You can view a single layer or many layers at one time. Simple drawings may be plotted on one layer. A simple plan for more complex drawings is to put construction lines on one layer and all the visible lines on a second layer; hidden lines may be drawn on a third layer. This allows you to keep the construction lines but gives you the option of omitting them from the printed or plotted drawing.

4.12.3 Selecting Pen Widths and Colors for a Plotter

You or your instructor will need to select the plotting colors, line styles, and line weights that will be used for a drawing. Normally a single color, such as dark blue or black, would be chosen to be used with the various line weights. This primary color can be represented on the screen by white if you are working on a monochrome CADD system. If colors are available, you should choose those that can be copied on a color copier.

Examples of colors, assuming a black background, that might be used and their accompanying line weights are:

Monochrome (Optional Weight)	Color	Use Application	Thickness (mm)
Solid/pattern (medium)	Green	Reference planes/lines, hidden lines, lettering	0.5
Solid (thin)	Red	Center, dimension, and extension lines	0.3
Solid (thick)	Blue	Visible lines	0.7
Dot	Yellow	Construction lines	0.3

4.12.4   Developing the Drawing on the Screen

After you have selected the origin, layers, line weights, and colors, you can start with the construction lines that define the location and overall size of the view on your construction layer. First, draw one horizontal line all the way across the screen to establish a Y coordinate for the lower limit of the front and profile views. Then repeat this line at the appropriate Y coordinates to establish the height of these views and the front and back of the top or bottom views.

Construction Lines

Construction is best done with a line type or color that will not be used for the finished drawing. You might choose a dot line pattern or the color yellow as outlined in the preceding. Yellow shows up well on a black monitor screen but would not be easy to read on white plotting paper. Save the high-visibility colors (black, blue, red, and green) for lines that must appear on the finished drawing.

It is efficient to locate the centers of circles and arcs on your construction layer (Figure 4.45a). Block in all of your views and then switch to your visible line layer for circles, arcs, and lines. When you locate tangent points, it is a good idea to bring them back to the construction layer. They are construction information and should not appear on the finished drawing.

Visible Lines

Circles and arcs should be drawn before the straight lines that are tangent to them (Figure 4.45b). This allows lines of the correct length to be added to the drawing. Otherwise, you may find yourself putting in lines followed by arcs and circles and then

Figure 4.45 (*a*) Construction lines and center lines should be located on construction layer. (*b*) Construct circles and arcs on visible layer before adding straight lines. (*c*) Add visible and hidden lines; locate any tangent points. (*d*) Lettering and text complete drawing, usually on separate layer. Drawing can be displayed or printed without construction lines. Computer generated image.

having to replace the original lines. Most CADD programs provide a continuous readout of the cursor position and display the length of a line or radius once the initial point or center has been selected. These features will help you locate the tangent points and draw the tangent lines. If the program you are using has a zoom feature, you can increase the scale of the tangent areas and locate the tangents accurately. At this time you should add any detailed construction that you could not determine before (Figure 4.45c) and carry it back to your construction layer.

Finishing the Drawing

The next step is to add remaining visible and hidden lines. Your last operation is to switch to a lettering or text mode and add the title block information and notes to your drawing (Figure 4.45d). You may choose to put text on the same layer as visible lines or reserve a layer just for text and notes.

Alternate Methods for Layering

If you decide to use more than one layer for the finished drawing, the following plan may be used. The first, or zero, layer can be used for construction. The second can be used for the visible lines. The third can be used for hidden lines and text material, and the fourth can be used for center lines, dimension lines, leaders, and extension lines. This plan allows a single pen per layer for the finished drawing exclusive of the construction, but it does assume that one plotting color with multiple line widths is available.

Another plan you may find efficient puts the construction on the first layer, the visible and hidden lines on the second layer, the center lines on the third layer, and the leaders, dimension, extension lines, and text on the fourth layer. This allows you to concentrate on the actual drawing on layers 2 and 3 and to add the size description (dimensioning) later or as needed.

If the steps outlined in the preceding have been followed, you should be able to tell the CADD system to plot the second and third layers or second, third, and fourth layers. This omits the first, or zeroth, layer where you did the construction so there is very little cleanup required. If lines have been drawn an incorrect length, CADD systems normally require you to erase the original line and replace it with a new line of the correct length. Some systems allow the line to be "shortened" or "lengthened." When using multiple layers, you should check your drawing a layer at a time to ensure that you have plotted the correct lines on the correct layer.

Use of layers is important for many types of construction or fabrication drawings. For example, an architectural/engineering firm may plan a building by putting the basic plan for each floor on one or two layers. They can add a layer for each of the following functions: heating and air conditioning, plumbing, electrical, and furniture layout. This technique allows them to provide the basic building plan to each specialty contractor with only the information that contractor needs.

4.13 *READING SURFACES OF OBJECTS*

You have read about what lines and points represent on a drawing and about creating an orthographic drawing manually or on a computer. As you practice with this information, you will become skilled at making drawings. You will also be called upon to read or interpret drawings in classes and in your professional career. One important part of reading orthographic projections is being able to recognize various types of surfaces.

4.13.1 Surfaces That Are True Size and Shape

The object in the next series should be an old friend. It is the L-shaped block you have seen before. Now all of the points on the surface of the object are labeled in all views. You can name every point, every line, and every plane surface on the object. Look at planes *VWXY* and *EFGH* in the top view (Figure 4.46). Do they appear in true size? Yes, they do because they are parallel to the horizontal plane. All the lines in the plane are horizontal lines. They are all true length in the top view, and the whole surface is true size. How do these horizontal surfaces appear in the front and side views? They appear as edges. You see a single line that is the edge view of the entire plane surface. *If a plane surface appears in true size in one view, it is parallel to the picture plane of that view and it will appear as an edge in adjacent views.*

Now examine the L-shaped plane *WXEGCA* (Figure 4.47). It is true size in the front view; it appears as an edge in the top view and in the right side view. Planes *GHCD* and *XYFE* are true size in the side view (Figure 4.48); they appear as edges in the top and front views. As you read drawings in the future, look for the surfaces that are parallel to the principal planes. These surfaces will be in true size in one view and will appear as edges in adjacent views.

Figure 4.46 Horizontal surfaces in true size on top view.

Figure 4.47 Frontal surface in true size on front view.

Figure 4.48 Profile surfaces in true size on side view.

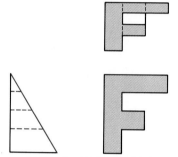

Figure 4.49 Inclined surface that is not true size in any principal view.

Figure 4.50 An *F*-block with inclined surface that shows its characteristic shape in two views and appears as edge in other view.

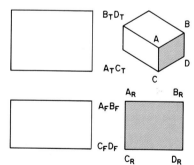

Figure 4.51 Rectangular prism with colored surface parallel to principal plane. Which plane?

4.13.2 Surfaces That Are Inclined

Figure 4.49 is another familiar shape, the inverted V. This time all of its points are named, so you can identify every point, line, and surface on the object. Surface *LNYX* appears as an edge in the front view, but it is neither an edge nor true size in adjacent views. It does appear as a rectangle in both the top and right side views. Plane surface *LYNX* is an inclined surface. It appears as an edge in one view and shows its characteristic shape in adjacent views. *It is not in true size in these views.* Now examine the views of the F-block (Figure 4.50). The inclined surface appears as an edge in the side view. You see the F shape in the front view and again in the top view. The characteristic shape is not in true size in either view, but you can identify the F in both views.

If a line represents the edge view of a plane surface, that surface will appear in true size in an adjacent view and as an edge again in the other principal view or that surface will appear in its characteristic shape in the other principal views.

4.13.3 Surfaces That Are Oblique

Look at Figure 4.51. First, you see a rectangular prism in part *a* that has a vertical plane on its right end. This surface, named *ABCD*, shows up as an edge in the top and front views. Because it is parallel to the profile plane, it shows up in true size in the right side view. In Figure 4.52 the right end of the block has been cut. Plane *EFGH* shows up as an edge in the front view and as a rectangle, but not true size, in the top and front views. Plane *EFGH* is an inclined surface. In Figure 4.53 the right end has again been cut. The cut is made so that no edge appears. Instead, we see the cut face in each of the principal views. Plane *JKL* is an **oblique surface** for it does not appear as an edge in any of the principal views. It does show up as a triangular shape in all views—none are in true size.

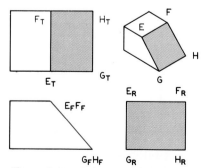

Figure 4.52 Prism with inclined colored surface.

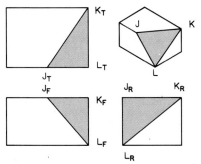

Figure 4.53 Prism with oblique colored surface.

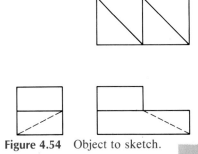

Figure 4.54 Object to sketch.

4.13.4 Reading by Sketching

The small pictorial sketches should help you to understand what has happened to the block each time. When you are reading orthographic drawings, it will help you to understand the object if you also make sketches of it. Start with the box that would contain the object and add lines to the sketch as you read each line of the drawing. You may try several positions for lines on the sketch before you get agreement with all the orthographic views. This is a part of the reading and visualizing process. Your eraser becomes as useful or more useful than your pencil when reading and sketching. Make a pictorial sketch of the object in Figure 4.54. If you start sketching it as the views are oriented in orthographic projection, you will find that the features at the right end of the object will not be visible in the sketch (Figure 4.55). At this time you would start a new sketch oriented similar to Figure 4.56. In this pictorial you can see what has happened to the right end of the block and still see most of the left end. When you brighten the lines of the sketch, you get Figure 4.57, which shows more of the object than the sketch started in Figure 4.55. Study this object. Can you find surfaces that are true size in the orthographic views? Can you find an inclined surface? An oblique surface?

Figure 4.55 First try: box that contains object and some of the lines illustrated.

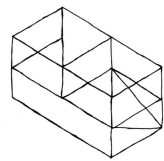

Figure 4.56 Second try: Turned around, it is easier to read.

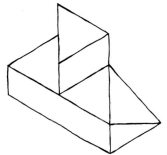

Figure 4.57 Completed sketch with all features visible.

Figure 4.58 Wedge with curved and tangent surfaces.

4.13.5 Curved and Tangent Surfaces

Some plane surfaces can also be identified as tangent to curved surfaces. Look at the wedge in Figure 4.58. The upper and lower surfaces can be described as being inclined and horizontal. They can also be described as being tangent to the cylindrical portion of the wedge. When a plane surface is tangent to a curved surface, you can locate the point of tangency in the circular view by drawing a perpendicular to the edge view of the plane from the center of the curve. This method will locate the intersection or tangent point for you, but *no line exists at this point*. The plane surface joins smoothly with the curved surface without a line of intersection. Look again at the top and side views of the wedge. There are no lines in these views in projection with the tangent point shown in the front view.

4.13.6 Conventional Practices

Figure 4.59 shows another kind of intersection with a curved surface. The vertical cylindrical part of the object meets a horizontal rectangular prism. The prism is as wide as the diameter of the cylinder, and the points of tangency can be seen in the top view where the vertical center line touches the visible edges of the prism. In the front view the edges of the prism end at the points of tangency. There are no lines of intersection; the contours blend smoothly. Now look at the small triangular part that joins the cylinder and the prism. It is not tangent, and you can see a visible line of intersection, $C_R D_R$, in the side view. You also see the intersection of the inclined face of the triangular block with the horizontal prism in the top view, $E_T F_T$. Note the colored lines on the front view. Think about what you have learned so far about orthographic projection. If the line *CB* were projected carefully from the top view to the front view, it would appear in the position of the colored line. It is drawn in black in projection with the limiting element of the cylinder. Is this a mistake? No, it is a **convention**. A convention is an irregular practice in the graphic language just as some verb forms are irregular in languages. Nothing would be gained by drawing the line in true projection in the front view, and it would take extra time of the drafter. Everyone familiar with the graphic language understands this convention, so you should use it when you make drawings.

4.13.7 Fillets, Rounds, and Runouts

The small triangular piece on the object in Figure 4.59 is called a **rib**. It is used to add strength to the object and to help maintain

Figure 4.59 Conventional treatment of rib intersecting cylinder.

Figure 4.60 Fillets, rounds, and runouts found on real objects make them stronger.

the vertical position of the cylinder. Sharp intersections on real objects are avoided when possible because forces that can cause fracture build up at sharp corners. If you were going to use this object in a machine, you could make it stronger by rounding the intersections of the rib with the cylinder and the prism. You would add **fillets** to these junctions—additional material that is rounded to make a smooth transition from one surface to another. When you do this, the views will change. Study Figure 4.60 and compare it with Figure 4.59. The sharp corners *CD* and *DE* are gone. Instead, you find smooth curves, fillets, joining the surfaces. The lines of intersection disappear in the top and side views. Instead, the parallel edges of the rib now end in **runouts,** which show that you have added fillets in the other view where you can see the curves. Another common practice used to reduce breakage is to round sharp outside corners. When you round these corners, the result is called a **round**. Note that in the new, improved version of the object one corner of the prism has been rounded. (The other corners have been left square because they will need to match with the surfaces of some other parts.)

4.14 *AUXILIARY VIEWS*

Many objects have surfaces that are not parallel to any of the principal planes. When you describe the shape of these objects with principal views, the surfaces not parallel to the principal planes will be distorted. They will not be in true size or shape. Some of the dimensions of the surface will appear shorter than they really are; they will not be in true length.

We can obtain a true-size view of such surfaces by assuming a point of view so that our line of sight and projectors are perpendicular or normal to the surface. The image can be captured

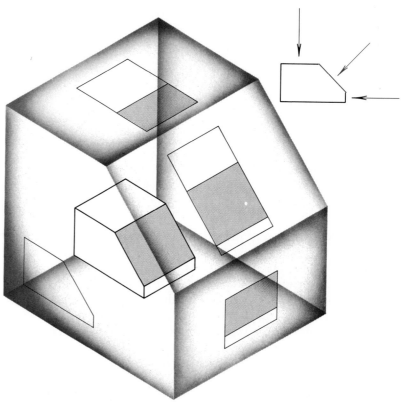

Figure 4.61 Projection box with inclined plane that captures image of inclined face of object in true size.

on a picture plane, which is not one of the principal planes. It will be parallel to the surface and perpendicular to the projectors. Such planes are called auxiliary planes. An infinite number of auxiliary planes are possible; in orthographic projection new planes must always be installed perpendicular to one of the existing planes.

Figure 4.61 shows an object with an inclined face in a projection box. Three of the planes are familiar to you. They are the principal planes: frontal, horizontal, and profile. One new plane has been added to the collection of planes, an auxiliary plane parallel to the inclined surface and perpendicular to the frontal plane. This plane is perpendicular to the line of sight and to the projectors that are normal to the inclined face. In the front view of the object you see the inclined face as an edge. You will also see the auxiliary plane as an edge in the front view, *parallel to the inclined face of the object*. Compare this picture to Figure 4.13, which introduced the projection box.

Figure 4.62 Projection box opened to show relationship of auxiliary view to other orthographic views.

When the collection of planes in three-dimensional space is opened up and laid flat on a two-dimensional surface, you see an arrangement such as Figure 4.62. The top view is still above and in projection with the front view. The side view is to the side of and in projection with the front view. The new auxiliary view is projected perpendicular to the edge view as the projection box opens up along the hinge line with the front view and the auxiliary plane lays out flat. Note that the front edge of the object is the same distance away from the hinge line in the auxiliary view as it is in the top and side views. The depth of the object is the same in all views. Compare the relationships in this view to Figure 4.14.

4.14.1 Preparing an Auxiliary View Drawing

Drawings with auxiliary views follow the same principles as orthographic drawings with principal views. The projectors are perpendicular to the picture planes, and the observer is assumed to

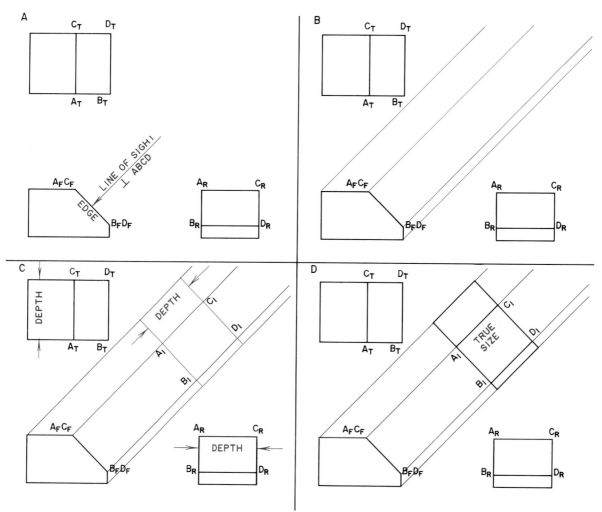

Figure 4.63 Construction of primary auxiliary view.

be at an infinite distance, so the projectors are all parallel for any one view. The picture planes for adjacent views (those views in projection with each other) are perpendicular.

Figure 4.63 shows the construction of an auxiliary view. You need at least two views to establish the size and location in space. In Figure 4.63 the surface we want to describe in true size appears as an edge in the front view, so we assume a line of sight perpendicular to this edge view. All the projectors for this auxiliary view will be parallel to this line of sight. The picture plane for the auxiliary view will be perpendicular to these projectors and parallel to the edge view of the surface we want to describe.

We select a convenient location for the front of the object in the auxiliary view just as you select a location for the top or side views at some convenient distance from the front view. Perpendicular to the projectors, we establish the line that represents the front of the object. Using dividers, we establish the rear of the object the same distance behind the front as it measured in the top or side views. This is the depth of the object. All points are located, and the visible lines are drawn as thick, black lines.

The relationships among the views are as shown in Figure 4.62, where the projection box was unfolded. When you prepare an auxiliary view, allow enough distance from the view from which you are projecting so that you do not overlap an existing view, such as the top or side view in this problem. A complete auxiliary view is not always needed; a partial view, showing only the surfaces or features that now appear in true size, will be adequate if other information is complete in the principal views. Auxiliary views may be projected from any existing view. You construct the view by projecting from an existing view and measuring from a view already in projection with the existing view.

4.14.2 Other Auxiliary Views

The views discussed in the preceding are primary auxiliary views. They are projected directly from principal views. We can also project auxiliary views from other auxiliary views as long as we follow the principles of orthographic projection. Successive auxiliary views are named in order as they are removed from principal views: primary, secondary, and tertiary. We can construct more and more views further removed from the principal views. However, it is difficult to maintain accuracy as we try to keep many sets of projectors parallel and to make perfect measurements. You will learn more about successive auxiliary views in Chapter 10, "Descriptive Geometry." This chapter will provide you with knowledge for solving problems in three-dimensional space.

4.15 *KEYS TO READING THREE-DIMENSIONAL OBJECTS*

Seeing a two-dimensional orthographic drawing with your eyes and reading this information in your mind as the three-dimensional object is a skill. People begin the study of graphics with different levels of this skill. Some people seem to see a three-dimensional object immediately when they look at an orthographic drawing

Figure 4.64 Visualization. Eye collects information; mind processes it to form three-dimensional image.

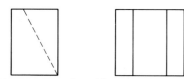

Figure 4.65 Can this object be visualized in three dimensions?

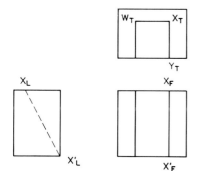

Figure 4.66 Read object by naming points and locating them in all views.

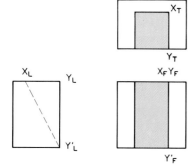

Figure 4.67 Read object by naming lines and locating them in all views.

(Figure 4.64). Others (most of us) have to practice this skill as they would any other, by starting with simple tasks and progressing to more difficult ones. The goal is to look at the multiview drawing and visualize the three-dimensional object in the mind. If this visualization does not occur immediately, there are methods and steps for organizing and interpreting the information that will lead you to the correct visualization of the object.

4.15.1 Identify and Locate Points on the Object

The first step is to identify *points* on the object and locate those points in all views. Look at the object in Figure 4.65. Perhaps you can visualize it right away, perhaps not. In either case, go through the steps of organizing the information in the given views. Start by naming points and locating them in all views. In Figure 4.66 point X is named in the top view. It is the intersection of lines WX and XY. Point X in the front view must be in projection with point X_T. It could be at X_F or at X'_F. Now look at the left side view. If you project from the front view, point X could be at X_L or at X'_L. Check the top view: X_T is not on the front of the object. It is about two-thirds of the distance from the front toward the rear. So X' cannot be the correct location for point X in the front or side views. You can name several points on the object and locate them by projection and measurement in the other views. Any locations that do not agree in all three views cannot be correct.

4.15.2 Identify and Locate Lines on the Object

A second step is to identify *lines* on the object and locate those lines in all views. Check Figure 4.67 and locate line XY in the top view. Line XY in the front view could be the point view $X_F Y_F$ or the line $X_F Y'_F$. Now look at the side view. Point Y is on the front of the object and line XY could be the horizontal line $X_L Y_L$ or the hidden, inclined line $X_L Y'_L$. Which is it? Either line could be the solution, for both points Y and Y' are in projection. How can that be? There are *two* lines that meet the requirements of the solution—there are *two* lines on the drawing, XY and XY'. In this example you followed a line on the object and found that you were locating two lines.

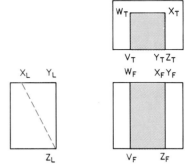

4.15.3 Identify and Locate Surfaces on the Object

A third method is to identify *planes* on the object and locate those planes in all views. Check Figures 4.68 and 4.69: point Y' has been renamed point Z and two other points on the colored plane named W and V, similar to points X and Z. These identify the surface you see in color in each view. Start with this surface in the top view. It resembles a small rectangle within the visible outline of the top view. Project this surface to the front view. It must lie between the two vertical lines inside the view, and it must include points below the top, horizontal surface of the object. If the colored plane were all on the top, horizontal surface, it would be one big surface because line XY and the other visible lines in the small rectangle would not exist. A line on a view of a three-dimensional object can represent only the edge view of a plane surface, the intersection of two surfaces, or the limiting element of a curved surface. Since the colored plane dips below the top plane, it must go all the way to the bottom. There are no breaks or lines between the top and the bottom. When you look at the side view, the only plane surface that could include points $VWXZ$ is the line XZ. This must be the edge view of the plane surface. (If you include point Y, you would have more than one surface or a warped, nonplanar surface.) In this drawing, line XZ in the side view represents the edge view of an inclined plane surface. This surface appears in its characteristic rectangular shape in both the top and front views. You can now identify more points in the side view, W_L and V_L.

You have seen Figure 4.70 before. It is an object with an oblique surface JKL. Note that JKL is a triangle in every view; look for such shape repetition to locate oblique as well as inclined surfaces on an object. Following plane surfaces on the views is a powerful method for reading a drawing. As you identify planes, you identify several points and lines at the same time.

Figure 4.68 Read object by identifying planes (coloring them works well) and locating them in all views.

Figure 4.69 Object in three-dimensional sketch with inclined plane in color.

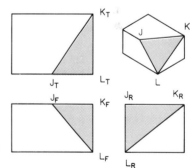

Figure 4.70 Shape repetition can help identify oblique surface.

4.15.4 Curved Surfaces and Limiting Elements

When you have curved objects, there is another type of line you can identify. This is the limiting element of the curved surface.

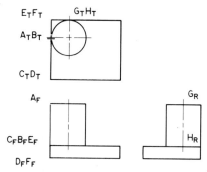

Figure 4.71 Limiting elements of cylinder; not same lines in every view.

Examine Figure 4.71. The colored line, line *AB*, appears in the front view. It is a limiting element of the cylinder. This line appears as a point in the top view. The line segment A_FB_F continues, and you see a line C_FD_F in the front view that resembles an extension of *AB*. It is not. These are two different lines. Line C_FD_F is the edge view of plane surface *CDEF*. Now check line *GH*. It is a limiting element of the cylinder in the side view and appears as a point in the top view. Where is *GH* in the front view? Where is *AB* in the side view? These lines do not appear in all views; they are elements of the cylinder and become visible lines only when they are the *limiting* elements or boundary lines of a view. Be careful when working with curved objects: The limiting elements are visible lines only in views where they describe the boundaries of the object.

4.15.5 Reading an Object by Sketching It

Another way to help visualize an object is to sketch a pictorial view of it. This method was described earlier in this chapter. It takes very little time and almost always helps you to understand a drawing. Begin by sketching lightly the pictorial "box" that would contain the principal dimensions of the object: width, height, and depth. Then sketch in parts of the object and remove those areas that are not a part of the finished object.

4.15.6 Reading by Modeling the Object

If you have used all of the methods described in the preceeding and still have difficulty visualizing an object, try modeling it in three dimensions. You can use plastic foam, soft wood, or modeling clay. The clay is recommended because you can use it over and over again. The principles are the same as sketching. Form a three-dimensional box that would contain the object, lightly scribe lines on the surfaces of the box to match lines on the drawing, and remove the pieces that are not a part of the object. If the object is mostly cylindrical, you can start with a cylindrical shape or join a cylindrical shape to another box. A pocket knife or putty knife and a pointed stick are the only tools you need to make readable models (Figure 4.72). Do not worry if your corners are not sharp and your plane surfaces not perfectly flat. These models should be aids to understanding; they need not be works of art (Figure 4.73).

Modeling on the Computer

You can also model the object on a computer. When you have learned to use your computer to make pictorial drawings, you

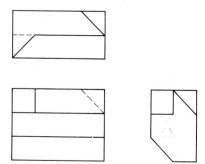

Figure 4.72 Object to model in clay, soap, or foam.

Figure 4.73 (*a*) Make block large enough to contain object. (*b*) Scribe lines from views on faces of block. (*c*) Cut along lines so that shapes on all faces remain same. (*d*) Clean up corners and surfaces with knife or pointed stick.

can "sketch" on the computer screen. Some programs will provide a pictorial grid so that you can easily construct the box that will contain the dimensions of width, height, and depth. As in manual sketching, you will want to place those lines on the pictorial drawing that match the multiview drawing and agree

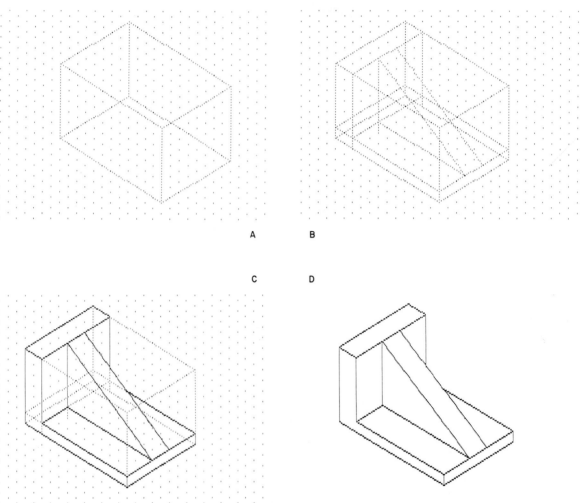

A   B

C   D

Figure 4.74 (*a*) Select grid for first layer and build box that contains object. (*b*) Use construction type lines to rough out shape.
(*c*) Switch to visible layer and add solid lines where should be; check agreement with views. (*d*) Turn off grid layer. You now have model of object on computer screen. Computer generated image.

with all of the given views. You can erase on the computer screen and try lines in a new position. It is a good idea to layer your pictorial drawing. The first layer could be the pictorial grid, the second layer the box that would contain the object and the trial lines, and the third layer the finished object (Figure 4.74).

4.16   *SUMMARY*

Orthographic projection is a fundamental part of the graphic language that all engineers and design professionals need to know. It is used to communicate with other technical people and with production and construction people. It is truly an international language.

Remember that the observer is an infinite distance away from the object; this makes all lines of sight (projectors) parallel. Projectors are all parallel to each other and perpendicular to the picture plane. The principal views include the front, top, and side; they may also include bottom and rear views. Auxiliary views may be needed to describe objects with surfaces not parallel to principal planes. Auxiliary views are always made with a line of sight perpendicular to the plane of the surface to be described. The new auxiliary picture plane will always be perpendicular to the line of sight and to one of the existing picture planes. It will be parallel to the surface to be described.

Visible features are shown with solid lines and hidden features with dashed lines. Center lines are shown for cylindrical objects if half a circle or more appears on one view. You should learn to read lines just as you learned to read the alphabet.

Creating a multiview drawing requires planning which views will best describe the object and how they will be positioned on the screen or paper. The overall dimensions of width, height, and depth should be blocked in as you start the drawing. All views should be developed together. Often, one view cannot be completed until intersections and other important information are plotted in another view.

Both reading drawings and making them require practice. Keys to understanding include visualization in three dimensions; identifying points, lines, and surfaces; and sketching in three dimensions with a pencil or on a computer screen. Modeling in clay or other material is also helpful. The problems at the end of this chapter will give you an opportunity to practice drawing and reading in orthographic projection.

PROBLEMS

Problems 4.1–4.12
Sketch the given views in each drawing. Sketch the lines that are missing in any of the views.

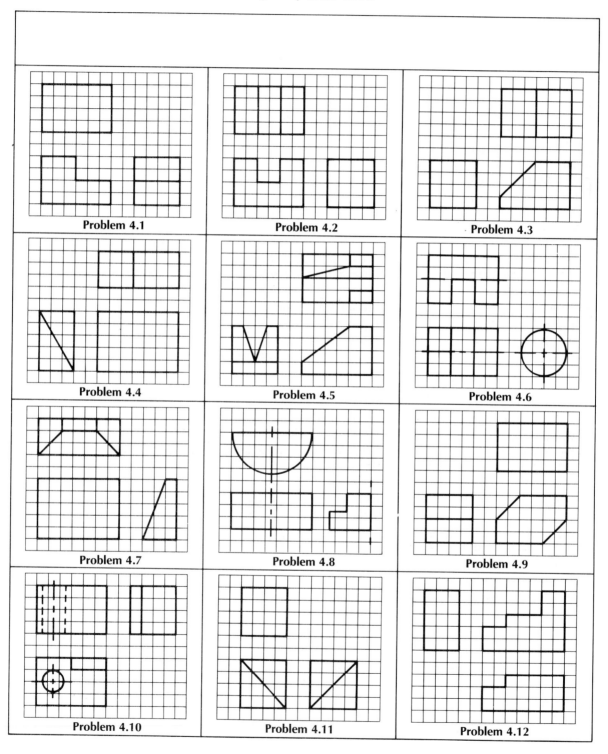

Problem 4.1

Problem 4.2

Problem 4.3

Problem 4.4

Problem 4.5

Problem 4.6

Problem 4.7

Problem 4.8

Problem 4.9

Problem 4.10

Problem 4.11

Problem 4.12

Problems 4.13–4.24

Sketch the given views in each drawing. Sketch the missing view as it
relates to the given views.

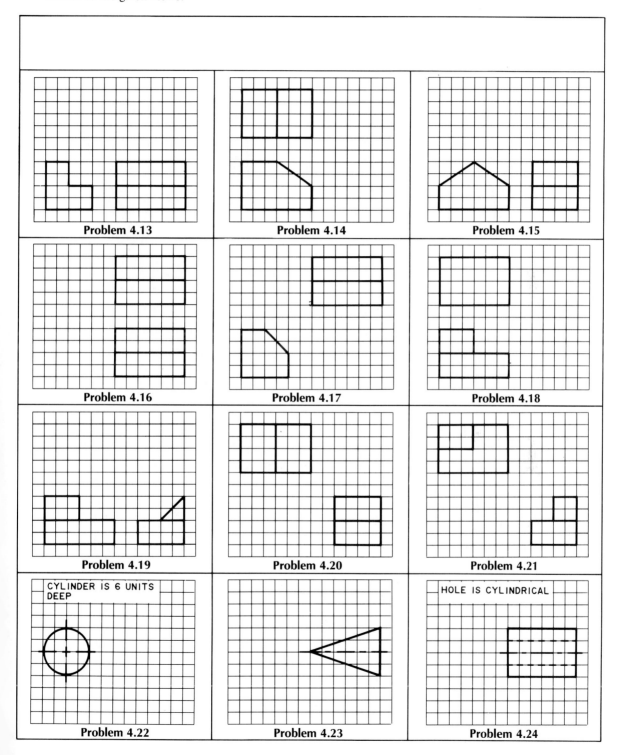

Problem 4.13	Problem 4.14	Problem 4.15
Problem 4.16	Problem 4.17	Problem 4.18
Problem 4.19	Problem 4.20	Problem 4.21
Problem 4.22	Problem 4.23	Problem 4.24

CYLINDER IS 6 UNITS DEEP

HOLE IS CYLINDRICAL

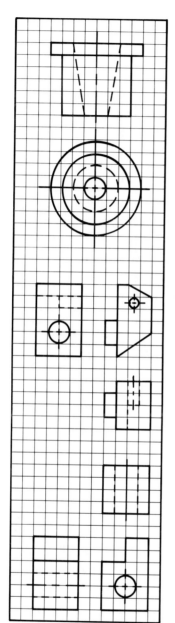

Problems 4.25–4.27
Read the orthographic drawings and sketch a pictorial of the object defined.

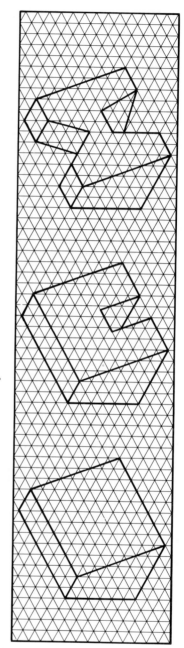

Problems 4.28–4.30
Sketch the top, front, and left side view of each problem.

Problems 4.31–4.32
Sketch a top, front, and side view of the object shown. Identify with color or code on each view the surfaces that are normal, inclined, and oblique.

Problems 4.33–4.41

Draw with instruments the given views. Complete the missing view as it relates to the given views.

Problem 4.33

Problem 4.34

Problem 4.35

13
25
69

.50
1.00
2.75

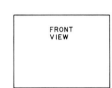

44
13 13

FRONT
VIEW

25

50

13

1.75
.50 .50

FRONT
VIEW

1.00

2.00

.50

Problem 4.36

13
25
44

.50
1.00
1.75

63
44
13

50

32
19
13

R.S.
VIEW

2.50
1.75
.50

2.00

1.25
.75
.50

R.S.
VIEW

Problem 4.37

TOP
VIEW

TOP
VIEW

100
32

63
32
20

100

75

32

4.00
1.25

2.50
1.25
.81

4.00

3.00

1.25

Problem 4.38

Problems 4.33–4.41
Draw with instruments the given views. Complete the missing view as it relates to the given views.

Problem 4.39

Problem 4.40

Problem 4.41

Problems 4.42–4.66

Draw three orthographic views of each of the pictorials shown. Specific problem sets are designed to give you practice with normal, inclined, and oblique surfaces as well as simple and complex curves.

Problem 4.42

Problem 4.43

Problem 4.44

Problems 4.42–4.66

Draw three orthographic views of each of the pictorials shown. Specific problem sets are designed to give you practice with normal, inclined, and oblique surfaces as well as simple and complex curves.

Problem 4.45

Problem 4.46

Problem 4.47

Problem 4.48

Problem 4.49

Problem 4.50

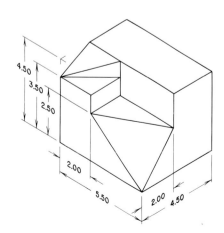

Problems 4.42–4.66
Draw three orthographic views of each of the pictorials shown. Specific problem sets are designed to give you practice with normal, inclined, and oblique surfaces as well as simple and complex curves.

Problem 4.51

Problem 4.52

Problem 4.53

Problem 4.54

Problem 4.55

Problem 4.56

Problems 4.42–4.66

Draw three orthographic views of each of the pictorials shown. Specific problem sets are designed to give you practice with normal, inclined, and oblique surfaces as well as simple and complex curves.

Problem 4.57

Problem 4.58

Problem 4.59

Problem 4.60

Problem 4.61

Problems 4.42–4.66

Draw three orthographic views of each of the pictorials shown. Specific problem sets are designed to give you practice with normal, inclined, and oblique surfaces as well as simple and complex curves.

Problem 4.62

Problem 4.63

Problem 4.64

Problem 4.65

Problem 4.66

Problems 4.67–4.70

Draw with instruments all the views required to describe the objects shown. An auxiliary view is required to describe the inclined surface.

Problem 4.67

Problem 4.68

Problem 4.69

Problem 4.70

TERMS YOU WILL SEE IN
THIS CHAPTER

PICTORIAL DRAWING   Graphic representation that shows features of width, height, and depth all in a single view.

ISOMETRIC DRAWING   Pictorial drawing that shows the horizontal, frontal, and profile planes in the same view, with the three planes 120° apart.

ELLIPSE   Closed symmetrical curve that does not have equal length radii.

FOUR-CENTER METHOD   Approximate method of constructing isometric ellipses.

IRREGULAR CURVE   Predetermined set of curves used to connect points on an irregular curve (curve that is not a function of a radius).

CAVALIER DRAWING   Oblique drawing where the width, height, and depth are all measured using full-length dimensions.

CABINET   Oblique drawing where the width and height are drawn at full length, but the depth measurements are drawn at half their dimensioned size.

PERSPECTIVE DRAWING   Pictorial view where width or depth lines are not parallel but converge at a vanishing point(s).

TWO-POINT EXTERIOR PERSPECTIVE   Specific type of perspective having two vanishing points that shows the exterior of an object.

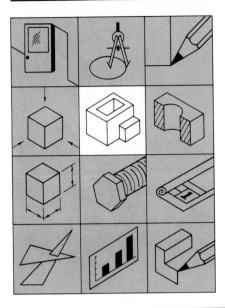

5

PICTORIAL DRAWINGS

5.1 INTRODUCTION

Pictorial drawings (Figure 5.1) make up a major portion of drawings used by the engineer. There are many types and variations of pictorial drawings since they must serve different purposes. Some of the major types of pictorials are axonometric drawings, oblique drawings, and perspectives. After you have learned to draw these pictorials, variations may be applied such as sectioning a pictorial or exploding an assembly.

ISOMETRIC

OBLIQUE

PERSPECTIVE

Figure 5.1 Samples of pictorial drawings.

Figure 5.2 Three visible normal surfaces have equal angles between them (120°) in an isometric drawing.

Figure 5.3 In dimetric drawings, two of the three visible normal surfaces have equal angles not equal to 120°.

Figure 5.4 Trimetric drawing has the three visible normal surfaces in a position where none of the angles between the three surfaces are equal.

5.2 *AXONOMETRIC DRAWINGS*

5.2.1 Types of Axonometric Drawings

The general term axonometric drawing is used within the technical drawing field to describe three types of pictorials: isometric, dimetric, and trimetric drawings. **Isometric drawings** are the quickest and easiest of all the pictorials to draw and the most commonly used. In an isometric drawing the three normal surfaces of a rectangular solid will have equal angles between them (120°) (Figure 5.2). In dimetric drawings, two of the normal surfaces will be equally spaced, but the third surface will have an angle of a different number of degrees (Figure 5.3). A trimetric drawing will have the three normal surfaces of the rectangular solid positioned so none of the three angles have the same number of degrees (Figure 5.4).

The drawing of dimetric and trimetric drawings takes more time because uncommon angles are often used. This means the axes for this type of pictorial drawing are not the same as those for isometric, and generally the three axes are a different number of degrees from each other. Surfaces at uncommon angles require you to construct all the ellipses for circular features or to have an expensive set of ellipse templates at hand. Since dimetric and trimetric drawings are seldom used in engineering work, they will not be discussed in detail. Instead, the isometric drawing will be covered in depth.

5.2.2 The Isometric Axes

There are four primary positions for placing the axes when starting an isometric drawing or a rectangular solid. You may elect to look down on the object (the most common) or elect to look up to the object (seldom drawn). Further, you may elect whether the long dimension will extend to the left or to the right of the near corner of the object (Figure 5.5).

The center axis is vertical in all cases, and the left or right axes vary from the horizontal by 30° in the chosen direction. One of the nice aspects of drawing an isometric drawing is that true length measurements may be made on or parallel to the three chosen isometric axes.

5.2.3 Normal Surfaces in Isometric

After the axes have been chosen and located, identify the overall width, depth, and height of the object to be drawn. Then measure these overall dimensions on the appropriate axes. From these

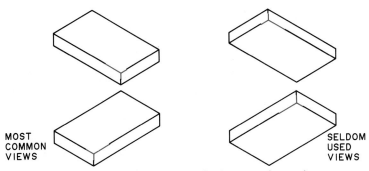

Figure 5.5 When drawing an isometric there are four primary positions for viewing the object. Looking down on the object with the long axis to the left or right are the most common views.

points you should be able to project the remaining corners of the rectangular solid that will contain the entire boundary surfaces of the object (Figure 5.6). It is important to first draw this box containing the extremes of the object and then remove desired portions in order to produce the pictorial representation of the object. You now have three normal surfaces on which measurements may be made, and you have the extremes within which the object must fit. Knowing this information, you should be able to draw objects having only normal surfaces by projecting and measuring on or parallel to the axes (Figure 5.7).

It should be remembered that with only rare exception, hidden lines are not used when drawing pictorials. You only draw the surfaces and edges of the object that your eye would see when viewing the object.

If normal surfaces are to be drawn within the box representing the extremes of the object, you will find it easy to measure on the box and project into the object. This is a good habit to get into because it will be an important key to drawing inclined and oblique surfaces later.

5.2.4 Inclined Surfaces in Isometric

If you need to draw inclined surfaces in isometric, you start by choosing and locating the axes. Then you measure the extreme width, height, and depth of the object and complete the rectangular solid representing the extremes of the object. Note that the entire box is drawn; then desired parts will be eliminated in order to produce the pictorial drawing of the object. It is best to measure the locations of the corners of the inclined surfaces on one of the outside three planes and then project and connect the points.

1. DRAW AXES
2. MEASURE W, D, & H
3. DRAW BOX VIEW LIGHTLY

Figure 5.6 Locate and measure the extreme width, depth, and height of the object; then project to complete the rectangular solid in which the extreme sizes of the object must fit.

Figure 5.7 Measure true length on one of the three original isometric axes or on any line parallel to these three lines.

Figure 5.8 Even though inclined surfaces to be drawn in isometric may be given in degrees, it is mandatory to locate the inclined surface by locating the points, not transferring the angle in degrees.

It is very important that you remember that angles given in degrees on an orthographic view may *not* be transferred to the isometric drawing by using the same number of degrees. This is because the surface of the pictorial is at a different angle to the picture plane than in the orthographic view; therefore, you must locate the points representing the corners of the inclined surface and then connect the points (Figure 5.8). In order to know the isometric line length representing the sides of the angle given in degrees in the orthographic view, you must first construct the angle in orthographic, measure the length of the lines in this construction, and then measure and locate the lines on the pictorial view.

When two inclined surfaces intersect, first draw one of the inclined surfaces using light lines. Then draw the second inclined surface by identifying points on the normal surfaces. The points

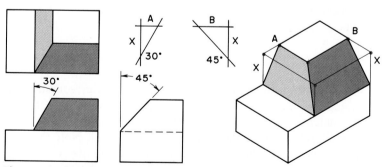

Figure 5.9 Intersecting inclined surfaces drawn in isometric must also be constructed by locating points from the orthographic views.

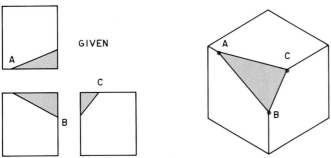

Figure 5.10 Oblique planes drawn in isometric must have their points located on normal surfaces or temporary construction on normal surfaces.

of the intersections can then be connected, and the correct lines may be darkened (Figure 5.9).

5.2.5 Oblique Surfaces in Isometric

Since an oblique surface is not parallel to any of the three normal surfaces drawn to outline the extremes of the object, all the points of the surface must be located on normal planes (or temporary construction of normal planes you must draw), and then the points must be connected (Figure 5.10).

To locate these points, you must make two measurements on each surface. On the frontal plane a height and width must be located. On the profile plane a height and a depth must be located, and on the horizontal plane the width and depth must be located. After the points have been located, then the lines outlining the oblique surface may be drawn by connecting the points (Figure 5.11).

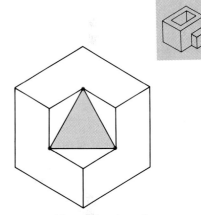

Figure 5.11 After points have been correctly located the oblique surface may be drawn.

5.2.6 Circles and Arcs Using an Isometric Template

When you draw a circle or arc on a normal surface of an orthographic drawing, you use a compass or a circle template because the normal surface is parallel to the plane of projection. In isometric drawing that plane of projection is not parallel to the three major planes of the rectanglar solid; therefore, the circle will not appear round. The circles will appear as **ellipses** (Figure 5.12). Although we draw the planes of the rectangular solid at 30° from the horizontal, these planes are technically 35° 16′ back from the horizontal. This means that you need a special template to draw isometric ellipses. It is called an isometric ellipse template, and

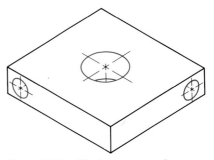

Figure 5.12 Circles on normal surface will appear as ellipses when normal plane is drawn in isometric.

Figure 5.13 Small isometric templates are easy to use and save time.

Figure 5.14 Set of large ellipse templates are more expensive and are used more as a specialty tool.

it is based on ellipses tilted back at an angle of 35° 16′ from the horizontal. The small isometric template will be helpful for drawing ellipses up to about 1.5 or 2 in. (38 or 50 mm) (Figure 5.13). Isometric ellipse templates are also available in metric sizes. If a larger ellipse is required, then a large ellipse template may be used (Figure 5.14).

The isometric ellipse template is easy to use, but you must orient the template correctly or the drawn pictorial will look drastically wrong. First look at the orientation of the ellipse on each of the three surfaces of the pictured rectangular solid (Figure 5.15). Note the direction of the long axis of the ellipse to the surface upon which it is pictured. You should be able to orient the template in the approximate position for drawing the ellipse on any of the three surfaces (Figure 5.16). Note also the position of the center lines for each of the three ellipses. The left and right planes have vertical center lines and one parallel to the 30° axis. The ellipse on the horizontal plane has two center lines parallel to the two 30° axes (Figure 5.17).

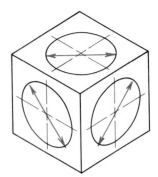

Figure 5.15 Note orientation of long axis for ellipses on each of the three normal surfaces of cube.

Figure 5.16 Very important to learn to correctly orient the isometric template in order that the ellipse will appear correctly on drawing.

Once you have oriented the template to the approximate position, then carefully line up the center line marks on the template with the center lines on the drawing. Finally draw the ellipse. Be careful to keep the pencil at a position perpendicular to the drawing surface or the ellipse will become distorted (Figure 5.18).

Arcs drawn on normal surfaces in isometric are simply a portion of the circumference of the ellipse. Therefore, you will first locate the center of the ellipse, draw the center lines, align the template, and then draw only the portion of the arc that is desired (Figure 5.19).

5.2.7 Circles and Arcs Using the Four-Center Method

To save time, use the ellipse template whenever possible to draw the isometric ellipses. However, sometimes a specific size ellipse

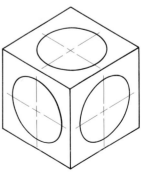

Figure 5.17 Note correct direction of center lines for ellipses that appear on the three visible normal surfaces.

Figure 5.18 To avoid distortion, pencil or pen should be perpendicular to drawing surface when using the ellipse template.

Figure 5.19 To draw an arc in isometric, follow same procedure as for drawing a circle but only darken the desired portion of the arc.

(one that does not appear on a standard template) will be required or a very large ellipse will be required. In either of these cases, you will have to construct an ellipse using a compass. The most efficient way is to use the **four-center method.**

To construct an ellipse on any of the three visible normal surfaces of the rectangular solid, complete the following steps:

1. Draw the center lines on the selected plane.
2. Measure the distance of the radius away from the center on all four center lines.
3. Draw an isometric box as shown.
4. Draw the extra construction lines to locate the centers for the compass.

5. Set the compass for the distance R and strike the two large arcs using the compass.

6. Set the compass for distance r and strike the two small arcs (Figures 5.20–5.22).

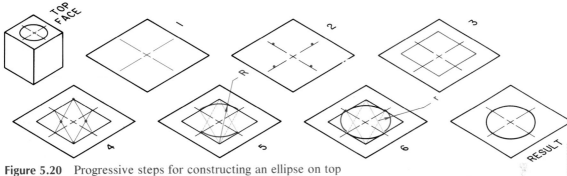

Figure 5.20 Progressive steps for constructing an ellipse on top surface using four-center method.

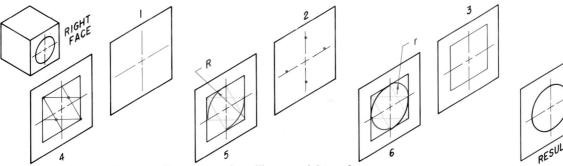

Figure 5.21 Progressive steps for constructing ellipse on right surface using four-center method.

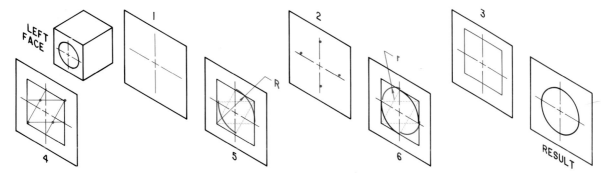

Figure 5.22 Progressive steps for constructing ellipse on left surface using four-center method.

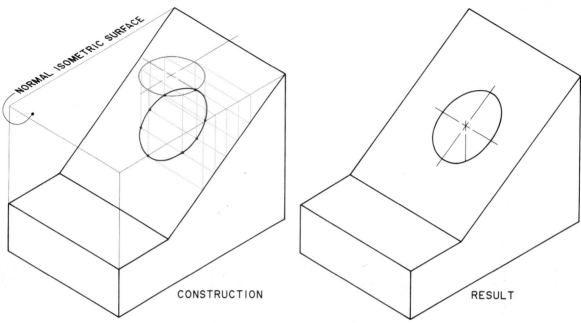

CONSTRUCTION

RESULT

Figure 5.23 Circle that appears on inclined isometric surface must first be constructed on normal plane and then projected to inclined isometric surface.

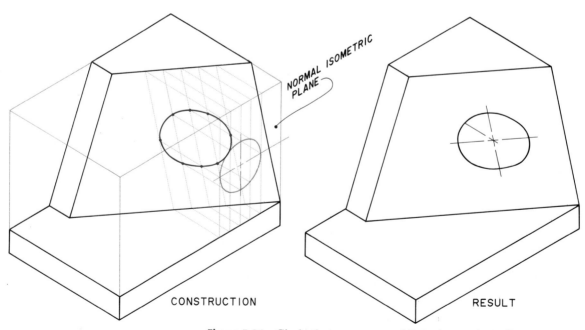

CONSTRUCTION

RESULT

Figure 5.24 Circle that appears on oblique isometric surface must first be constructed on normal plane and then projected to oblique isometric surface.

5.2.8 Drawing Ellipses on Nonisometric Surfaces

You will occasionally have to draw an ellipse on an inclined surface that appears in an isometric drawing. When this happens, you must first plot the circle (as an ellipse) on one of the two adjacent normal isometric surfaces. Then plot a grid starting at any point on the ellipse. Project this point to the adjacent isometric surface. Then project to the inclined edge and finally along the inclined surface. Repeat this procedure until enough points have been plotted to construct the ellipse on the inclined plane. You will have to use the irregular curve to draw the ellipse on the inclined surface (Figure 5.23).

A circle appearing on an oblique plane will appear as an ellipse. You must first locate and draw the isometric ellipse on a normal isometric plane. Then project selected points from the isometric ellipse to the oblique surface until they intersect each other creating a grid network. When enough points are located on the oblique surface, then the irregular curve can be used to connect the points that produces the correctly shaped ellipse on the oblique surface (Figure 5.24).

5.2.9 Using the Irregular Curve

The **irregular curve** is a difficult tool to master until you have had some practice using it. There are many sizes and shapes of irregular curves. You must select an appropriate size and shape to connect all the identified points (Figure 5.25).

Figure 5.25 Irregular curves are available in a large number of sizes and shapes.

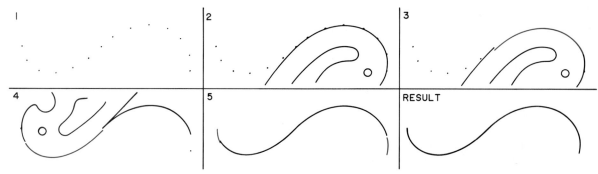

Figure 5.26 To use irregular curve correctly, align curve with as many locator dots as possible (minimum four dots except for end spaces). Then draw connecting line through all but the last space. When drawing next segment of curve, overlap last space.

Follow this procedure to use this tool:

1. Plot all the points you wish to connect.
2. Place the curve so that you can align as many points as possible (at least four dots must be used except for the end spaces).
3. Connect all spaces between the points except for the space at each end.
4. Now repostion the irregular curve so the first space aligned overlaps the end space drawn last. Continue this procedure until the curve has been drawn as a continuous curved line.
5. Note that you cannot overlap the end two spaces; you will have to do the best you can with these two spaces.
6. If necessary, clean up the irregular curve so it looks like a continuous curve (Figure 5.26).

5.2.10   Exploded Views

Once you have mastered the drawing of objects in isometric, you can begin making exploded views. These views consist of isometric drawings of several parts that are displayed along their axis of assembly. This type of drawing allows the untrained eye to immediately visualize how several parts of an assembly fit together. These drawings are particularly valuable when the viewer must either assemble or disassemble a number of parts that are manufactured to fit together. Exploded views are often used in maintenance manuals (Figure 5.27).

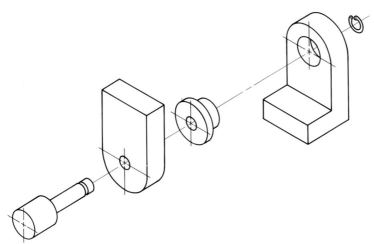

Figure 5.27 Exploded view shows parts of assembly separated along center line. This makes each part visible to viewer.

The easiest way to produce one of these drawings on the drawing board is to make rough drawings first. Then cut the pieces apart and reposition them with the spacing between the views being equal or pleasing to the eye. Then trace the original rough drawings to make the final presentation.

5.2.11 Isometric Assembly Drawings

There are times when a pictorial drawing is required to show the outside of an assembly made up of several parts. This type of drawing is known as an isometric assembly drawing. Depending on the complexity of the drawing, you will either make the drawing part by part in its assembled position, or draw the objects individually and trace them; locating them correctly. These views tend to show how the entire assembly appears from the outside after the parts have beem assembled (Figure 5.28).

5.2.12 Sectioned Isometric Drawings

Sometimes it is important to look inside a part drawn in pictorial. When a drawing is made of one part it is a sectioned isometric drawing. If an assembly drawing is sectioned it is called an isometric sectioned assembly. This might be used when the need arises to show how the internal parts in an assembly work together. Cylindrical and spherical parts such as shafts, bearings, bolts, and keys are not generally sectioned in this type of drawing (Figure 5.29).

Figure 5.28 Assembly drawing is a pictorial view that shows visible features of group of parts after they have been assembled.

Figure 5.29 Sectioned isometric drawing shows how the parts of assembly fit together by employing sectioning practices. Note that connective shafts, rods, bolts, and so on are not sectioned.

5.2.13 Isometric Projections

Up to this point you have been making isometric drawings. Now you should be exposed to a little pictorial theory and try to understand the difference between an isometric drawing and an isometric projection. This is particularly important since most computer-generated pictorials are actually projections.

The biggest visual difference between isometric drawings and isometric projections is the size of the two images. The isometric drawing is drawn using 100% true length measurements on the height, width, and depth axes (Figure 5.30). However, in isometric projection the height, width, and depth measurements are displayed at 82% of their true length (Figure 5.31). For the purpose of projection the X, Y, and Z axes, remain fixed, and we think about the object rotating in space about the axes.

To understand why these are different you must first visualize the axes as they relate to (1) the drawing board or the computer screen or (2) the axes of the object. Note that the X axis goes to

Figure 5.30 On isometric drawing the height, width, and depth measurements are made at 100%.

Figure 5.31 On isometric projection the height, width, and depth measurements are made at 82%.

Figure 5.32 Note orientation of the X, Y, and Z axes as they relate to observer at drawing board or at computer.

the right, the Y axis is vertical, and the Z axis goes toward the observer (Figure 5.32).

In isometric projections the object is first rotated about the Y axis by $-45°$ (Figure 5.33). Then the object is rotated about the X axis by $35°$ $16'$ (Figure 5.34). Because the normal surfaces of the object are no longer parallel or perpendicular to the picture plane, the image edges will appear foreshortened on each axis by 18%. The foreshortened view is called an isometric projection.

When you draw a pictorial on a computer, the image that appears on the screen is a projection, and therefore it is foreshortened. The big advantage of drawing these pictorials on the screen is that the object may easily be rotated about the axes in a countless number of positions once the data for the object's features have been entered. This allows the observer to view the object from many positions while the computer redraws the object. Copies of the rotated object can be printed or plotted and used for a set of working drawings. The printout can also be the basis for a technical illustration of the object or assembly (Figure 5.35).

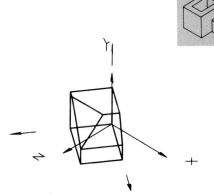

Figure 5.33 Object is first revolved around Y axis. Computer generated image.

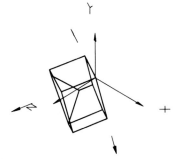

Figure 5.34 Object is then revolved about X axis. Computer generated image.

Figure 5.35 Computer-generated isometric pictorials may easily be revolved to produce various revolved views. Computer generated image.

ISOMETRIC
DRAWING

ISOMETRIC
PROJECTION

Figure 5.36 Resulting difference between isometric drawing and isometric projection.

If an isometric drawing is made on the drawing board, you only get a view from one vantage point. To show the object rotated, it would have to be completely redrawn. In order to compensate for the difficulty of creating manually drawn isometric projections, drafters decided that, if each of the isometric axes was foreshortened the same amount, then a standard scale could be used for isometric drawings. For this procedure to work, it was assumed that any measurements made parallel to the object axes would be whatever was on the orthographic drawings of the principal views. This had the effect of creating a proportional but "larger-than-life" drawing. This results in the need for an 82% scale for making measurements on the X, Y, or Z axes if an isometric projection is created on the drawing board (Figure 5.36).

CAD software packages, including the one available with this book, will allow you to rotate the object so you may view the object from several vantage points and allow you to get a hard copy of the rotated object.

5.3 OBLIQUE DRAWINGS

5.3.1 Oblique Views

Oblique drawings are very similar to isometric drawings, but there are some noticeable differences. The object has one face of the rectangular solid parallel to the picture plane. Two other faces do appear, but since the angle of the depth axis may vary, the angles of the other two surfaces may vary from drawing to drawing. The axes of an oblique drawing are horizontal, vertical, and a depth axis. You may select any angle for the depth axis but a 45° angle is usually preferred. Notice how much change there is in the appearance of the object when the depth axis is altered (Figure 5.37). The main advantage of drawing an oblique drawing is that you may use the circle template on the plane that is parallel to the picture plane. This is possible because the circles will appear as true circles and not as ellipses on this one surface only. Remember, any plane parallel to the picture plane is true size and shape (Figure 5.38).

The computer is used quite often on a commercial basis for making oblique drawings of large exploded assemblies. The express purpose is for making parts catalogs for assemblies where repair on the assembly might be necessary. The parts are identified with a part number for easy, quick identification.

PREFERRED   SPECIAL USE ONLY

Figure 5.37 Oblique drawings generally have a front view that shows true width and height measured on horizontal and vertical axes. Depth measurements are generally on axis that recedes at 45° (although another angle may be selected).

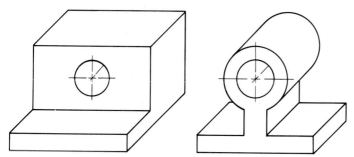

Figure 5.38 Circles appearing on the front surface of an oblique drawing will be drawn as circles not as ellipses.

CAVALIER DRAWING

FULL HEIGHT

FULL WIDTH

FULL DEPTH

SELECTED ANGLE

Figure 5.39 Cavalier drawings are oblique drawings using full-size height, width, and depth measurements. With this type of drawing, depth appears visually to be exaggerated.

5.3.2 Cabinet and Cavalier Oblique Drawings

Oblique drawings of objects drawn with a 45° depth axis that have a long depth measurement do not appear visually correct to the eye. These drawings are known as **cavalier drawings.** Notice that the cube, when drawn in cavalier, with true height, width, and depth, visually appears much too deep (Figure 5.39).

It is common for the person making an oblique drawing to take some ''artist's liberties.'' The furniture industry uses a type of oblique drawing that somewhat remedies this depth problem. They make a **cabinet** type of oblique drawing. For this type of drawing, the width and height measurements are made at full dimension. However, the depth axis is shortened and is measured at half dimension along the depth axis. This produces a completely different and more visually pleasing drawing (Figure 5.40).

CABINET DRAWING

FULL HEIGHT

FULL WIDTH

HALF DEPTH

SELECTED ANGLE

Figure 5.40 Cabinet drawings are oblique drawings that use full width and height measurements but only half the size of the depth measurement.

Figure 5.41 Ellipse templates for making oblique drawings range in increments from 10 to 80°. Note how shape of ellipses changes as degree increment changes.

5.3.3 Special Ellipse Templates

Those persons who frequently make oblique drawings will certainly have a complete set of ellipse templates that are commonly identified by 5° increments from 10 to 80°. Because these sets are expensive, every effort should be made to keep the sets intact. These ellipse templates are used to draw ellipses on the depth axis when the given depth axis angle equals one of the templates in the set. These templates are usually purchased in small, medium, and large size sets and range from $\frac{1}{2}$ to 8 in. (13 to 203 mm) or more in size (Figure 5.41).

5.4 TYPES OF PERSPECTIVE DRAWINGS

The engineer will hardly ever make a **perspective drawing** because it is very time consuming to draw. The engineer's concern is to communicate not to spend time making a perspective drawing. However, the engineers should know what a perspective drawing looks like so they may request to have one made if that type of communication is necessary.

The most common perspective is a **two-point exterior perspec-**

Figure 5.42 With two-point perspective, visual rays converge at either left or right vanishing points.

Figure 5.43 Image shows what perspective view looks like without all the construction lines.

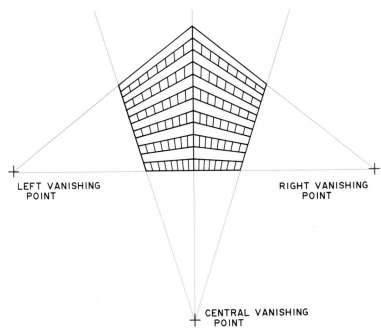

Figure 5.44 Note that vertical lines on view also converge at vanishing point in a similar manner at the visual rays to the left and right vanishing points.

tive, which is usually drawn for architectural renderings. In this type of perspective, the visual rays are not parallel as with the other drawings that have been discussed, but they converge at two vanishing points (Figures 5.42 and 5.43). A three-point perspective is used to emphasize some features or to emphasize the view from the location of the observer's eye. They are almost exclusively used for special architectural renderings but are seldom drawn due to the distortion caused. They are used as "attention getters." Note that this type of drawing has three vanishing points (Figures 5.44 and 5.45).

A one-point perspective is used almost exclusively for interior-room views. They give the observer the illusion of looking into the room. In this case there is only one vanishing point (Figures 5.46 and 5.47).

Perspective grids are commercially available. These printed grid patterns serve as time savers. The drafter places a piece of vellum or film over the grid pattern and then sketches or draws the perspective using the lines on the grid as a guide.

5.5 COMPUTER APPLICATIONS

There are professional firms that specialize in producing perspective line drawings for architects and engineers. When the need for such a service occurs, you send the elevations and plan views to the firm; they digitize the information and print out a line illustration in the type of perspective requested. Once the information has been digitized, it is easy to turn out different types of perspectives or different views based on different locations of the observer's eye (Figures 5.48 and 5.49).

Persons having access to computers that accommodate computer graphics can make a perspective drawing of sorts. However, it is much easier if the rubber banding feature is available. This allows the person creating the drawing to start a line at the

Figure 5.45 Three-point perspective portrays an image that is not like the eye would view the object, but architects and others often use these views to emphasize certain features for presentation purposes.

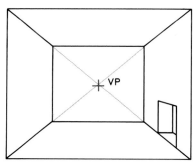

Figure 5.46 One-point perspective is used for interior views. Note that vanishing point is located at center of drawing.

Figure 5.47 One-point interior perspective results in view that closely resembles what eye would actually see.

Figure 5.48 Computer-generated perspective view is generally used to save time on complicated perspectives. Computer generated image. Courtesy of Ohio State University School of Architecture

Figure 5.49 Computer-generated perspective view. Computer generated image. Courtesy of Ohio State University School of Architecture

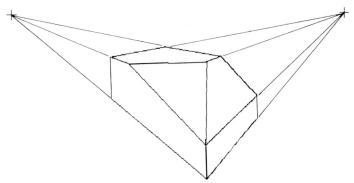

Figure 5.50 Computer software that allows for rubber banding can be used to draw perspective views on computer screen. Computer generated image.

vanishing point and connect the other end of the line on the object where desired. This is a great time saver (Figure 5.50).

5.6 LINE APPLICATIONS

Very light construction lines are used when making a pictorial drawing. You will find that some drawings requiring many sets of center lines will result in a cluttered drawing during the layout process. If the drawing can easily be cleaned up, then use that

CONSTRUCTION   RESULT

Figure 5.51 If pictorial view construction requires many lines, it may be smarter to trace lines that make up final pictorial drawing on separate sheet.

piece of paper for the final drawing. If not, it is suggested you trace the final drawing on a piece of vellum or film using the original construction and center lines as guides for the tracing process (Figure 5.51).

Since hidden lines are hardly ever used on pictorial drawings, the only lines commonly drawn are object lines and some center lines. The object lines are wide lines, and the center lines are thin lines. The exception is when some dimensions are added to the pictorial view in which case some additional thin lines are used.

5.7 SUMMARY

Pictorial drawings are drawn to improve the communication process. They are likely to be less detailed than the orthographic or working drawings. Visualization for both the trained and untrained eye will be quicker when pictorial drawings are presented.

PROBLEMS

Problems 5.1–5.60

The following problems are given in one-, two-, or three-view orthographic drawings. You are to draw these objects as isometric drawings. Where three orthographic views are given, these three views will become the three visible planes in the isometric drawing.

Problem 5.1

Problem 5.2

Problem 5.3

NOTE: ALL FILLETS AND ROUNDS R 3

ø 44 BOSS
ø 22

NOTE: ALL FILLETS AND ROUNDS R.13

ø1.75 BOSS
ø.88

Problem 5.4

R 19   R 13

NOTE:
ALL FILLETS AND ROUNDS R 3

R 0.75   R 0.50

NOTE:
ALL FILLETS AND ROUNDS R.125

Problems 5.1–5.60

The following problems are given in one-, two-, or three-view orthographic drawings. You are to draw these objects as isometric drawings. Where three orthographic views are given, these three views will become the three visible planes in the isometric drawing.

Problem 5.5

Problem 5.6

Problem 5.7

Problem 5.8   Computer generated image.

Problem 5.9

Problems 5.1–5.60

The following problems are given in one-, two-, or three-view orthographic drawings. You are to draw these objects as isometric drawings. Where three orthographic views are given, these three views will become the three visible planes in the isometric drawing.

Problem 5.10

Problem 5.11

Problem 5.12

Problem 5.13

Problems 5.1–5.60

The following problems are given in one-, two-, or three-view orthographic drawings. You are to draw these objects as isometric drawings. Where three orthographic views are given, these three views will become the three visible planes in the isometric drawing.

Problem 5.14

Problem 5.15

THE CENTER OF
EACH ANGLE IS
10 DEEP

90

125

50

Problem 5.16

3.50

5.00

2.00

50
19
6
6
31
15°

25
19
3
15°

.25
1.25

2.00
.75
.25
15°

1.00
.75
.13
15°

Problem 5.17   Computer generated image.

18

.75

20

36

R 20, TYP.

.88

1.38

R .88, TYP.

Problem 5.18   Computer generated image.

Problems 5.1–5.60

The following problems are given in one-, two-, or three-view orthographic drawings. You are to draw these objects as isometric drawings. Where three orthographic views are given, these three views will become the three visible planes in the isometric drawing.

Problem 5.19

Problem 5.20

Problem 5.21

Problem 5.22

Problem 5.23

Problems 5.1–5.60

The following problems are given in one-, two-, or three-view orthographic drawings. You are to draw these objects as isometric drawings. Where three orthographic views are given, these three views will become the three visible planes in the isometric drawing.

Problem 5.24

Problem 5.25

62
25
100
50
100
50

ø 90

2.50
1.00
4.00
2.00
4.00
2.00

ø 3.50

Problem 5.26

ø 50
90
38
50
100

ø 2.00
3.50
1.50
2.00
4.00

12
38
64

.50
1.50
2.50

Problem 5.27

Problems 5.1–5.60

The following problems are given in one-, two-, or three-view orthographic drawings. You are to draw these objects as isometric drawings. Where three orthographic views are given, these three views will become the three visible planes in the isometric drawing.

Problem 5.28

Problem 5.29

Problem 5.30 Computer generated image.

Ø 13
Ø 44 x 82° CSK

Ø .50
Ø 1.75 x 82° CSK

Problem 5.31 Computer generated image.

Ø 13
Ø 31 x 90° CSK

Ø .50
Ø 1.25 x 90° CSK

Problem 5.32 Computer generated image.

Ø 6

Ø 6   R 19

R 9

Ø .25

Ø .25   R .75

R .38

Ø .13

Problem 5.33 Computer generated image.

Problems 5.1–5.60

The following problems are given in one-, two-, or three-view orthographic drawings. You are to draw these objects as isometric drawings. Where three orthographic views are given, these three views will become the three visible planes in the isometric drawing.

Problem 5.34   Computer generated image.

Problem 5.35   Computer generated image.

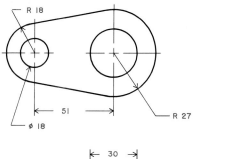

Problem 5.36 Computer generated image.

Problem 5.37 Computer generated image.

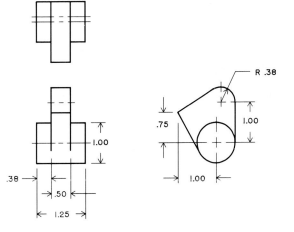

Problem 5.38 Computer generated image.

Problems 5.1–5.60

The following problems are given in one-, two-, or three-view orthographic drawings. You are to draw these objects as isometric drawings. Where three orthographic views are given, these three views will become the three visible planes in the isometric drawing.

Problem 5.39 Computer generated image.

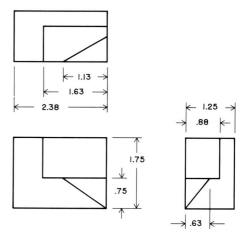

Problem 5.40 Computer generated image.

Problem 5.41 Computer generated image.

Problem 5.42 Computer generated image.

Problems 5.1–5.60

The following problems are given in one-, two-, or three-view orthographic drawings. You are to draw these objects as isometric drawings. Where three orthographic views are given, these three views will become the three visible planes in the isometric drawing.

Problem 5.43 Computer generated image.

 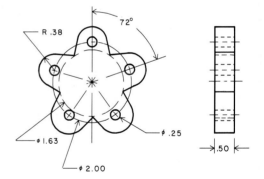

Problem 5.44 Computer generated image.

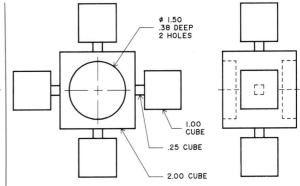

Problem 5.45 Computer generated image.

Problem 5.46 Computer generated image.

Problem 5.47 Computer generated image.

Problems 5.1–5.60

The following problems are given in one-, two-, or three-view orthographic drawings. You are to draw these objects as isometric drawings. Where three orthographic views are given, these three views will become the three visible planes in the isometric drawing.

Problem 5.48 Computer generated image.

Problem 5.49

Problem 5.50 Computer generated image.

Problem 5.51 Computer generated image.

NOTE:
FRONT, TOP, AND
RIGHT SIDE VIEW

ⵁ 39 TYP

6 TYP

Problem 5.52 Computer generated image.

NOTE:
FRONT, TOP, AND
RIGHT SIDE VIEW

ⵁ 1.63 TYP

.25 TYP

Problems 5.1–5.60

The following problems are given in one-, two-, or three-view orthographic drawings. You are to draw these objects as isometric drawings. Where three orthographic views are given, these three views will become the three visible planes in the isometric drawing.

Problem 5.53 Computer generated image.

Problem 5.54 Computer generated image.

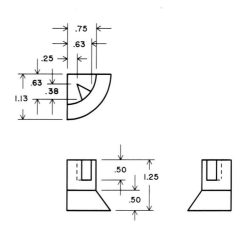

Problem 5.55 Computer generated image.

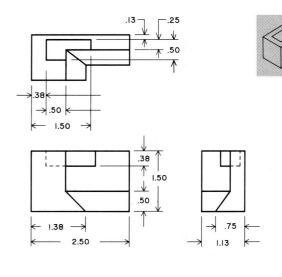

Problem 5.56 Computer generated image.

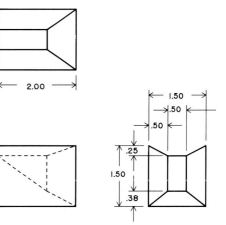

Problem 5.57 Computer generated image.

Problems 5.1–5.60

The following problems are given in one-, two-, or three-view orthographic drawings. You are to draw these objects as isometric drawings. Where three orthographic views are given, these three views will become the three visible planes in the isometric drawing.

Problem 5.58

Problem 5.59 Computer generated image.

Problem 5.60 Computer generated image.

TERMS YOU WILL SEE IN THIS CHAPTER

CUTTING PLANE Imaginary surface used to cut through an object to reveal its interior.

SECTION VIEW Orthographic view that shows the interior features of an object as visible lines.

SECTION LINES Lines drawn on the cut surfaces of an interior portion of an object that has been cut by a cutting plane also called crosshatch lines.

FULL SECTION Section view formed by passing a cutting plane entirely through an object to reveal its interior features.

HALF SECTION Section view of symmetrical objects formed by passing the cutting plane in such a manner so as to show the interior of half the object.

OFFSET SECTION Section view created with multiple cutting planes and used to show interior features that cannot be located with a single cutting plane.

REVOLVED SECTION Section view of an elongated symmetrical feature where the cutting plane is passed perpendicular to the axis of symmetry and the interior feature is revolved 90° into the plane of the drawing.

REMOVED SECTION Section view similar to a revolved section except that the section view is not drawn within the view containing the cutting plane.

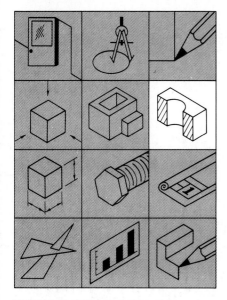

6

SECTIONS AND CONVENTIONS

6.1 INTRODUCTION

Standard orthographic views show all features of objects. Interior features may be very complex, and standard orthographic views may lead to confusion when several hidden lines are present. The confusion associated with these hidden lines may be reduced by cutting away part of the object and looking at the interior as a section view (see Figure 6.1 for samples).

In this chapter you will learn the concepts and techniques of sectioning. Also, there are conventional practices and standards used in making engineering drawings that you will learn and become familiar with so that you are able to read and produce engineering drawings.

Figure 6.1 Sectioning samples.

Figure 6.2 (*a*) Imaginary cutting plane passing through object. (*b*) Interior features of object are shown along line where cutting plane passed.

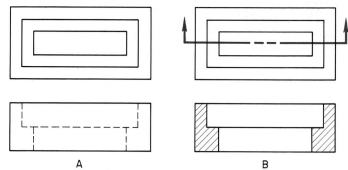

Figure 6.3 (*a*) Normal front and top orthographic views of object showing hidden features in front view. (*b*) Same object showing cutting plane in top view and resulting sectional view.

6.2   *CUTTING PLANE*

Figure 6.2*a* is a pictorial drawing with an imaginary **cutting plane** passing through an object. This cutting plane is similar to a knife or saw blade cutting the object. After the object is cut, a portion is removed to reveal the interior features as shown in Figure 6.2*b*. The object shown pictorially in Figure 6.2 is shown orthographically in Figure 6.3. The interior features are represented by dashed lines in the front view. These interior features become visible and are shown more clearly in Figure 6.3*b*, which is a section view of the object. The **section view** is created by passing a cutting plane through the object and imagining that it is "cut" along this plane and a portion of the object is removed. The hidden features are now shown as visible lines, and the part of the object in contact with the cutting plane is indicated with crosshatching. When draw-

Figure 6.4 (*a*) Cutting plane and front section view. (*b*) Cutting plane and top or horizontal section view. (*c*) Cutting plane and profile or right side section view.

ing section views, the edge view of the cutting plane is shown by a thick line as shown in the top view of Figure 6.3*b*. Avoid hidden lines in section views. A second section view or additional orthographic view may be used to reveal additional interior or exterior details.

Figure 6.4 shows front, top, and right side section views. Each cutting plane line has arrows that are perpendicular to the cutting plane and point in the direction of the line of sight for the section view. Also, the cutting plane often is identified with alphabet characters such as *A–A* or *B–B*. When the cutting plane is identified in this manner, the view is called section *A–A* or section *B–B*. This identification technique is especially useful when showing removed sections (discussed in Section 6.8). Two types of cutting plane lines used in section views are shown in Figure 6.5. The line shown in Figure 6.5 (top marked *A–A*) is typically used when the cutting plane line is long. The cutting plane line shown in Figure 6.5 (bottom marked *B–B*) shows the line style that is used for short cutting plane lines.

Figure 6.5 (*Top*) Example of type of cutting plane line used when cutting plane is long. (*Bottom*) Example of type of cutting plane line used when cutting plane is short.

6.3 *SECTION LINES*

Section lines, or crosshatch lines, are shown in the section view. Generally, the portion of the object that is in contact with the cutting plane is crosshatched to show the features that were once hidden but are now visible.

Historically, the section line pattern indicated a specific type of material to be used to make the object. For example, Figure 6.6 shows the patterns traditionally used for many fairly common materials. The pattern of uniformly spaced lines shown for cast iron in Figure 6.6 is now accepted as the standard for section lines regardless of the material from which the object is made; however, use of the other material patterns is optional. This standard is established by The American National Standards Institute (ANSI) in ANSI Y14.2M-1979.

Section lines are normally spaced approximately 0.06–0.12 in. (2–4 mm) apart by eye although some triangles have a line etched on them to ensure uniform spacing of section lines. A line can be scribed on the triangle using the point of the dividers or compass (Figure 6.7). Simply place the triangle over a previously drawn section line, with the etched line covering the section line, to draw another section line at the correct spacing (Figure 6.8).

Section lines should be thin and dark, never varying in thickness. There should be a distinct contrast between the thickness of object (or visible) lines and the thickness of section lines. The 1982 ANSI standards call for object lines to be 0.7 mm thick and section lines 0.35 mm thick. Section lines must end exactly at the object outline, never extending beyond the object outline or stopping short of the object outline.

Section lines are normally slanted at a 45° angle, but any standard angle, such as 30 or 60°, can be used. Choose this angle so that the section lines will not be parallel to the object outlines of the part. Section lines that are perpendicular to the object outlines of the part are acceptable but not preferred. Horizontal and vertical section lines are not preferred as well (Figure 6.9).

Drawing section lines on a computer screen using a computer-aided design and drafting (CADD) program can be done with ease and precision. Most CADD programs allow the user to define a boundary area, and then by using a fill or hatch command the area is filled with a pattern selected by the user. The user also selects the angle and spacing of the section lines. The more sophisticated CADD programs offer the user a wide range

Figure 6.6 Section line patterns for various engineering materials. Courtesy of ANSI

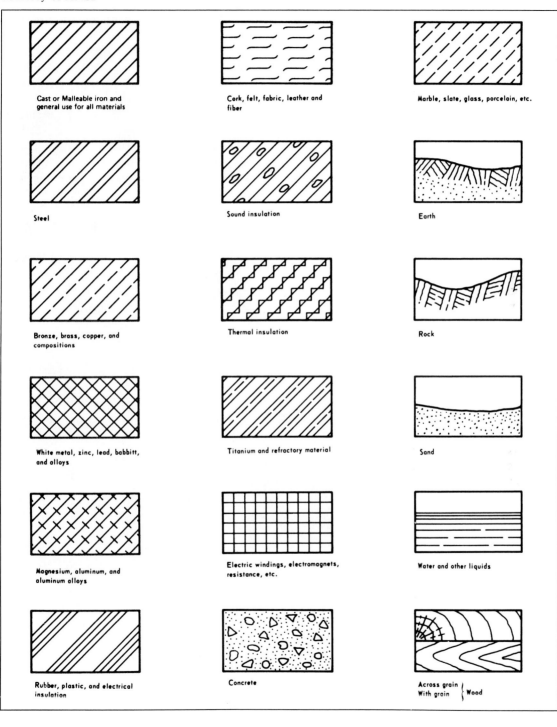

Figure 6.7 Etching triangle with line to aid in drawing uniform section lines.

of hatch styles to accommodate engineering, architectural, and various other applications.

When using a CADD program that does not have an automatic fill or hatch command, a section view can still be drawn by carefully laying out a line at a 45° angle on the area to be sectioned. Using the repeat command, you can completely cross-hatch the section view by repeating the line at the desired spacing. The crosshatching will not only cover the desired area but also areas that you do not want to show as sectioned. By deleting the portion of the section lines that are not part of the section view, using the line removal feature, a good section view can be made.

Figure 6.8 Using etched triangle to properly space section lines.

Figure 6.9 Choosing correct angle for section lines.

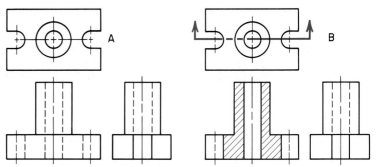

Figure 6.10 (*a*) Normal orthographic drawing of object. (*b*) Orthographic drawing of same object showing front view in full section (half the object is removed).

6.4 *FULL SECTION*

A **full section** is a section view formed by passing a cutting plane entirely through an object to reveal all of its interior features. An entire view (top, front, right side, etc.) is shown in section, and this view is said to be a full section. A full-section view generally shows half the object removed.

Figure 6.10 shows the top, front, and side views of an object. The front view can be drawn as a full section simply by passing a cutting plane fully through the top view and removing the front portion of the object. This procedure yields a full-section view in the front view as shown in Figure 6.10*b*. The position of the cutting plane indicates the exact location of the section view, and the arrows on the cutting plane indicate the viewing direction. The area that is cut by the cutting plane is crosshatched with uniform section lines. For simple full sections, the cutting plane may be omitted when its position is obvious.

A full section through a cylindrical object introduces some common problems with sectioning. Figure 6.11 shows a cylindrical

Figure 6.11 Cylindrical object in full section.

part in full section. A common error that should be avoided is to omit visible lines behind the cutting plane (Figure 6.11).

Hidden lines in full sections should be omitted unless they are necessary to provide a clear understanding of the sectioned view. Including hidden lines in the section view is orthographically correct but may lead to confusion rather than clarity.

6.5 HALF SECTIONS

Symmetrical objects often can best be shown using **half-section** views. This method of sectioning simply requires that a quarter of the object be removed to expose the interior features of the part. Figure 6.12 shows an example of a half-section view. The

Figure 6.12 Half section: (*a*) cutting planes passing through object and (*b*) half-section view of the object.

cutting plane is really two cutting planes meeting perpendicular to one another at the axis of symmetry of the object. Removing one quarter of the object reveals the internal features of the object (Figure 6.12).

A half-section view has the advantage of exposing the interior features of half of the object and keeping the exterior features displayed on the other half. It is extremely important to keep in mind that one side of the half section shows the interior while the other side shows the exterior. The dividing line between the interior and exterior is not a solid line. It is a center line that denotes the axis of symmetry of the object. A common error made in drawing half sections is to include hidden lines on the half showing the exterior. This is wrong.

Since a half-section view shows both the interior and exterior in a single view, its usefulness is largely limited to symmetrical objects. Half-section views are very common in assembly drawings. Assembly drawings often require that both the interior and exterior features of the object be shown. Therefore half-section views are very useful under these circumstances.

6.6 OFFSET SECTIONS

Many times interior features are not located within a single plane. In order to include such features in a section view, the cutting plane may be stepped, or offset, to pass through all these features. The cutting plane is offset at right angles. The section view is then drawn as if the internal features were in a single plane or in a straight line. This type of section view is called an **offset section** and can be considered a special type of full section. Note that the offsets, or 90° turns, that the cutting plane makes to pass through each feature are not shown in the section view. Figure 6.13 shows an example of an offset section.

NONSECTIONED VIEW

OFFSET SECTIONAL VIEW

Figure 6.13 Offset section. Computer generated image.

6.7 REVOLVED SECTIONS

The shape of a bar, arm, spoke, or other elongated symmetrical feature may be shown in the longitudinal view by means of a

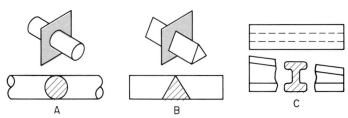

Figure 6.14 Revolved sections showing (*a*) solid cylinder, (*b*) solid triangular prism, and (*c*) conventional break to show section.

Figure 6.15 Revolved sections. Note that line covered by the section view must be removed.

revolved section (Figure 6.14). The cutting plane in a revolved section is passed perpendicular to the axis of the elongated symmetrical feature and then revolved 90° into the plane of the drawing. A conventional break (discussed in Section 6.11) should be used to provide room for the revolved section if it conflicts with the original view (Figure 6.14*c*).

The revolved section is superimposed on the drawing. This means that lines that are significant in describing that view may interfere with the section view (Figure 6.15). The lines covered by the revolved section must be removed.

6.8 *REMOVED SECTIONS*

A **removed section** is similar to a revolved section but is not drawn within the view containing the cutting plane line. It is displaced from its normal projection position. The displacement from the normal projection position should be made without turning the section from its normal orientation. Figure 6.16 shows three examples of removed sections. In Figure 6.16, each removed section is labeled with letters and designated section *A–A*, section *B–B*, and section *C–C*. The section labels correspond to letters identifying the cutting plane that defines the exact location from which the removed section was taken.

Figure 6.16 Removed sections. Section view always goes behind cutting plane or away from direction of view as defined by the arrows on cutting plane.

Figure 6.17 Broken-out section.

6.9 BROKEN-OUT SECTIONS

Many times only a partial-section view is needed to expose the interior features of an object. This section, called a **broken-out section,** is limited by a break line. Figure 6.17 shows a broken-out section. A full or half section is not needed to explain the design. A small broken-out section is sufficient and takes less time to construct. Cutting planes are not shown for broken-out section views.

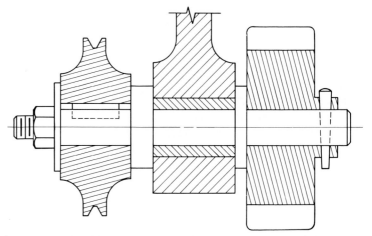

Figure 6.18 Sectioned assembly. Section lines vary in angle and spacing in adjacent parts. Standard parts such as nuts, bolts, keys, and washers (see Sections 8.4 and 8.6) do not get sectioned. Also, solid or hollow shafts do not get sectioned.

6.10 SECTIONED ASSEMBLIES

An assembly drawing showing several parts is sometimes sectioned to show the relationship of the parts to each other. When sectioning an assembly, it is important to separate the parts by varying the angle of the section lines. Figure 6.18 shows a typical sectioned assembly and the importance of distinguishing between the individual adjacent parts by varying the angle, direction, and spacing of the section lines. Small parts and fasteners that are not sectioned many times have their outline shown as hidden lines in the sectioned assembly.

6.11 *CONVENTIONAL BREAKS*

Conventional breaks are usually used to describe a long piece that has a uniform cross section. Figure 6.19 shows a typical conventional break. The shaft need not be drawn in true length; instead a conventional break can be used, and the dimension showing the true length can be placed on the drawing. This procedure is a time and material saver.

A conventional break is shown in different ways depending on the geometry of the cross section. Figure 6.20 shows various cross-sectional geometries as conventional breaks.

Of the conventional breaks shown in Figure 6.20, three are most often used: the solid cylinder, the cylindrical tube, and the rectangular cross section. The conventional break for a solid cylinder is shown in more detail in Figure 6.21*a*. Generally the curved portion of the break is drawn carefully freehand; however, instruments can be used, particularly when the diameter is large.

The conventional break for a cylindrical tube is shown in more

Figure 6.19 Typical conventional break.

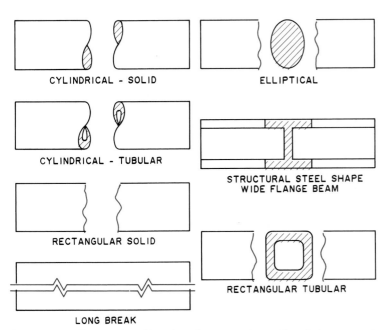

CYLINDRICAL - SOLID

ELLIPTICAL

CYLINDRICAL - TUBULAR

STRUCTURAL STEEL SHAPE
WIDE FLANGE BEAM

RECTANGULAR SOLID

RECTANGULAR TUBULAR

LONG BREAK

Figure 6.20 Conventional breaks of some cross-sectional shapes used in engineering.

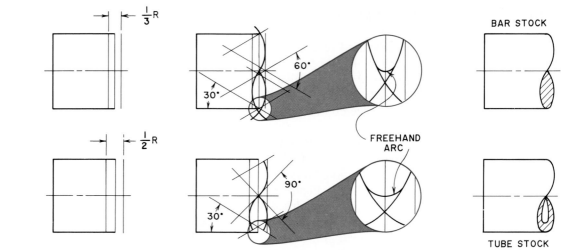

Figure 6.21 Cylindrical break: (*a*) solid construction, (*b*) tubular construction.

Figure 6.22 Rectangular break.

detail in Figure 6.21*b*. This conventional break resembles the conventional break used to show a cylindrical solid, except the interior space of the tube is shown without section lines. Again, the curved portions of the break are drawn freehand unless the diameter is especially large. In the case of both the conventional break for the cylindrical solid and the cylindrical tube, the area that represents the break is crosshatched with uniform section lines.

The conventional break for a rectangular uniform cross section is not as complex to draw as the cylindrical breaks. Two "squiggle" lines approximately 1 in. apart are drawn transverse to the long dimension, and the area between these squiggles is erased (Figure 6.22).

6.11.1 Conventional Breaks by Computer

The computer can be used to do the construction of a conventional break. A CADD system may or may not have a primitive or a symbol to create the conventional break. A method that will work on most CADD systems involves the use of construction circles, the arc function, straight lines, and, if available, the crosshatching or hatching function. The method is discussed in what follows.

Begin by locating the position of the break on the object (Figure 6.23a). You may want to zoom in on this area until it fills the screen to make the construction easy to see. Add a grid to aid in the construction of the conventional break. Draw two dotted lines perpendicular to the object outlines to locate the center of the break symbols (Figure 6.23b). Then choose a circle with a diameter approximately one-half of the radius of the object. Draw the circles in the dotted line style so that they are just tangent to the upper and lower surfaces of the object (Figure 6.23b). It may help to do this construction on a separate drawing layer to aid in separating the construction lines and the finished symbol.

If the CADD system in use has a tangent function, draw two construction lines tangent to the two circles in dot style. They will cross in the middle of the object (Figure 6.23b). If a tangent function is not available, you can use the rubber band line mode introduced in Chapter 2 to draw the tangents. Start lines where the vertical marking line crosses the center line of the object. Locate the estimated tangent point on the circle and draw in a construction line. Zoom in further and check the tangency (Figure 6.23b). If the line made in rubber band mode does not touch the arc at one specific point, the tangency is not good; therefore, you should zoom out and try again. When your construction is complete for both tangents on one side of the break, repeat the procedure for the other two tangents.

Use the tangent points as the starting and ending points for the arc of the full symbol. Go back to the original drawing layer and draw in the solid arcs and half arcs first and then draw in the solid tangents (Figure 6.23c).

Select the crosshatch function if it is available and mark the arc and the tangent lines as the limits on the crosshatched area. Then using the crosshatch function, fill the area with the proper crosshatch symbol.

If a crosshatch function is not available, use the grid dots and the repeat line function to locate some evenly spaced, construction lines that cross the boundaries of the break symbol (Figure 6.23d). Do this on the construction layer. Go back to the drawing layer and zoom in on the portion of the break to be crosshatched. Select the proper line thickness and use intersections of the construction lines as the beginning and ending points for the crosshatching lines (Figure 6.23e).

Zoom in on the other half of the symbol and crosshatch the bottom portion of that half. Zoom back out and redisplay without the construction layer to verify the construction (Figure 23f).

Figure 6.23 (*a*) Construction of conventional cylindrical break using computer. Locate the position of the conventional break. (*b*) Draw construction lines perpendicular to object outlines to locate center of break symbol. Draw two construction circles with radius of one half the radius just tangent to upper and lower surfaces of object. Draw two tangents to construction circles so they cross in center of object. (*c*) Darken in half arcs and tangents on drawing layer to complete boundary of conventional break. (*d*) Locate evenly spaced construction lines on construction layer to crosshatch break if crosshatch function is not available on your system. (*e*) Zoom in on drawing layer and draw in crosshatch lines using arc and tangents as boundaries. (*f*) Zoom out and redisplay conventional break without construction layer to verify. Computer generated image.

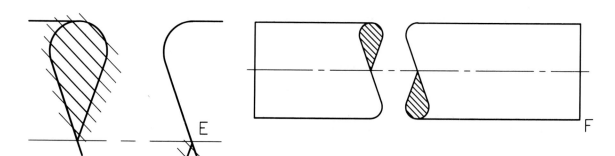

If the system being used allows the user to create, enlarge or shrink, and copy symbols, then the construction may be saved for future use on other breaks. It is probably not worth saving as a picture file only.

6.12 *CONVENTIONAL REVOLUTIONS*

Conventional revolutions are used to make section views clear and avoid confusion in interpretation. Conventional revolutions are associated with objects that have holes, ribs, spokes, or lugs in any combination equally spaced in circular fashion. The holes, ribs, spokes, or lugs may not project orthographically to the section view, but the convention allows for this practice and it is accepted as standard.

6.12.1 Conventional Hole Revolutions

Figure 6.24 shows three examples of holes in a circular plate. The first example shows four holes equally spaced around the circular

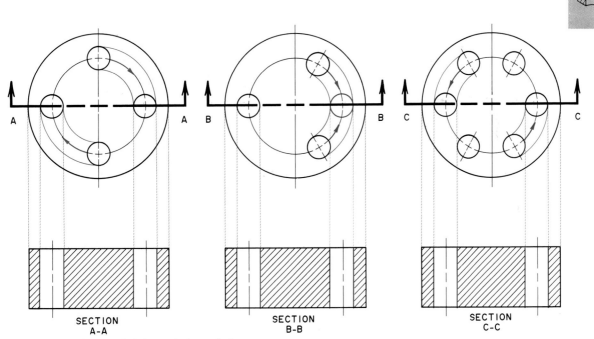

Figure 6.24 Conventional revolutions, holes.

plate. The cutting plane passes through two of the holes and a correct full-section view is shown. The holes that are not cut by the cutting plane are not shown. These irregular features are revolved into the cutting plane and then projected to the sectioned view. They are not shown in the section view with hidden lines. The second example shows a circular plate with three holes equally spaced. The cutting plane is shown going through only one of these holes; however, the other two holes are revolved into the cutting plane before they are projected to the section view. When comparing the section view of the first example with the section view of the second example, the views are exactly the same. The difference in the orthographic drawings is shown in the top views, which show the arrangement of the holes. The section view would look the same if there were six holes equally spaced in the circular plate as shown in the third example.

6.12.2 Conventional Rib Revolutions

Ribs are simply stiffeners that anchor geometric objects to a supporting plate or surface. Ribs, like holes, may be equally spaced about a circular plate base. Figure 6.25 shows three examples of ribs equally spaced on a circular plate securing a cylinder on top

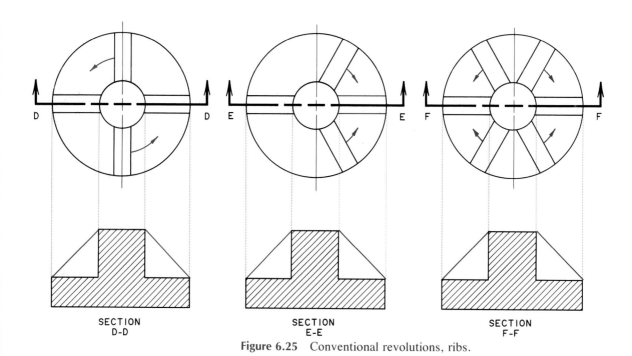

Figure 6.25 Conventional revolutions, ribs.

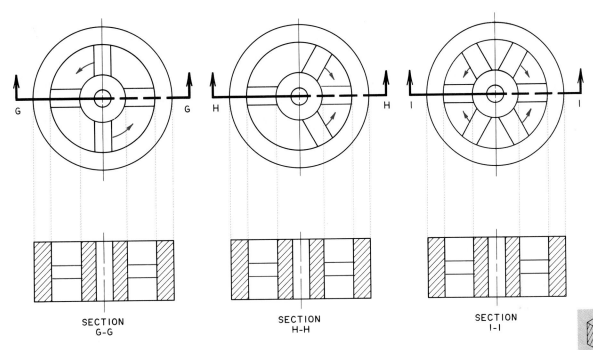

Figure 6.26 Conventional revolutions, spokes.

of the plate. Like the holes in the previous examples, the ribs are revolved into the cutting plane and projected to the section view. Since the ribs are not a continuous feature around the circular base, the section view shows them without section lines. This is an important convention used in all engineering drawing.

6.12.3 Conventional Spoke Revolutions

Spokes are treated in exactly the same manner as ribs. Figure 6.26 shows three examples of a mechanical wheel with equally spaced spokes arranged in various configurations. The spokes are revolved into the cutting plane in each case, and the section views are shown to be identical.

6.12.4 Conventional Lug Revolutions

Lugs are similar to ribs except that they do not rest on a circular base. They are arranged about a cylinder. Figure 6.27 shows three examples of lugs and the section view that would be drawn for each case.

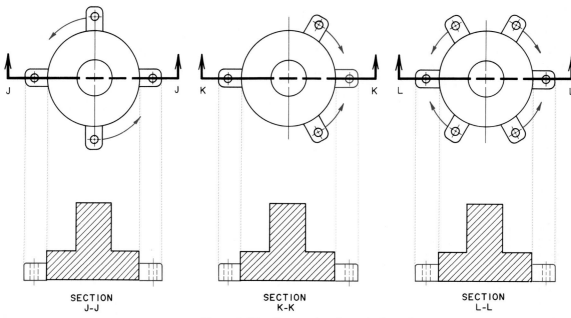

SECTION
J-J

SECTION
K-K

SECTION
L-L

Figure 6.27 Conventional revolutions, lugs.

SECTION
M-M

SECTION
N-N

Figure 6.28 Conventional revolutions, holes and ribs in combination.

6.12.5   Hole, Rib, Spoke, and Lug Combination

When holes, ribs, spokes, and lugs are shown in combination, each is revolved into the cutting plane and projected to the section view. Figure 6.28 shows an example of holes and ribs occurring together and the resulting section view.

6.12.6   Alternate Crosshatching

Omitting crosshatching on ribs and similar features sometimes can lead to confusion and an unclear drawing. For example, Figures 6.29*a* and *b* show two full section views that look identical. Figure 6.29*a* is a part that does not have ribs, while Figure 6.29*b* shows the same part that includes ribs that are flush with the top portion. If you look at the full section views, you can see that the ribs are not identified in Figure 6.29*b*. This is not clear. Greater clarity is achieved by the practice of alternate crosshatching (Figure 6.29*c*). Section lines (alternating) are drawn through the rib identifying it as a solid part of the object rather than air space. The lines separating the rib and the solid portions of the part are dashed.

6.13   *SUMMARY*

Standard orthographic views show all features of objects. Interior features shown as dashed lines may confuse the person interpreting the drawing. Sectioning cuts the object with an imaginary plane, making interior features, which were hidden to the viewer, visible. Sectioning makes the drawing more clear.

    Many types of section views can be drawn depending on the complexity of the object. A full-section view shows the entire

Figure 6.29   (*a*) Section view of part without ribs. (*b*) Section view of part with ribs that are flush with the top. (*c*) Section view with ribs identified with alternate crosshatching.

orthographic view as a section view, with half the object removed. A half-section view shows one-half of the orthographic view as a section view; however, a half-section view requires that the object have an axis of symmetry. An offset-section view is a type of full section in which several cutting planes meet at 90° angles to show the interior features of the object. A revolved-section view is a section shown within the longitudinal view of a long object. The cutting plane is located perpendicular to the long axis of the object, and the revolved section is then rotated 90° into the plane of the paper or computer screen to show the section view of the object.

Conventional breaks are usually used to describe a long piece with a uniform cross section. The long piece, such as a structural steel beam or a solid cylindrical shaft does not need to be drawn in true length. A conventional break can be used to save time and drawing materials. The true length dimension can be shown on the conventional break drawing.

Conventional revolutions are used to make section views clear and avoid confusion in interpretation. Conventional revolutions are associated with objects that have holes, ribs, spokes, or lugs in any combination symmetrically spaced in a circular fashion. When a section view of an object with these features is drawn, many times the cutting plane will not pass through one or more of the symmetrically spaced holes, ribs, spokes, or lugs. The features that are missed by the cutting plane are revolved into the cutting plane and projected to the section view from the revolved position. Conventional revolutions make the drawing clear and easily interpreted.

This chapter has presented techniques used in drawing section views as well as conventional practices and standards used in making engineering drawings. Mastery of these concepts will enable you to read and produce engineering drawings and become literate in the language of the engineer.

PROBLEMS

Problems 6.1–6.12

Sketch the given views and the indicated section for each problem.
Note that the cutting plane in Problems 6.5 and 6.6 indicates a broken-out section.

Problem 6.1

Problem 6.2

Problem 6.3

Problem 6.4

Problem 6.5

Problem 6.6

Problem 6.7

Problem 6.8

Problem 6.9

Problem 6.10

Problem 6.11

Problem 6.12

Problems 6.13 and 6.14
Given the top and front views of the object, draw the top view as shown and the front view in full section.

Problem 6.13

Problem 6.14

Problem 6.15
Given the top view and right side view (alternate position) of the object, draw the top view as shown and a front view in full section.

Problem 6.15

Problems 6.16–6.18
Draw the views that completely describe the object. One of the views must be a full section.

Problem 6.16

Problem 6.17

Problem 6.18

Problems 6.19–6.21

Given the two views of the object, draw the appropriate half section view.

Problem 6.19

Problem 6.20

Problem 6.21

Problem 6.22
Draw the orthographic views that best describe the object. One of the views must be a half section.

Problem 6.22

Problem 6.23
Draw the orthographic views that best describe the object shown for Problem 6.17. One of the views must be a half section.

Problem 6.23

Problems 6.24 and 6.25
Draw the orthographic views that best describe the object shown. One of the views must be a full section.

Problem 6.24

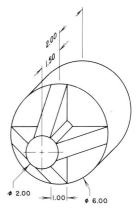

Problem 6.25

Problems 6.26–6.28
Given the front and right side views of the object, draw the front view as shown and a right side view as a full section.

ø 6
ø 13×13 DEEP
3 HOLES EQ. SP.
ON ø 56

ø 25
ø 75 ø 33
31
38

ø .25
ø .50×.50 DEEP
3 HOLES EQ. SP.
ON ø 2.25

ø 1.00
ø 3.00 ø 1.32
1.25
1.50

Problem 6.26

ø 16
R 6
2 PLACES
ø 30
ø 76
45×2
BOTH ENDS
ø 52
ø 8
5 HOLES
EQ SPACED
13
32

ø .63
R .25
2 PLACES
ø 1.19
ø 3.00
45×.09
BOTH ENDS
ø 2.06
ø .31
5 HOLES
EQ SPACED
.25
1.25

Problem 6.27

19
43
11
ø 102
ø 79 ø 38
ø 16
ø 25 CBORE
8 DEEP
6
25
R 6
2 PLACES

.75
1.69
.44
ø 4.00
ø 3.12 ø 1.50
ø .62
ø 1.00 CBORE
.31 DEEP
.25
1.00
R .25
2 PLACES

Problem 6.28

Problem 6.29
Draw the orthographic views required to best describe the object shown. One of the views must be a full section.

Problem 6.29

Problems 6.30 and 6.31
Draw the orthographic views required to best describe the object shown. One of the views must be a full section.

Problem 6.30

Problem 6.31

Problems 6.32–6.36

Given two views of the object, draw both views with the appropriate view showing the object in full section.

Problem 6.32

Problem 6.33

Problem 6.34

Problems 6.32–6.36
Given two views of the object, draw both views with the appropriate view showing the object in full section.

Problem 6.35

Problem 6.36

Problems 6.37 and 6.38
Select the most appropriate views to describe the object shown.

Problem 6.37

Problem 6.38

Problems 6.39 and 6.40
Draw the given views of the object shown. At the indicated location, draw a revolved section to indicate the interior shape.

Problem 6.39

Problems 6.39 and 6.40

Draw the given views of the object shown. At the indicated location, draw a revolved section to indicate the interior shape.

Problem 6.40

Problem 6.41

Complete the two views of the object shown. The top and bottom of the right arm are semicircular in shape while the top and bottom of the left arm are rectangular. The top view will show the conventional representations for runouts for these shapes.

Problem 6.41

Problems 6.42–6.44

Draw the indicated views of the object shown.

FILLETS AND ROUNDS R 3

SECTION A-A

Problem 6.42

FILLETS AND ROUNDS R .125

SECTION A-A

SECTION B-B φ 12

FILLETS AND ROUNDS R 3

Problem 6.43

SECTION B-B φ .50

FILLETS AND ROUNDS R .125

FILLETS AND ROUNDS R 3

SECTION C-C

Problem 6.44

FILLETS AND ROUNDS R .125

SECTION C-C

Problems 6.45–6.47
Complete the two views shown of the object. The cutting plane indicates that a broken-out section is required.

Problem 6.45

Problem 6.46

Problem 6.47

Problems 6.48–6.52

Draw the views that best describe the objects shown. One of the views must be an offset section.

Problem 6.48

Problem 6.49

Problem 6.50

Problems 6.48–6.52
Draw the views that best describe the objects shown. One of the views must be an offset section.

Problem 6.51

Problem 6.52

TERMS YOU WILL SEE IN THIS CHAPTER

DIMENSION   Numerical value that defines the size of an object or a part of an object.

DIMENSION LINE   Thin solid line used to show the extent and direction of a dimension.

EXTENSION LINE   Thin lines drawn perpendicular to dimension lines to indicate the feature of the object to which the dimension refers; also known as projection lines or witness lines.

LEADER LINE   Thin solid line used to direct dimensions or notes to the feature being described.

DATUM LINE   Thin solid line that establishes a reference for all dimensions on a drawing.

ARROWHEADS   Placed at the ends of dimension lines to show the limits of the dimension.

NOTES   Placed on a drawing to clarify or define a dimension or manufacturing process.

LIMIT   Maximum or minimum size allowed for a particular dimension.

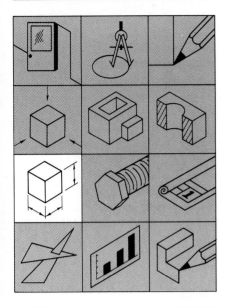

7

DIMENSIONS AND TOLERANCES

7.1 INTRODUCTION

In earlier chapters you learned how to describe completely the shapes of objects on drawings following the principles of orthographic projection. Before the object can be built, however, the size of the object must also be completely described. Adding information to a drawing about the size of an object is known as *dimensioning* a drawing (Figure 7.1).

Dimensioning a drawing is not a simple task. In this chapter you will learn the standard vocabulary used, how drawing scale is specified, techniques for dimensioning basic geometrical shapes, appropriate placement of dimensions on drawings, and standard dimensioning practices. This chapter also includes information regarding the proper specification of tolerances, the principles of good dimensioning, and techniques for checking a completed drawing for dimensioning accuracy.

Figure 7.1   Adding size description to familiar shapes tells how big they are.

7.1.1 Dimensioning Standards

It is important that all persons reading a drawing interpret it in exactly the same way. This is why the views in an orthographic drawing are always placed in the same relationship to each other. When reading size information (**dimensions**) on a drawing, it is also important that there be only one way to interpret the information. Misunderstanding will result in parts that may fail. This could cost loss of reputation, time and money, or even human life. To increase the clarity of dimensioning, standard practices have been developed.

The American National Standards Institute (known as ANSI) has established standards for dimensioning mechanical drawings. The material in this chapter (except where otherwise noted) conforms to the latest ANSI dimensioning standards known as ANSI Y14.5M.

7.1.2 Dimensioning Practices

You, as a beginning student, should note that there is more to describing the size (dimensioning) of an object than simply associating numbers with the geometry of the object. First, it should be remembered that there is no such thing as an "exact" measurement of a dimension. A carpenter installing the framework for the walls of a house may be satisfied if the wall is within a quarter of an inch of the specified measurement. A jet aircraft engine component, however, might have to be located to within less than 0.0001 in. of the specified measurement,* and the location of conductor paths in a large-scale integrated circuit in a computer must be held to within fractions of a micron. Therefore, every dimension on a drawing will have a **tolerance,** or allowable variance, associated with it.

The selection and placement of dimensions will be influenced by the methods used to produce the object and the relationships between various surfaces of the object. The functional relationship between an object and other parts in an assembly (how parts fit together to do their job) will also influence selection and placement.

---

*ANSI Standard Y14.5M specifies that where a dimension is less than 1 millimeter, a zero precedes the decimal point. It also specifies that a zero is not used before the decimal point for values less than 1 inch. Your editors have included the zero before the decimal point for values less than 1 inch to improve the readability of this text for the student not familiar with standard practices. You may include the zero in decimal inch calculations but you should not include the zero in decimal inch dimensions on your drawings.

Figure 7.2 Pencil drawn and dimensioned to three different scales.

7.2 SCALING AND DIMENSIONS

More often than not it is impossible to make a drawing the same size as the finished product. For example, you would not expect the floor plan for your house to be drawn full scale (or life-size), but the dimensions (numerical values) on a drawing are always given as full size.

Let's suppose that you are making a half-scale drawing of an object that is 6.5 in. wide. When you lay out the width on your drawing, you will measure 3.25 in., or one-half of full size. Your half-size scale will enable you to do this with no calculations. When you place the dimensions on your drawing, however, you will dimension the width as 6.5 in., or actual size (Figure 7.2). The title block of the drawing will indicate the scale to which the drawing was made.

7.3 DIMENSIONING BASIC SHAPES

There are a number of standard practices used in dimensioning basic shapes. The practices discussed in this section apply to mechanical drawings used in the manufacturing industries. Different practices are used in architectural drawings, structural drawings, or printed circuit board design.

7.3.1 Assumptions

A number of assumptions are made by those interpreting engineering drawings. If there are exceptions to these assumptions, they must be clearly stated on the drawing. Two important ones concern perpendicularity and symmetry.

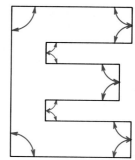

Figure 7.3 Right angles are assumed to be 90° and are not dimensioned.

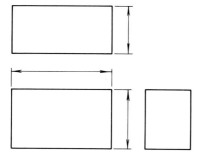

Figure 7.4 Three dimensions describe size of rectangular prism.

Figure 7.5 Positive cylinders with diameters and heights dimensioned.

Perpendicularity

All lines on a drawing that appear to be perpendicular (at a 90° angle) are assumed to be perpendicular to within the specified tolerance (Figure 7.3). All angles other than 90° must be dimensioned.

Symmetry

Objects, patterns of holes, or other features that appear to be symmetrical are assumed to be symmetrical unless otherwise specified. If there is any chance of confusion by the user of the drawing, symmetry (or the lack of symmetry) should be noted clearly on the drawing.

7.3.2 Simple Shapes

Even the most complex parts can be broken down into a collection of simple shapes. When you can dimension the simple shapes on a complex object, you will have dimensioned most of the object. Three shapes are most common: the rectangular prism, cylinder, and cone (or frustum of a cone).

Rectangular Prisms

The height, width, and depth of a prism must be specified (Figure 7.4).

Cylinders

There are two types of cylinders to dimension: positive and negative. Positive cylinders are solid objects. Positive cylinders are dimensioned by giving the length and outside diameter of the cylinder. The numerical value of the diameter is always preceded by the diameter symbol (ϕ). It is usually placed on the longitudinal, or noncircular, view (Figure 7.5). A negative cylinder is a circular cavity (a hole). Negative cylinders are usually dimensioned on the circular view. When the hole goes all the way through an object, the depth is not given [the shape description shows this fact (Figure 7.6)]. In the case of a blind hole (a negative cylinder that does not pass completely through an object), the depth of the hole refers to the depth of the full diameter and may be given in a note (Figure 7.7).

Figure 7.6 Positive cylinder with negative cylinder or hole through it.

Figure 7.7 Cylindrical object with properly dimensioned blind hole.

Cone and Frustum

A cone may be dimensioned by giving the diameter of the base and the height (Figure 7.8). Another method of describing a cone is to give the height and the included angle of the cone. Still another method is to specify the diameter of the base and the taper per foot of the sides. A frustum of a cone is a truncated cone, that is, a cone with its point cut off. It may be dimensioned by specifying the diameter of base, the included angle, and the height; the diameters of both ends; and the height or the diameter of the base and height of the original cone and the height of the frustum (Figure 7.9).

Figure 7.8 Size of cone totally described by its base diameter and height.

Figure 7.9 Frustum of cone may be described by three different sets of dimensions.

ALL FILLETS AND
ROUNDS R .08

Figure 7.10 All fillets and rounds of same size can be described in single note.

Figure 7.11 Slots with round ends may be dimensioned in three different ways. Courtesy of ANSI

7.3.3 Simple Curved Shapes

In Chapter 4 you read about fillets and rounds, which are used at the intersection of surfaces to provide a curved contour and improve strength. You could dimension each of these arcs separately and would if they were all different sizes. Small arcs used for fillets and rounds are often the same size on a part. These are usually dimensioned with a single note (Figure 7.10). This note, similar to one of the following, would be placed in a prominent place on the drawing:

All Fillets and Rounds R .08

All Undimensioned Radii .08

We often find slots with rounded ends on mechanical parts. There are several accepted ways to describe the dimensions of these holes as shown in Figure 7.11. Note that the curved ends have radii noted, but the radii are not dimensioned.

7.3.4 Circular Center Lines

Features on round objects are often located on circular center lines. Ribs for reinforcement and holes for connecting parts are two types of features that may be located on circular center lines. A group of holes arranged on a circular center line are often used for bolting two parts together; these center lines are called bolt circles. The diameter of a bolt circle or other circular center line is dimensioned on the circular view. Three acceptable ways to dimension them are shown in Figure 7.12. You may find the letters ''B.C.D.'' included with the size of the diameter on some drawings.

7.4 PLACEMENT OF DIMENSIONS

The proper placement of dimensions is important. If dimensions are placed in a haphazard manner, confusion and misunderstanding may result. Some general guidelines for proper placement of dimensions follow.

7.4.1 Characteristic View

Dimensions for a particular feature of an object should be placed in the most characteristic view of that feature. In the example shown in Figure 7.13, the notch shows most clearly in the front view. If you look at the top view, it is hard to tell if there is a notch or a series of steps. The front view, however, clearly shows the notch. Therefore the width and height of the notch should be dimensioned in the front view rather than in the top or side view.

7.4.2   Dimensions Should Be Grouped

You should plan to group the dimensions on a drawing in such a way that the reader can locate them with a minimum of effort. It is also a good idea to start dimensions from a *common corner*. Study Figure 7.13 and you will note that all height dimensions are measured from the bottom and all width dimensions from the right side of the front view. Imagine trying to make these measurements on a block of metal—it is much easier if all measurements read from a common corner. Note how much harder it is to read the dimensions in Figure 7.14 compared to Figure 7.13. This problem gets worse as the complexity of the object increases.

7.4.3   Preferred Locations

The preferred locations for dimensions are, first, between the views; second, to the right of or below a view; and third, to the left of or above a view. The view should be one that describes the surface in true size and shape. Dimensions should not be located within the boundaries of an object unless other choices are less clear. Hidden lines should not be dimensioned since a hidden view of a feature is usually not the most characteristic view. It may be clearer if you construct a section view to expose hidden features for dimensioning.

Figure 7.13   Dimensions placed on characteristic view and located from common corner.

Figure 7.14   Dimensions placed haphazardly are hard to find and understand.

Figure 7.12   Diameter of circular center lines should be given by one of three methods shown.

7.5 DIMENSIONING SYSTEMS

A number of different dimensioning systems have been developed, and many are still in use. You should be aware of these systems in addition to the standards used in the classroom.

7.5.1 Alignment of Dimension Numbers

Older drawings were frequently dimensioned with the aligned dimensioning system. In the aligned system, the dimensions were placed in alignment with the dimension lines and were oriented to be read from either the bottom or right of the drawing. Current ANSI standards call for use of the unidirectional system where all dimensions are read from the bottom of the drawing (Figure 7.15).

Figure 7.15 Aligned and unidirectional placement of dimensions; note height dimensions.

7.5.2 Units of Measurement

The two most common systems standards are those published by the International Standards Organization (ISO) and the American National Standards Institute (ANSI Y14.5M). Drawings that follow ISO standards have the designation SI. The commonly used unit of measurement on SI drawings is the millimeter. The ANSI standards specify the decimal inch or the millimeter.

Dimensions on drawings made in the United States are usually in decimal inches unless otherwise specified on the drawing. The ANSI standards specify that a note of the following form should be used if there is any doubt about which units are used: UNLESS OTHERWISE SPECIFIED, ALL DIMENSIONS ARE IN MILLIMETERS (OR INCHES).

A convenient way to specify which system is used is to put the word *metric* or *inches* in a rectangle near the title block (Figure 7.7). Fractional dimensions are mostly found on older drawings, although fractional dimensions are sometimes used on new drawings. You should be aware that although the ANSI Y14.5M standards are the most commonly used, many companies maintain their own drawing and dimensioning standards. It is important, when starting a new job, to become familiar with the standards and practices in use in your new organization.

7.6 STANDARD PRACTICES

The purpose of engineering drawings is to communicate information. As in any system of communication, consistency and clarity are important. There are several standard practices relative to lines, **arrowheads,** and notes with which you should be familiar.

7.6.1 Lines for Dimensioning

Relative line weights are important for improving the readability of drawings. The ANSI standards call for object lines to be 0.7 mm thick. **Dimension, extension, leader,** and center lines are to be 0.5 mm thick (your authors believe that the use of three line weights are preferable—the 0.5 mm thickness can be reserved for hidden lines and 0.35 mm used for dimension, extension, leader, and center lines). Computer-generated drawings may also be drawn with three weights of lines as suggested in Chapter 4. All lines must be dark for proper reproduction. Remember, there are no light lines on finished copies of drawings—only thick, medium, or thin dark lines.

7.6.2 Arrowheads

Well-made arrowheads will give a drawing a professional appearance and make the drawing easier to read. Arrowheads should be ~$\frac{1}{8}$ in. (3 mm) long, and the width should be about one-third of the length (Figure 7.16*a*). They may be a little longer on long dimension lines but should not be shorter. If too short, they become just tiny blobs on the drawing. Arrowheads may be either open, closed, or filled (Figure 7.16*b*). Do not attempt to "dress up" your drawing with fancy arrowheads—they reduce clarity rather than helping it (Figure 7.16*c*).

7.6.3 Lettering and Notes

All of the numbers, letters, and words on your drawing must be readable. As you learned in Chapter 2, the minimum size for

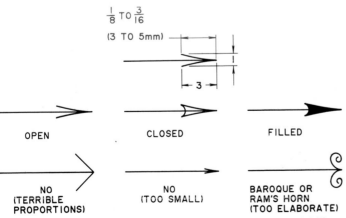

Figure 7.16 (*a*) Well-proportioned arrowheads improve readability of drawing. (*b*) Acceptable arrowheads may be open, closed, or filled. (*c*) Wide, tiny, or ornate arrowheads are unacceptable.

Figure 7.17 Leader lines for note are made up of straight lines; leader always goes to beginning or end of note, never in middle.

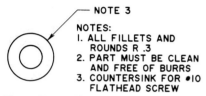

Figure 7.18 If there are many notes on a drawing, they may be lettered on a note list.

readable letters is $\frac{1}{8}$ in. (3 mm) high. None of your dimension numbers or letters should be smaller than this; guidelines should be used for all lettering.

Notes are placed on drawings to indicate the size or depth of holes, the number of holes or spaces in a repeating pattern, or other information required in manufacturing an object. Notes may be placed at the end of a leader line, which has a short horizontal flag (Figure 7.17). If a note is too large to fit comfortably in the body of a drawing, it may be placed in a note list. In this case the leader line will have an indication as to which note applies (Figure 7.18). Leader lines should go to the beginning or the end of a note, never to a line in the middle of the note (Figure 7.17).

7.6.4 Readability

The principles and methods outlined in the preceding sections are not rules—they are the results of years of practical experience. Original drawings never leave the engineering office—only copies are sent to production or construction sites. These sites may be open to wind or rain; they may be using oily cutting fluids; and there may be no clean surface on which to lay the drawing for study. So drawings are folded to fit hip pockets, splashed with oil, and splattered with rain—after this treatment, *they must still be readable*. The principles outlined will help to make your drawings or those done under your supervision readable under shop and field conditions.

7.7 PLANNING AND LAYOUT

You will need to plan a dimensioned drawing more carefully than you planned only the views (Chapter 4). Keep in mind that most dimensions should be attached to the characteristic view and should be placed between the views. If you don't provide enough space for dimensions, the drawing will become crowded and hard to read. A good approach is to make a sketch of the object with all the dimensions before you begin the formal drawing. The views can then be placed to allow proper placement of dimensions. Review Chapter 4 for planning a drawing and now include space between the views and around the views for the dimensions and notes.

7.8 FUNCTIONAL RELATIONSHIPS

Selection of dimensions involves knowledge of the function as well as the shape of an object. Since no feature of an object can

Figure 7.19 Total length of (*a*) has a tolerance of 0.020; total length of (*b*) has a tolerance of 0.080 (0.020 × 4).

be made to an exact size, every dimension will have an associated tolerance. Some times a general tolerance note will be given that will apply to all dimensions on a drawing except those having a more precise tolerance. When an object is part of an assembly of parts, the tolerances must be such that the parts will fit together. For example, the largest permitted shaft must fit properly into the smallest permitted hole. Individual tolerances, more precise than general tolerances, would be specified for these parts.

7.8.1 Tolerance Accumulation

When a series of dimensions is involved, an accumulation of tolerances may cause problems. For example, if there were four uniform steps on an object and each step were dimensioned from the next step with a tolerance of 0.020, then there would be a tolerance accumulation of 0.080 in. (4 × 0.020) across the series of steps. This tolerance buildup could be avoided by dimensioning each step from one edge of the object. Look at Figure 7.19 to see how this is done. The tighter the tolerance, the more expensive it will be to manufacture the part. Therefore a good designer will select the method of dimensioning that will minimize tolerance accumulation.

7.8.2 Dimensioning to Finished Surfaces

You need to remember when dimensioning that finished surfaces are more important than unfinished surfaces. A finished surface has been machined (turned, milled, ground, or polished). Finished surfaces are shown with a small V, which resembles a cutting tool touching the surface. Because they have been cut, finished corners are shown as sharp corners where they intersect unfinished surfaces or other finished surfaces (Figure 7.20). Parts for machinery or buildings are made by forging, casting, rolling, welding, and other methods that result in a rough exterior surface.

Figure 7.20 Finished surfaces (note square corners) and finish marks—new and old.

When parts must fit together or be precise, we invest extra time, machinery, and people to make the rough surfaces smoother and more exact.

Figure 7.21 shows two methods for dimensioning the height of the same object. Figure 7.21*a* shows dimensioning the height as a chain of dimensions, while Figure 7.21*b* shows an overall height dimension with the upper and lower protrusions dimensioned separately. At first glance the two methods might appear equal since a simple calculation indicates the same overall height. When tolerances are considered, however, there is a significant difference.

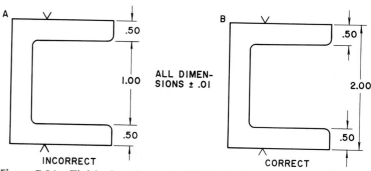

Figure 7.21 Finished surfaces should always be used to locate features on drawing.

If the designer has specified a general tolerance for dimensions of 0.01 in., then the maximum and minimum permissible heights for the object in Figure 7.21*a* would be 2.03 and 1.97 in. The maximum and minimum for Figure 7.21*b* would be 2.01 and 1.99 in. Since there are finish marks on the upper and lower surfaces of the object, we must assume the distance between these surfaces is more important than the size of the cavity. You should select the method shown in Figure 7.21*b*.

7.8.3 Dimensions for Locating Features on an Object

Finished surfaces have some important function (otherwise why go to the additional expense of finishing the surface?). When you locate the centers of holes such as those found in the locator block in Figure 7.22, the center should be dimensioned to the finished rather than the unfinished surfaces.

A pattern of holes, such as the counterbored holes in Figure 7.22, usually will be matched to similar holes in an assembly. Therefore, the distance between holes should be specified. The reason for this is the same as that discussed in Section 7.8.2. Although simple arithmetic might lead you to believe that the results would be the same if we dimensioned from the outside

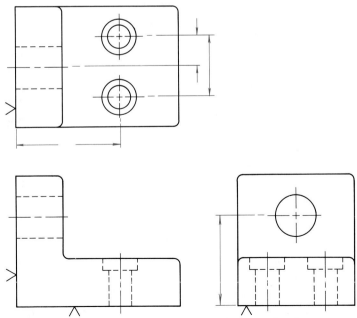

Figure 7.22 Holes need to be located relative to each other or to finished surfaces.

edges to the centers of the holes, the tolerance accumulation will either cause the parts to not fit together or will require specification of tighter tolerances, which will add to costs.

7.8.4 Procedure for Functional Dimensioning

A part that has finished surfaces or holes should be dimensioned so that these features are located with respect to each other. Unfinished surfaces are less important and may be located relative to finished surfaces or other unfinished surfaces. A procedure for dimensioning that has worked for many designers is outlined:

1. Dimension the basic shapes
2. Dimension the cavities
3. Locate the basic shapes and cavities
   a. Finished surface to finished surface
   b. Finished surface to center line
   c. Center line to center line
   d. Other locations
4. Add the notes
5. Check your work
   a. Are all dimensions given that are needed?
   b. Are the numbers correct?
   c. Do the notes describe all special needs?
   d. Will this part fit with matching or mating parts?

7.9 DIMENSIONING ON THE COMPUTER SCREEN

You can use the principles outlined for dimensioning drawings made on a computer. When you plan the drawing, you will need to leave room between the views for the dimensions. If you don't leave enough, it is easier to redraw or move a view on the computer screen and make more space for the dimensions than it is on paper.

7.9.1 A Layer for Dimensions

It is a good idea to put the dimensions and notes on a different layer or level than the views or construction lines. This was mentioned in Chapter 4. If you need to make several changes, you can change dimensions without the danger of changing your views or you can modify the views and then overlay the dimensions and change only the affected values.

7.9.2 Automatic Dimensioning

Some graphics programs have "automatic" dimensioning. This means that the length of the lines you specified on your orthographic views will be lettered on the drawing if you use this feature. If you were a little sloppy when making the views, the dimensions could be in error. You may get the dimensions you drew rather than the ones you wanted.

7.9.3 Commercial Software and Educational Software

Graphics packages designed for professional drafters have many useful features—these are often specified by picking options from a series of menus. This can mean finding your way through a series of four or five menus. A professional will soon become used to the options he or she needs for their type of work. Some programs will allow them to specify a path that automatically selects the needed option from consecutive menus.

There are other CADD programs designed for classroom use that do not have so many automatic features and options. They require that you draw each line and select each number and letter as you would when making a drawing by hand. This might prove faster for a student who does not spend every day with a graphics package. Try the packages you have available and become as familiar with the computer screen as you have with a drawing board or sketch pad. The problems at the end of this chapter are designed to be done on the computer screen or manually.

7.10 OTHER STANDARD DIMENSIONING SYSTEMS

New manufacturing methods, including computer-controlled machine tools, have created the need for additional dimensioning systems. Computer-controlled machines sometimes allow us to transfer information directly from the data base for computer-aided design and drafting systems (CADD) to computer-aided manufacturing systems (CAM). Engineering disciplines other than mechanical design have evolved their own particular dimensioning systems. You need to be aware of the existence of these different systems.

7.10.1 Datum Lines and Surfaces

Numerically controlled machine tools and computer-aided machining frequently employ a system of dimensioning using rec-

Figure 7.23 Rectangular coordinates measured from datum lines are used to dimension parts for computer-aided manufacturing.

tangular coordinates. Figure 7.23 shows an example of a drawing using rectangular coordinates. In this drawing horizontal and vertical **datum lines** are established. Sometimes the datum lines will correspond to the edge of the object as in Figure 7.23, and sometimes the center lines of the first hole are used as datum lines. The datum lines are labeled 0, and all other numbers represent the distance in the horizontal or vertical direction from the datum lines. Datum lines or surfaces may be established for all the principal directions—width, height, and depth. In the example shown in Figure 7.23, the sizes of the holes are shown in a hole table. In some drawings, the hole table may also show the required number of each size hole.

7.10.2 Tabular Dimensioning

Tabular dimensioning is used when several parts have the same general shape, but a dimension, or series of dimensions, changes. Figure 7.24 is an example of a drawing with tabular dimensions. In this case there are three different washers required. Rather than make three different drawings, one drawing is made and a table is used to indicate the numerical values for each dimension. Tabular dimensioning is used to reduce the number of drawings required without reducing the amount or clarity of information.

7.10.3 Other Design Disciplines

The dimensioning systems and examples given have all been taken from mechanical engineering practices used for producing machine parts. Different dimensioning systems and practices are used in other types of drawings such as architectural, structural, or electrical. Figures 7.25–7.27 show examples of dimensioning practices for these types of drawings.

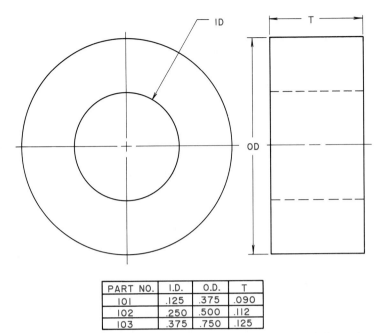

PART NO.	I.D.	O.D.	T
101	.125	.375	.090
102	.250	.500	.112
103	.375	.750	.125

Figure 7.24 Tabular drawings allow description of several sizes of same shape on one drawing.

Figure 7.25 Some architects use slashes or large dots in place of arrowheads. They don't break dimension lines and specify feet and inches rather than just inches. Courtesy of McDonald-Cassel & Bassett, Inc.

Figure 7.26 Steel designers also use feet and inches and don't break dimension lines. Slashes or double slashes may be used rather than arrowheads. Courtesy of McDonald-Cassel & Bassett, Inc.

Figure 7.27 Drawings for electronic printed circuits have no dimensions but are drawn many times full size with great precision and copied photographically.

7.11 DIMENSIONING FOR INTERCHANGEABLE PARTS

Many industries rely on manufacturing large numbers of items that consist of assemblies of parts. (Automobiles, home appliances, and weapons are examples.) To be cost effective, mass production requires the use of interchangeable parts. No object can be made exactly to a given dimension. Therefore, tolerances are specified to indicate the permissible variation from the given dimension. The smaller the tolerance, the more expensive the object is to manufacture. Dimensions with greater tolerances are less expensive to produce but may result in parts that cannot be assembled or will not perform the function for which they were intended. There is a need, therefore, for dimensioning and tolerancing systems for interchangeable assembly. Eli Whitney, the inventor of the cotton gin, is given credit for the first application of interchangeable parts when his shop produced weapons for the U.S. government before the Civil War.

7.11.1 Types of Fits

There are three types of fits of mating parts. They are clearance, interference, and transition fits. Each of these fits will be illustrated using the groove block shown in Figure 7.28. The slot in the block has a width with a tolerance of 0.020 in. Remember, the tolerance is the difference between the minimum and maximum permitted sizes (**limits**). Three different sliders are shown in Figure 7.28: A, B, and C. Figure 7.29 shows the slider in place in the groove block. It is possible to write an expression to calculate the maximum and minimum clearance between the slider and the slot (Figure 7.30). Maximum clearance will occur when

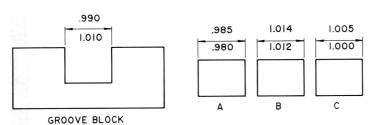

GROOVE BLOCK

Figure 7.28 Grooved block with three different sliders that may fit in it.

Figure 7.29 Slider in place in groove.

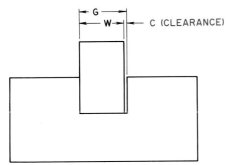

Figure 7.30 When slider fits in groove, space left is called clearance.

the slider is small and the slot is large. Minimum clearance will occur when the slider is large and the slot is small. The relationships can be expressed as follows:

$$C_{max} = G_{max} - W_{min}$$

Where C is the clearance, G is the groove size, and W is the width of the slider.

Clearance Fits

If we calculate the C_{max} and C_{min} for slider A and the block, the results are as follows:

$$C_{max} = 1.010 - 0.980$$
$$= 0.030$$
$$C_{min} = 0.990 - 0.985$$
$$= 0.005$$

Therefore the clearance will vary between 0.005 and 0.030 in.

When slider A is assembled with the groove block, we have determined that there will be clearance no matter what combination of permissible sizes of slider and block we choose. This type of fit is known as a clearance fit. Some clearance fits are called running or sliding fits.

Checking Fit Calculations

Slider A has a tolerance of 0.005 in., while the tolerance on the slot is 0.020. Note that the difference between the maximum and minimum clearance in this case is 0.025 in. $(0.030 - 0.005)$, which is equal to the sum of the tolerances of the two parts generating the clearance. It will always be true that the difference between the maximum and minimum clearance will equal the sum of the tolerances of the parts generating the clearance. This provides a quick way to check your work when calculating fits. Simply compare the sum of the tolerances with the difference between the maximum and minimum clearance. If they are not equal, there is an error. The tolerance of the system may be expressed mathematically as

$$T_s = \Sigma \, (T_1 + T_2 + \cdots T_n) = C_{max} - C_{min}$$

We calculate or examine the maximum and minimum conditions of a fit (the extremes). If these values are acceptable, we know that every value between the extremes will be acceptable.

Interference Fits

Now consider the case of assembling slider B with the block. You can use the same expression to calculate the maximum and minimum clearance:

$$C_{max} = 1.010 - 1.012$$
$$= -0.002$$
$$C_{min} = 0.990 - 1.014$$
$$= -0.024$$

In both extremes we get negative clearance. Negative clearance is another way of saying interference. This type of fit is called an interference fit.

Check the calculations as follows:

$$\text{Tolerance on slot} = 0.20$$
$$\text{Tolerance on slider} = 0.002$$
$$T_s = \Sigma (T_1 + T_2) = C_{max} - C_{min}$$
$$\text{Sum of tolerances } T_1 + T_2 = 0.020 + 0.002 = 0.022$$
$$\text{Difference between } C_{max} \text{ and } C_{min} = (-0.002) - (-0.024)$$
$$= 0.022$$

Interference fits are used to permanently (or semipermanently) assemble parts. Some interference fits are called force or shrink fits.

Transition Fits

Now assemble slider C with the block. The maximum and minimum clearances are

$$C_{max} = 1.010 - 1.000$$
$$= 0.010$$
$$C_{min} = 0.990 - 1.005$$
$$= -0.015$$

In this case the clearance varies from 0.010 to -0.015. At one extreme we have clearance and at the other we have interference. This type of fit is known as a transition fit. Transition fits are sometimes called locational fits.

Checking calculations you see the following:

$$\text{Tolerance on slot} = 0.020$$
$$\text{Tolerance on slider} = 0.005$$
$$\text{Sum of the tolerances } T_1 + T_2 = 0.025$$
$$\text{Difference between } C_{max} \text{ and } C_{min} = 0.010 - (-0.015)$$
$$= 0.025$$

7.11.2 Standard Fits

In the examples presented, the objects were dimensioned and the clearances were calculated. In most cases, however, the clearances will be specified and the dimensions must be calculated. The important thing to recognize is that the mathematical relationship between the clearance and the dimensions remains constant. By first writing a mathematical expression and then solving that expression for the required variables, clearance calculation problems can be solved with a minimum of confusion. You should be able to analyze an assembly, write an expression for the relationship, determine maximum and minimum clearances, and determine the type of fit.

Since many people are concerned with specifying types of fits, standard fits have been developed by various organizations such as ISO and ANSI. Tables for standard clearance, interference, and transition fits are available from the standards organizations and included in many engineering handbooks. Samples of the ANSI tables for fits are included in the Appendix.

7.11.3 Cylindrical Fits

There are two systems for determining the dimensions for cylindrical fits—the fit between round holes and shafts (Figure 7.31). These systems are known as the basic hole system and the basic shaft system. To understand both systems you must be familiar with the terms *allowance* and *nominal*. Allowance is defined as the difference in size between the hole and shaft for the tightest fit. The allowance always equals the minimum hole minus the maximum shaft. Nominal refers to the general size of a dimension independent of the tolerance. For example in Figue 7.28 the slot in the groove block and the width of the sliders all would be considered to have a nominal size of one inch.

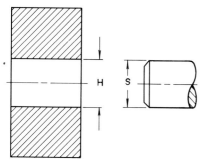

Figure 7.31 Cylindrical fit system with a round shaft fitting round hole, shown in section.

Basic Hole System

The basic hole system is used when the size of the hole is considered more important or more expensive to produce than the size of the shaft. If a machine shop only has certain sizes of reamers (tools for finishing round holes very accurately), it would be desirable to adjust the shaft size to fit the hole. In the basic hole system, the minimum hole diameter is taken as the basic hole size. For example, for a nominal $\frac{1}{2}$-in. hole and shaft, a $\frac{1}{2}$-in. reamer might be used. Because reamers and other tools cannot be operated perfectly, the minimum hole size would be 0.500 in., and the largest hole likely to be produced would be a little larger. Let's calculate the limit dimensions for the hole and shaft with a nominal size of $\frac{1}{2}$ in.

Assume that the following design specifications are to be met:

Nominal size	$\frac{1}{2}$ in.
Allowance	0.005
Hole tolerance	0.003
Shaft tolerance	0.001

To find the limit dimensions for the hole:

Basic size (minimum hole diameter)	0.500
Assigned hole tolerance	0.003
Maximum hole (minimum hole plus tolerance)	0.503

To find limit dimensions for the shaft:

Basic size	0.500
Allowance	0.005
Maximum shaft (basic size minus allowance)	0.495
Assigned shaft tolerance	0.001
Minimum shaft (maximum shaft minus tolerance)	0.494

Basic Shaft System

Sometimes the diameter of the shaft may be more important than the diameter of the hole. For example, it may be desirable to use standard size stock shafts, which can be purchased with carefully controlled, precise diameters. In these situations, the basic shaft system would be used. The basic shaft system is the least common of the two systems. The basic shaft system is similar to the basic hole system except that the maximum shaft diameter is used as the basic size. Here is an example calculation of maximum hole and shaft diameters using the basic shaft system.

Assume the following design specifications:

Nominal size	$\frac{1}{2}$ in.
Allowance	0.010
Shaft tolerance	0.002
Hole tolerance	0.006

To find the limit dimensions for the shaft:

Basic size (maximum shaft diameter)	0.500
Assigned shaft tolerance	0.002
Minimum shaft diameter (maximum minus tolerance)	0.498

To find the limit dimensions for the hole:

Basic size	0.500
Allowance	0.010
Minimum hole diameter (minimum hole plus allowance)	0.510
Assigned hole tolerance	0.006
Maximum hole diameter (minimum hole plus tolerance)	0.516

7.12 ADVANCED DIMENSIONING TOPICS

The information you have studied about dimensioning is only a small part of the total you would need to do a professional job of specifying parts for manufacture. What you have studied is basic to all size descriptions, however, there are several other topics with which you need to become familiar to be a professional designer.

7.12.1 Metric Dimensioning with International Standards

All of the limit dimension examples given can be done using metric values. However, a standard for metric tolerances exists that considers tolerances in a different way: a system of tolerances has been established that uses code letters (Hole: 20H8, Shaft: 20f7). This system is described in ANSI standard B4.2. The standard and its application are discussed in Appendix 1.

7.12.2 Surface Control

Another area of interest is surface control. When we design parts for interchangeable assembly, we need to be concerned with not only the size but also the surface roughness and quality. Imagine trying to measure the height of course sandpaper to a thousandth of an inch! We cannot achieve precise tolerances unless the surfaces are smooth, and sometimes the nature of the tool marks or roughness affects the function of the part. ANSI standards B46.1 and Y14.36 provide information about surface texture and specification for controlling it. Some of this information is presented in Appendix 1.

7.12.3 Geometric Tolerancing

Geometric tolerancing is another advanced topic. It concerns the control of factors affecting parts to be assembled such as straightness, flatness, concentricity, parallelism, and others. There is a complete set of symbols available to describe on the drawing what is required for a part to function as needed. The use and application of these symbols is a complete course in itself. Some of this information is given in Appendix 1 so that you may recognize the symbols when you find them on a drawing.

None of the topics mentioned here can be included in an introductory course in graphics. Your instructor may have you study Appendix 1 to become aware of these methods and symbols. As you specialize you may study them in detail in advanced courses.

7.13 *PRINCIPLES OF GOOD DIMENSIONING*

A number of fundamental principles have been developed that will help you in selection and placement of dimensions. These principles, understood by engineers and craftsmen, will avoid possible misinterpretations. They have been summarized below for easy reference. Remember, however, that the overriding principle is that of *clarity*. Whenever it seems that the fundamental principles are in conflict, the conflict should always be resolved in a way that promotes maximum clarity.

1. *Each feature of an object is dimensioned once and only once.* Remember that every dimension has a tolerance associated with it. In Figure 7.32a, the overall height, the height of the step, and the depth of the notch are given. Assume for the moment that

Figure 7.32 Each feature of object is dimensioned only once, with the dimensions selected to suit function.

the tolerance on all three dimensions is ±0.010 in. This leads to a dilemma relative to the permissible variation in the height of the object. If the height is determined to be the sum of the notch and the step, the permissible variation would be ±0.020 in. If the height dimension is used, however, the variation could only be ±0.010. To resolve this problem, one of the dimensions must be left out. In practice, the least important dimension is the one left out, and error is allowed to accumulate in a part of the object that is not critical. Unless there is some functional requirement, it is usually better to dimension material rather than space. Figure 7.32*b* shows the preferred way to dimension the object.

2. *Dimensions should be selected to suit the function of the object.* If, as in the case of the object in Figure 7.32*b*, the object were used in an assembly where the depth of the slot were more important to the assembly than the height of the step, then the step height dimension would be replaced by the slot depth dimension.

3. *Dimensions should be attached to the most descriptive view of the feature being dimensioned.* In Figure 7.33 the most descriptive view of the notch is the front view. It is incorrect, therefore, to dimension this notch in the less descriptive top or side views.

4. *Dimensions should not specify the manufacturing method.* The manufacturing method should only be specified if it is a mandatory design requirement. A common practice found in older drawings is to see hole diameters specified by the drill size as in Figure 7.34*a*. The preferred method is shown in Figure 7.34*b*. The only time a manufacturing method should be specified is when it is the only acceptable way to produce the object. If, for example, the tolerance could be met by either punching, drilling, laser machining, chemical milling, or electrical discharge machining, the production department should have the flexibility to select the most economical manufacturing process. On the other hand, if a particular process is important to the function of the object e.g., the surface obtained by electric discharge machining rather than drilling, the process should be specified on the drawing.

5. *Angles shown on drawings as right angles are assumed to be 90° unless otherwise specified.*

6. *Dimensions should be located outside the boundaries of the object whenever possible.* When dimensions are located within the boundary of an object, such as in Figure 7.35*a*, they are difficult to find and read. Figure 7.35*b* shows the preferred way to locate them.

7. *Dimension lines should be aligned and grouped to promote uniform appearance.* Unnecessary crossing of object and exten-

Figure 7.33 Dimensions should be placed on most descriptive view.

A. AVOID   B. PREFERRED
Figure 7.34 Dimensions should specify only size, not process.

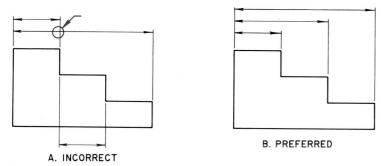

Figure 7.35 Dimensions and dimension lines should not be placed on views.

Figure 7.36 Dimensions should be aligned and grouped. Dimension lines should never be crossed.

sion lines as shown in Figure 7.36*a* should be avoided. By aligning and grouping the dimensions, as in Figure 7.36*b*, the drawing not only looks neater, it is much easier to read.

8. *Dimension lines should be unbroken except for the number between the arrowheads.* It is very easy to misread a dimension if the dimension line is crossed by any line other than the extension lines at the ends (Figure 7.36*a*).

9. *The space between the first-dimension line and the object should be at least $\frac{3}{8}$ in. (10 mm). The space between dimension lines should be at least $\frac{1}{4}$ in. (6 mm).* Proper spacing is shown in Figure 7.37. Remember, the spacing shown is minimum. The spacing can be increased to improve readability.

10. *There should be a visible gap between the object and the origin of an extension line.* This gap of about $\frac{1}{16}$ in. (2 mm) is shown in Figure 7.37.

11. *Extension lines should extend $\frac{1}{8}$ in. (3 mm) beyond the last dimension line* (Figure 7.37).

Figure 7.37 Dimension lines should be spaced for readability; extension lines should have visible gap at origin and extend beyond last dimension line.

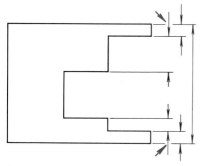

Figure 7.38 Extension lines should be broken if they cross arrowheads.

12. *Extension lines should be broken if they cross or are close to arrowheads* (Figure 7.38).

13. *Leader lines used to dimension circles or arcs should be radial.* Radial lines are those that, if extended, pass through the center of a circle or arc. Figure 7.39*a* shows a radial line; Figure 7.39*b* shows a nonradial line.

14. *Dimensions should be oriented to be read from the bottom of the drawing.* This orientation is known as the unidirectional dimensioning system. Older drawings frequently used the aligned system. These systems were shown earlier in this chapter in Figure 7.15.

15. *Diameters are dimensioned with a numerical value preceded by the diameter symbol (ϕ).* You may find many drawing examples that were produced before publication of the 1982 ANSI standards that do not conform to this practice. All new drawings should use the diameter symbol as shown in Figures 7.34*b*, 7.39*b*, and 7.40.

16. *Positive cylinders should be dimensioned in the longitudinal view.* Figure 7.40 shows the proper dimensioning of concentric cylinders. The hole was not dimensioned in the longitudinal view because this would have resulted in dimensioning to a hidden line. Hidden lines are not considered to be the most descriptive view of features. If a section view of the hole in Figure 7.40 were shown instead of a hidden view, the hole could be dimensioned in the longitudinal view.

Even more important is the manner in which a machinist usually makes a circular hole. It is most often made with a tool, like a drill, that requires the machinist to locate the center as a starting point. Therefore, the machinist needs to know the location of the center point, which is shown on the characteristic, circular view of the hole. It is here that the size should be given, so that all information about the hole is grouped in one place. Positive cylinders are most frequently produced by turning on a lathe. The machinist is looking at the work turning between centers in a longitudinal position. It is in this position that measurements of diameters are taken, so it is in this position on the drawing that these diameters should be given.

A. CORRECT   B. INCORRECT

Figure 7.39 Leaders for notes on cylindrical parts should be radial; diameters should always be preceded by symbol (ϕ).

Figure 7.41 Internal and external radii should be dimensioned with numerical value preceded by symbol R.

Figure 7.40 Positive cylinders should be dimensioned in longitudinal view; depth of blind hole is given in note with diameter.

17. *Radii are dimensioned with a numerical value preceded by the radius symbol (R).* Older drawings have the R following rather than preceding the numerical value. All new drawings should conform to the current standards. Figure 7.41 shows how to dimension a radius.

18. *When a dimension is given to the center of an arc or radius, a small cross is shown at the center.* Figure 7.41b shows a radius with a located center. In Figure 7.41A it is not necessary to locate the center of the radius since it can be located from the horizontal and vertical lines to which it is tangent. It is not necessary to locate the centers for small fillets and rounds.

19. *The depth of a blind hole may be specified in a note, and is the depth of the full diameter from the surface of the object* (Figure 7.40).

20. *Counterbored, spotfaced, or countersunk holes should be specified in a note.* Figures 7.42–7.44 show dimensioning of counterbored, countersunk, and spotfaced holes. All are concentric with a starting hole. A counterbored hole is usually intended to provide space for the head of a socket head or fillister head screw; a countersunk hole is the negative frustum of a cone and provides space for the head of a flat head screw. A spotface is a very shallow cavity intended only to provide a smooth bearing area for a fastener. You will learn more about fasteners in Chapter 8.

Figure 7.42 Counterbored holes: depth may be given with diameter or to visible line on longitudinal view, when sectioned. Courtesy of ANSI

Figure 7.43 Countersunk and counterdrilled holes. Courtesy of ANSI

7.14 *CHECKING A DIMENSIONED DRAWING*

Dimensions on drawings are very important for communicating technical information. Incorrect or missing dimensions could result in faulty parts. This can be costly in terms of time, money, or professional reputation. Therefore you should check all dimensions before completing a drawing.

You should first check to see that every feature has been di-

Figure 7.44 Spotfaced holes: either depth or remaining thickness of material may be specified. Courtesy of ANSI

mensioned once and only once. This can be done by moving a straightedge from left to right and from top to bottom across the drawing, stopping at each feature and making sure it has been properly dimensioned.

Next make sure that all numerical values are correct. Take a scale and lay it on every dimension, checking for accuracy. These two steps will only take a couple of extra minutes in preparing a drawing, and by getting into the habit of checking your dimensions, you will avoid many embarassing and costly mistakes.

7.15 *SUMMARY*

Describing the size of objects is important. Without size information, nothing could be built. Without the concept of tolerances, there could be no interchangeable parts, and manufactured products would be prohibitively expensive. Because of the importance of dimensioning drawings, standard practices have been developed. Since the primary reason for producing drawings is that of communicating information, these standard practices must be learned and followed. Proper attention to line weights and selection and placement of dimensions will result in drawings that communicate clearly and accurately the requirements of a design. If you will take a few minutes to review the principles of good dimensioning before making a drawing and then check each completed drawing, you will soon be making clear, professional-quality, engineering drawings.

PROBLEMS

Problem 7.1

Problem 7.2

Problem 7.3

Problem 7.4

Problem 7.5   HOLE IS 3 UNITS DEEP

Problem 7.6   ECCENTRIC BUSHING THROUGH
HOLE IS OFFSET $\frac{1}{2}$ UNIT

Problem 7.7

LARGE CYLINDER IS
IO UNITS ø,
3 UNITS DEEP
SMALL CYLINDER IS
4 UNITS ø
2 UNITS DEEP
ALL HOLES ARE
2 UNITS ø
AND THRU
CIRCULAR CENTER
LINE IS 7 UNITS ø

Problem 7.8

FLANGE PLATE
IS 2 UNITS DEEP

SMALL HOLES
ARE $1\frac{1}{2}$ UNITS ø

ODD HOLE IS
DISPLACED 30°

Problems 7.1–7.8
Prepare the needed orthographic views of each assigned problem on $8\frac{1}{2} \times 11$ paper or on your computer.
Let each unit on the sketch equal $\frac{1}{4}$ in., or 6 mm, on your drawing, as designated by your instructor.
Completely dimension the object so that it could be built from your drawing. Be sure to allow enough space
around the views for good size description.

Problem 7.9

Problem 7.10

FINISH ALL OVER

Problem 7.11

HOLES
EQUALLY SPACED

Problem 7.12

Problem 7.13

Problem 7.14

IS THIS THE
BEST CHOICE
OF VIEWS?

Problems 7.9–7.14

The views of each of these objects should be drawn on 8½ × 11 paper or on your computer. Determine dimensions with your dividers and the scale printed in color on this page. Dimension your drawing in one-place millimeters or two-place inches as specified by your instructor. Include a note on each drawing that states which system is used.

Problem 7.15

COMPLETE THE TABLE BELOW FOR CONDITIONS AT A, B, & C

TOLERANCE-SHAFT	——	——	——
TOLERANCE-HOLE	——	——	——
TOLERANCE-SYSTEM	——	——	——
TIGHTEST FIT	——	——	——
LOOSEST FIT	——	——	——
TYPE OF FIT	——	——	——

Problem 7.16

DETERMINE THE MINIMUM AND MAXIMUM CLEARANCE
AND THE TOLERANCE OF THE SYSTEM

TOLERANCE OF SYSTEM _____

$$Cl = \frac{MAX}{MIN}$$

Problem 7.17

DETERMINE THE TOLERANCE OF THE SYSTEM
AND THE LIMITS FOR PART "X"

TOLERANCE OF SYSTEM _____

LIMITS OF CAVITY "X" _____

Problem 7.18

USE THE ANSI FIT TABLES TO DETERMINE
THE CORRECT HOLE AND SHAFT SIZE FOR
CONDITIONS A, B, C & D.

DIMENSION TOLERANCES	NOMINAL	CLASS	HOLE	SHAFT
A	$\frac{1}{2}$	FN4		
B	$1\frac{1}{4}$	LT5		
C	$\frac{3}{8}$	RC8		
D	2	LC9		

Problems 7.15–7.21

Complete the required information for each tolerance problem assigned by your instructor. You may be asked to letter the information neatly on a drawing sheet or a sheet of note paper or to reproduce the drawing manually or on the computer and complete the information on your drawing.

Problem 7.19

DETERMINE LIMITS OF INSIDE DIAMETER AND LENGTH OF WHEEL.

$\dfrac{3.009}{2.997}$

$\dfrac{.010}{.030}$

$\dfrac{2.000}{2.010}$

∅ 1.125

FIT OF COPPER WHEEL ON SHAFT SHOULD BE RC8

Problem 7.20

A STEEL TIRE IS TO BE FORCE FIT ON AN IRON WHEEL. THE WHEEL HAS A BRONZE BEARING IN IT TO RECEIVE THE SHAFT. COMPLETE THE FOLLOWING INFORMATION:

WHEEL
TIRE

∅ 10.00
(FN5)

∅ .750
(RC7)

O.D. OF WHEEL _____

I.D. OF TIRE _____

I.D. OF BUSHING _____

O.D. OF SHAFT _____

Problem 7.21

∅ 1"

$\dfrac{.7500}{.7508}$

1.25

COUPLING

BEARING

IMPELLER

LENGTH OF ∅$\frac{3}{4}$ SHAFT _____

LIMITS OF ∅$\frac{3}{4}$ SHAFT _____

TOLERANCE OF SYSTEM _____

BEARING:
LIMITS OF SHAFT _____

LIMITS OF BEARING _____

TOLERANCE OF SYSTEM _____

IMPELLER:
LIMITS OF SHAFT _____

LIMITS OF BEARING _____

TOLERANCE OF SYSTEM _____

GENERAL TOLERANCE ± .03
FITS REQUIRED:

   SHAFT W/COUPLING - FN2
   SHAFT W/BEARING - RC3
   SHAFT W/IMPELLER - FN4

DIMENSIONS OF COUPLING ARE SHOWN. SHAFT IS NEVER TO EXTEND BEYOND END OF COUPLING

Problems 7.15–7.21

Complete the required information for each tolerance problem assigned by your instructor. You may be asked to letter the information neatly on a drawing sheet or a sheet of note paper or to reproduce the drawing manually or on the computer and complete the information on your drawing.

Problems 7.22–7.26
Refer to the problems at the end of Chapter 4. Draw the views necessary to describe the object and dimension them for production. Use decimal inches or millimeters as specified by your instructor; note on your drawing which system is used. Remember that the dimensions used to describe an object in a pictorial drawing may not be the dimensions most appropriate for an orthographic multi-view drawing.

Problem	Reference
7.22	4.43
7.23	4.50
7.24	4.54
7.25	4.61
7.26	4.66

TERMS YOU WILL SEE IN
THIS CHAPTER

 HELIX Curve traced on a cylinder by the rotation of a point crossing its right section at a constant oblique angle.

THREAD Inclined plane wrapped around a cylinder in a helical pattern.

SCREW Fastener having an external thread.

 NUT Fastener having an internal thread.

 WELD Joint produced in materials by heat or pressure, often with the addition of a filler material.

 ADHESIVE Material applied in thin films that bonds parts together by sticking to both.

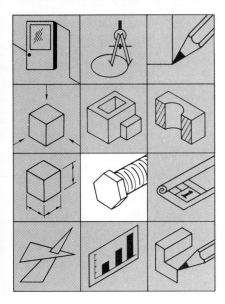

8

FASTENING, JOINING, AND STANDARD PARTS

8.1 INTRODUCTION

Imagine a world without fasteners! Automobiles, airplanes and appliances would be only piles of loose parts; your clothes would be "swaddling clothes"—just wrapped around you; and your typewriter, telephone, and television could no longer communicate. Fasteners of many types hold the tools of our civilization together (Figure 8.1). The most common fasteners are threaded fasteners such as bolts, nuts, and screws. Other mechanical fasteners include pins, rivets, and keys. Joining processes you will learn about in this text are welding and adhesion. New designs of fasteners and new methods for joining parts together are constantly being developed. This chapter will introduce you to the basics of fastening and joining; as you pursue your professional career, you will need

Figure 8.1 Fasteners are all around us.

to learn more about existing fasteners and new developments that may solve the fastening and joining problems you encounter.

8.2   SCREW-THREAD TERMS

Screw **threads** look like the pictures in Figure 8.2. In order for them to function as fasteners, or adjustment devices, you need two parts: an external thread and internal thread. The external thread looks like a V-shaped strip around a shaft in the form of a **helix**. (The **screw** is one of the basic machines described in the history of technology and was conceived by the Greek mathematician Archimedes as an inclined plane wrapped around a cylinder.) The internal thread is the complement of the external thread—a cavity with the screw thread removed from the material.

Other important terms are:

Axis - Longitudinal center line of screw or threaded cavity.

Crest - Outer surface of thread helix.

Figure 8.2   Screw-thread terms.

Root - Inner surface of thread helix.

Major Diameter - Largest diameter of screw thread, measured crests to crests (external and internal).

Minor Diameter - Smallest diameter of screw thread, measured roots to roots (external and internal).

Pitch - Distance between similar points on adjacent threads, measured parallel to the axis.

Pitch Diameter - Diameter of an imaginary cylinder that cuts the threads so that the distance to the crest equals the distance to the root.

Depth of Thread - Distance between crests and roots, measured normal to the axis.

Side - Surface of the helix connecting crest to root.

Thread Angle - Angle included between sides of thread, measured in a plane through axis of thread.

8.3 OTHER PROPERTIES OF THREADS

There are three other properties of threads with which you should be familiar. They include hand, form, and multiple threads. You will find a sketch for each property and an explanation in the following paragraphs.

8.3.1 Right-hand and Left-hand Threads

Threads can be made to advance when the screw is turned to the right (clockwise) or to the left (counterclockwise) (Figure 8.3). Common fasteners are all made right hand; if you need a left-hand fastener you must specify LH. One common example of a

Figure 8.3 Right-hand and left-hand threads.

left-hand thread is the outlet of a propane tank. Safety regulations require that this thread be different from standard threads, that is, left hand. The old adage, "Turn it right to make it tight," does not apply when you attach a regulator to a propane tank. Look at the threads on your bow compass—one side of the screw is right hand, the other is left hand.

8.3.2 Screw-Thread Forms

Not all screw threads look like V's when sectioned. The form of a thread is also called the profile. There are several types of threads, developed at different times for different purposes. The sharp V thread with a 60° thread angle was the U.S. standard thread for many years (Figure 8.4). It was replaced by the American National form, which has flattened crests and roots and is stronger than the sharp V. While the United States flattened the crests and roots, the British rounded them. Their standard thread form for many years was the Whitworth standard. Cooperation during World War II led the United States, Great Britain, and Canada to adopt a common thread form, the Unified, which includes features of the Whitworth and the American National. (This was the ABC accord.) The angle and proportions are from the American National, and the rounded roots and optional rounded crests on the external thread are from the Whitworth.

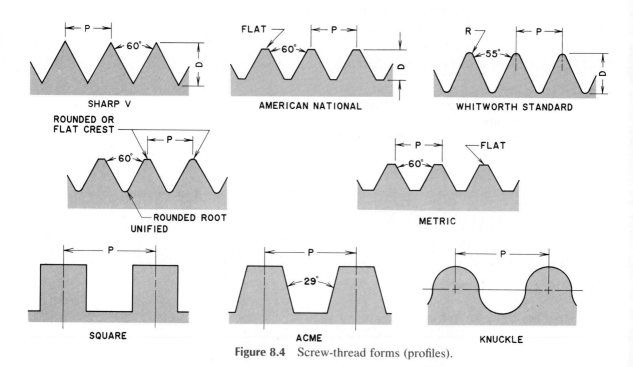

Figure 8.4 Screw-thread forms (profiles).

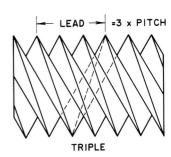

Figure 8.5 Single and multiple threads.

International commerce includes metric measurements and metric fasteners. Metric threads are similar in form to the American Standard or Unified; however, they are not cut as deep as ABC threads. Metric threads may have flat or rounded crests and roots, and they are measured in millimeters rather than in inches.

Threads may be used for fasteners, adjustment, or power transmission. Two forms used for power transmission are the square thread and the Acme thread. The square thread is theoretically ideal for power transmission, but it is hard to produce. A stronger thread form, easier to manufacture, is the Acme thread. An automobile jack or the lead screw on a lathe uses an Acme thread.

Most threads are cut with tools called taps and dies or rolled into the metal with hard metal-rolling tools powered by hydraulic cylinders. Another form of thread, which can be cast or rolled into the material, is the knuckle thread. If you examine a glass jar with a threaded top you will see a form of the knuckle thread. It can be cast with fully rounded crests and roots. It can also be rolled into thin metal parts such as lids for glass jars and bases for light bulbs.

8.3.3 Multiple Threads

So far, you have been reading about single threads in which a single helix is wrapped around the cylinder. Sometimes we need to have the thread advance farther for each turn; in this case we wrap more than one helix about the cylinder and have multiple threads (Figure 8.5). The lead of a thread is the distance it advances in one revolution. The pitch is equal to the lead for a single thread. With a double thread (two parallel helices on the shaft), the lead is twice the pitch; with a triple thread (three parallel helices), the lead is three times the pitch.

Multiple threads advance more rapidly than single threads but are not as strong. They are used for speed and convenience. If you have a large bow compass among your drawing instruments, examine the threads. They will be multiple threads for quick operation.

Figure 8.6   Profiles of course and fine threads for same diameter.

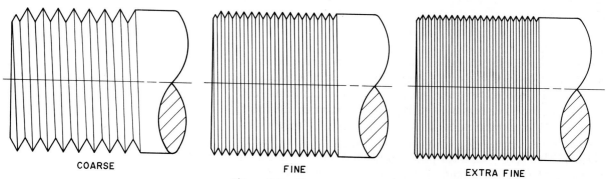

COARSE   FINE   EXTRA FINE

Figure 8.7   Three standard series of threads for same-diameter screw.

8.3.4   Thread Series

Series indicates the number of threads per unit of length for a given diameter. When threads were handmade, more than a century ago, each craftsman made the number of threads on a shaft that he felt were appropriate for the job. Over the years the number of threads per unit length has been standardized for different applications. There are several different series in the Unified system and in the metric or the International Standards Organization (ISO) system.

The two most common series in the inch system are the coarse series and the fine series. The Coarse series is used for most fasteners; when greater strength is needed or the length of thread engagement is short, the Fine series would be used. For the same diameter the Fine series has about 50% more threads per inch than the Coarse. The pitch is smaller for the fine thread for there are more threads per unit length, which puts them closer to each other. The profile, too, is smaller (Figure 8.6).

The National thread form is also available in the Extra Fine series (UNEF or NEF). This thread is used for fine adjustments or very short length of thread engagement. There are also three fixed-pitch series available: 8-pitch (8UN or 8N), 12-pitch (12UN or 12N), and 16-pitch (16UN or 16N) (Figure 8.7). Fixed pitch means that the number of threads per inch is the same for a group of fasteners regardless of diameter. Fixed-pitch threads are used mostly on large-diameter fasteners and for adjustment devices.

Threads have been discussed so far as applied to cylindrical surfaces. They also apply to tapered surfaces—frustums of cones. The National Pipe Thread (NPT) series is used for connecting pipes. The tapered threads seal securely to prevent leakage in piping systems. These threads are described in ANSI B2.1, which also includes straight pipe threads.

8.3.5 Specifying a Thread

When you specify a thread on a drawing or on a purchase order, you do not need to write out a complete description of all the features you need for that thread. There is an accepted shorthand way to describe what is needed. Unified (inch system) threads are specified as shown in Figure 8.8. The first number is the nominal or major diameter. Current standards call for the diameter to be expressed as a decimal inch figure; most fastener catalogs and older drawings will have a fractional inch size. The next number (after the dash) is the number of threads per inch. This will always be an integer. It is followed by letters that specify the thread form (UN or N) and the series, (C for coarse, in this example).

After the next dash you will find a number and, sometimes, a letter. The number indicates the class of fit. You do not need to specify tolerances on standard fasteners—the one number covers a paragraph of specifications in ANSI standard B1.1, which describes the tolerance to which the thread should be made. Class 2 is most common; class 1 is for quick assembly of noncritical parts, and class 3 is used for precision threads and where close adjustment may be needed. The letter following the number (A

Figure 8.8 Unified thread specification.

Figure 8.9 Metric thread specification (ISO).

or B) specifies an external thread (A) or an internal thread (B). Standard B1.1 describes both internal and external specifications for each class.

If the thread is not right hand, the class would be followed by a dash and the letters LH. In the case of a multiple thread the word DOUBLE or TRIPLE would be added to the thread specification.

Metric (ISO) threads are specified in a different manner (Figure 8.9). They always start with the letter M to indicate immediately that this is a metric thread conforming to the standards published by the ISO. The next number if the major diameter in millimeters. Mostly these are integers, but in smaller sizes they may be a one-place decimal such as 1.6 or 2.5. The diameter is usually followed by an × and another number that is the pitch of the thread in millimeters. Since coarse threads are most common, the pitch may be omitted for coarse threads. If the fastener is a general-purpose one, the total specification could be M16.

Classes of tolerance may be specified after a dash. The first number refers to the tolerance grade for the pitch diameter—the smaller the number, the tighter the tolerance. Recommended tolerance grades are 4, 6, and 8. A tolerance grade of 5 represents good commercial practice. The lowercase letter following this number is the tolerance position or allowance for the pitch diameter. The next number is the tolerance for the major diameter, followed by the allowance for the major diameter. Tolerance positions are specified by lower case letters for external threads and by capital letters for internal threads. The codes follow:

External threads: e, large allowance; g, small allowance; h, zero allowance.

Internal threads: G, small allowance; H, zero allowance.

BOLT   CAP SCREW   STUD

Figure 8.10 Common threaded fasteners—bolt, cap screw, and stud—as they appear on drawings.

8.4 *THREADED FASTENERS*

The concept of screw threads becomes useful when it is applied to fasteners. Three of the most common threaded fasteners are the bolt, the cap screw, and the stud as drawn in Figure 8.10. Photographs of actual fasteners are shown in Figure 8.11. These are the most common fasteners: many other types exist and hundreds of new, special fasteners are designed each year.

Figure 8.11 Bolt, cap screw, and stud photographed not holding parts together.

Figure 8.12 American standard cap screws with five types of heads.

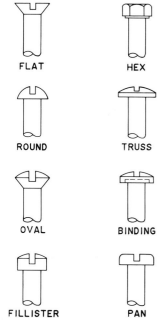

Figure 8.13 Head styles of American standard machine screws.

8.4.1 Screws

Screws with a major diameter of $\frac{1}{4}$ in. or more are called cap screws. They may have several types of heads as shown in Figure 8.12. Standard diameters in the U.S. system go from $\frac{1}{4}$ up to 4 in. Standard diameters in the ISO system go from 1.6 to 100 mm. There are many lengths available in each diameter: The larger the diameter, the longer the lengths you can get. Screws are available in coarse, fine, or extra fine threads with the coarse series most readily available. Useful information about cap screws is given in the Appendix.

Small screws are called machine screws. Sometimes they are used with small nuts and act as little bolts. The types of heads available are pictured in Figure 8.13. They are available in the U.S. system from size 0 (0.060) to size 12 (0.216) and overlap cap screw sizes in some head styles from $\frac{1}{4}$ up to $\frac{3}{4}$. Where they have similar heads and the same diameters as cap screws, the machine screws will usually be threaded almost all the way to the head while cap screws are threaded for a length of a little over twice the diameter. Metric machine screws are available as small as 1 mm diameter and go up to 20 mm diameter. Lengths are available proportional to diameters, with the smallest screws produced as short as 2 mm and the largest as long as 100 mm. Metric machine screws mostly have coarse threads. You will find more data about machine screws in the Appendixc.

Set screws are used to hold parts in place on a shaft. They pass through a threaded hole in the part and bear firmly on the shaft (Figure 8.14). Sometimes a slot or dimple is formed in the shaft to make a receptacle for the end of the set screw. Common set screw heads and points are shown in Figure 8.15. The headless screws are produced with diameters from 0 (0.060) to 2 in. Square-head set screws are available from $\frac{1}{4}$ to $1\frac{1}{2}$ in. diameter. Lengths run from $\frac{1}{16}$ to 6 in. for the headless varieties and up to 12 in. for square-head screws. They can be ordered with coarse or fine threads.

Metric set screws range in diameter from M1 to M20 in lengths as small as 2 mm for the small-diameter screws to 80 mm for the larger diameters. Commercial standard threads are coarse threads.

Set screws are available with ends of soft material such as nylon or brass to bear on the shaft. This feature reduces damage to the shaft—desirable if the shaft is threaded or polished. They may also have locking devices built into the threads so they do not become loose from vibration. Nylon plugs imbedded in the thread are often used.

Another useful type of screw is the shoulder screw, or stripper bolt (Figure 8.16). Originally developed for holding movable parts of metal-forming dies together, it has a smooth section of constant diameter next to the head and a threaded section on the end. The smooth section is precisely machined with a close tolerance on the diameter. These fasteners can be used as guide pins or as axles in mechanisms. They are relatively inexpensive and can save time and cost over custom designs. Shoulder screws are available in inch and metric sizes.

Several other varieties of screws have been standardized. Wood screws are similar to machine screws, but the ends are tapered so that they start easily into wood and form their own threads in the wood. Lag screws are large wood screws with square or hex heads. Sheet metal screws are also tapered on the starting end. They are of hardened metal and have large heads for easy driving. Other threaded fasteners made of hardened metal can cut or form their own threads as they are installed.

There is an infinite variety of specialized fasteners. They are

Figure 8.14 Set screws prevent relative motion between shaft and part mounted on it.

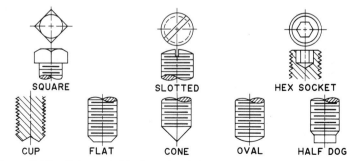

Figure 8.15 Set screw heads and points.

Figure 8.16 Shoulder screw (stripper bolt) has smooth, precisely finished shoulder.

Figure 8.17 Some of the many specialized fasteners available.

made for various industries having specialized needs, to combine functions, and to meet special fastening requirements. Screws are available that cut their own threads: They are tapered on the end, hardened, and act like a tapping tool to form threads. Others have integral locking devices such as plastic inserts in the thread, deformed threads, or one-way drives in the heads that allow the fastener to be installed but require special tools for removal. A few such fasteners are pictured in Figure 8.17.

8.4.2 Bolts, Nuts, Studs, and Washers

Bolts look like cap screws (Figure 8.18). They may have hexagonal or square heads or special-purpose heads. The length of the

HEX HEAD BOLT AND NUT WITH JAM NUT

SQUARE HEAD BOLT AND NUT WITH WASHER

Figure 8.18 Bolts are designed to be used with nuts.

thread on bolts may be less than the thread length on cap screws of the same size. Specifications for cap screws and finished hex head bolts are combined. An unfinished fastener is machined only in the threads. Most hexagonal-head bolts and **nuts** have a smooth bearing face, called a washer face, machined on the surface that contacts the parts being fastened. These are known as semifinished. Finished nuts have a washer face on both sides. The tolerances for finished fasteners are more strict than those for semifinished. Square-head bolts and nuts are not finished except in the threads.

Bolts and nuts may be grouped according to use: Regular are for normal service; heavy are for harder service or for easier wrenching because the nuts and bolt heads are larger than regular head and nuts. Square-head bolts are only available as regular. A special type of nut may be used with a standard nut to provide locking action. These thin nuts are called jam nuts.

Inch-size bolts and nuts are produced with diameters from $\frac{1}{4}$ to 4. Nuts are available down to size 0 for use with small machine screws. U.S. bolts and nuts are produced with coarse or fine threads.

Metric bolts and nuts are hexagonal. They can have coarse or fine threads extending the full length of the shaft or partly threaded. They are all semifinished or finished. Metric bolts and cap screws are considered to be the same and are known as metric hexagon-head screws. They come in diameters from 5 to 100 mm. Nuts are available in sizes from 1 to 100 mm. A special series of metric fasteners is standardized for structural applications. These are called high-strength structural bolts and nuts and are produced in diameters from 16 to 36 mm appropriate for fastening the framing of buildings.

Threaded fasteners hold parts together by squeezing or gripping. Washers are used to increase the gripping area, to provide a smooth surface for wrenching the screw heads and nuts, to act as a locking agent, and to provide thickness adjustments. The simplest form of washer is a circular disc with a hole through it. Some are not simple circles: They may have cuts or teeth to provide a locking action (Figure 8.19).

Washers are specified by the diameter of the fastener with which they are used. U.S. washers have two designations in many sizes: narrow, based on specifications of the Society of Automotive Engineers, and wide, based on structural needs. These flat washers are specified as type A, plain, N, or W. The tolerances on these washers are very broad for they are stamped from metal sheets at high speed.

Figure 8.19 Washers come in different thicknesses and may be smooth or have locking features.

8.4.3 Materials for Fasteners

Steel is the most common material for fasteners. Steel is a generic name that includes many alloys of iron and carbon with small quantities of other materials that add special properties. Controlled heating and cooling of steels is called heat treatment and can significantly change physical properties. Carbon steels are often heat treated to improve the tensile strength of the material.

Heads of U.S. hex-head bolts and screws are code marked to indicate their grade (strength). They may bear a number indicating grade or the specification number that they meet. Radius marks are also used to show the grade: no radius marks, tensile strength = 60,000 psi; three radius marks = 120,000 psi; and six radius marks = 150,000 psi. Metric screws with 5 mm or larger diameter, both hex head and socket head, are marked with a property class number that indicates tensile strength and the ratio of yield strength to tensile strength. These numbers appear as 8.8, 9.8, 10.9, or 12.9. The higher integer number denotes a higher tensile strength. The higher decimal number denotes a higher ratio of yield strength to tensile strength.

Fasteners for corrosive environments may be made of stainless steel, brass, bronze, Monel metal, copper, nickel, or plastics. Nylon is often used for fasteners in chemical equipment. Aircraft fasteners may be made of high-strength aluminum or titanium. Fasteners of different materials are shown in Figure 8.20.

Figure 8.20 Many different materials are used to make fasteners.

8.5 *REPRESENTING THREADS AND FASTENERS ON DRAWINGS*

Screw threads are helices on positive, negative, or tapered cylinders. Drawing them as continuous curves requires careful plotting of points located in two views of the screw thread. We seldom do this, except for presentation or catalog illustrations, because it is very time consuming.

8.5.1 Detailed Thread Symbols

The representation that looks most like the true helices is the detailed symbol (Figure 8.21). The diameter is the actual diameter

Figure 8.21 Detailed thread representation.

of the thread. The pitch can be read from thread tables taken from ANSI or ISO standards (reproduced in the Appendix) and plotted to an approximate, convenient value. Remember, this is a symbol and 11 threads per inch can be plotted as 10 threads per inch with no loss in meaning (and 10 per inch can be measured directly with a standard scale). The actual value of the pitch will be given in a note. The pitch should be laid out on the major diameter line on the side of the thread and offset one half the pitch distance when laid out on the opposite side. You locate the minor diameter or root lines by drawing 60° angles between adjacent crests. Sharp V's are drawn for each pitch interval on each side of the thread. Crests should be connected with medium weight (0.5 mm) lines. The crest lines will not be perpendicular to the axis, but will slope. A right-hand, external thread will slope to the left (think: turn it clockwise to make it advance.). A right-hand, internal thread will slope to the right; you are looking at the far side of the thread. You should also connect the roots with medium-weight lines. Note: root lines with *not* be parallel to crest lines. If you are drawing a left-hand thread, the external thread lines will slope to the right and the internal ones to the left. Crest and root lines may be omitted on detailed internal thread symbols. The ends of external threads are usually tapered or chamfered for easier starting—the chamfer is normally drawn at 45°.

Internally threaded holes that are through holes are drawn the same from one end to the other. Some threaded holes are blind holes, which do not go all the way through. To produce such a threaded hole requires that the tap drill that removes material to the diameter of the root of the thread go further into the part than the desired depth of the thread. The tap (thread-cutting tool) is tapered on the end so that it can start cutting in a hole the size of the minor diameter. As it continues to cut, the full diameter of the tap does the cutting and produces full threads, but the tapered end never cuts full threads at the bottom of the hole. Full threads can be cut to the bottom of the hole with special "bottom" taps—this process requires a second tapping operation and more time. For this reason the tap drill hole is made a little deeper than the needed depth for the threaded hole. The end of the drilled hole is drawn with a 120° angle to show the conical shape made by the conical end of the drill. This is shown in one of the internal threads in Figures 8.21–8.23. Blind holes should be avoided when possible.

The end view of a detailed symbol for an external thread shows a solid circle for the major-diameter distance and a dashed circle for the minor diameter (you can't see the roots looking at the end of an external thread). If the end of the thread is chamfered, the minor-diameter circle will be shown as a solid line. The end view of an internal thread will have a solid circle for the minor diameter and a dashed circle for the major diameter. This part of the thread is hidden inside the part, but you can see a hole the size of the minor diameter.

8.5.2 Schematic Thread Symbols

Detailed symbols are seldom used on drawings for production. Although they are quicker and easier to draw than true helices, they are time consuming. Schematic symbols have been developed that convey the appearance of screw threads; they can be drawn more quickly than detailed symbols (Figure 8.22).

The major diameter of the thread is drawn and an approximate pitch scaled for an external symbol. A chamfer is usually shown at the end of the thread. Crest lines are drawn without slope as medium-weight lines (0.5 mm). One sharp V should be constructed to locate the root lines. Light construction lines should be drawn for the root diameter and the root lines drawn as thick lines (0.7 mm).

Internal threads, in section, are drawn similar to external threads. Internal threads, not in section, appear as four parallel dashed lines—two major-diameter lines and two minor-diameter lines. The circular view is the same as a detailed symbol.

8.5.3 Simplified Thread Symbols

The exact requirements for a thread are specified in the thread note that gives the diameter, the pitch, the thread form, and the tolerances. Simplified symbols that can be drawn quickly indicate that a thread is present but leave all specifications to the note (Figure 8.23). They do not show pitch or thread form. The external symbol is simply the outline of the threaded shaft, with chamfer, and a pair of dashed lines indicating the root diameter of the thread. If the thread runs less than full length, the symbol terminates at a visible line across the shaft.

The internal symbol is the same as the schematic symbol if the thread is not sectioned. If it is sectioned, the root lines are shown as visible lines, and the thread cavity between the root lines is not crosshatched. The major-diameter lines are shown as dashed lines on the crosshatching and extend as far as the length of thread. The termination of threads is a line perpendicular to the axis of the thread, hidden in the sectioned portion and visible in the unsectioned cavity.

Figure 8.22 Schematic thread representation.

Figure 8.23 Simplified thread representation.

8.5.4 Drawing Screw Heads and Nuts

You have learned how to represent screw threads on drawings with detailed, schematic, and simplified symbols. Threads by themselves are not of great value—they are useful as part of a fastener. The proportions of several types of fasteners are given in the Appendices. These are taken directly from the standards for such fasteners. If you buy a hex-head cap screw of a given size, it will always have the same proportions unless you request a special type of screw.

To draw a bolt and nut, you need to know where the center line of the fastener will be. You need to know where the contact surfaces are—where does the head of the bolt touch the parts—where does the nut touch? You need to know the size (diameter) of the fastener. It is a hex-head bolt or a square-head bolt? When you have this information and the specification tables for standard bolts, you can draw the fastener. Figure 8.24 outlines the steps for drawing a hex-head bolt and nut. Bolt and screw heads and nuts are usually drawn across corners to show the maximum room that the fastener will take as it is turned to tighten or loosen it.

Figure 8.24 Drawing hex-head bolts and nuts (semifinished).

Figure 8.25 Drawing square-head bolts and nuts (Unfinished).

If you don't have the standards table available, these proportions for regular, finished bolts and nuts will be helpful:

Diameter of head or nut	$1\frac{1}{2}$ × thread diameter
Thickness of bolt head	$\frac{2}{3}$ × thread diameter
Thickness of nut	$\frac{7}{8}$ × thread diameter
Length of thread	
< 6 in.	2 × thread diameter + $\frac{1}{4}$ in.
≥ 6 in.	2 × thread diameter + $\frac{1}{2}$ in.

The steps for drawing a square-head bolt and nut are similar and are outlined in Figure 8.25. The proportions for other fasteners are given in the standards copied in the Appendix.

8.5.5 Drawing Fasteners with a Computer

Computer graphics is a great tool for drawing fasteners. Many advanced CADD packages have a library of symbols and patterns that can be called from memory and displayed on the screen. A

Figure 8.26 Using computer graphics to generate fastener drawings.

threaded fastener symbol, once brought to your screen, can be enlarged, reduced, shortened, lengthened, moved, mirrored, copied, or otherwise modified to meet your needs. If you do not have an advanced CADD package available, you can create your own pattern or "block." You create a "generic" threaded fastener by zooming in to draw to a large scale, so you can get the details right. You would use the same procedure on the screen as shown in Figure 8.25 or 8.26. Perhaps you would create a hex-head bolt and nut and store it in memory at some convenient scale. If you need a hex cap screw, call up your bolt and nut and delete the nut (Figure 8.26). If you need a headless set screw, delete the nut and the bolt head and modify the tip. Imagine two circular disks to be bolted together—you retrieve your end view of the bolt head and duplicate it wherever there is a center for a bolt.

The power in a computer system lies in its ability to make changes rapidly. With a few fastener patterns in memory you can modify or combine them to produce almost any kind of fastener in a very short time. When making a pictorial drawing, you can enter the three-dimensional information—the data base for a fastener. You can then call the data base and show the fastener in any position or orientation on your drawing. You may also modify it as you would a two-dimensional drawing.

8.5.6 Specifying Fasteners

The symbol for the fastener alerts the reader of your drawing that a fastener is used at that location. The symbol by itself does not give the reader enough information to obtain a specific fastener. You need to provide the following information:

1. Thread specification (discussed in Section 8.3.5)
2. Fastener length
3. Fastener series (only for bolts: regular or heavy)
4. Type of finish (unfinished, semifinished, or finished)
5. Material (other than steel)
6. Head style or shape
7. Point style (only for set screws)
8. Name of fastener family
9. Special requirements (coatings, specifications to meet)

The information is grouped differently for U.S. and ISO fasteners. U.S. fasteners are described with the thread specification first and the length, followed by the other information in the order already listed. Metric fasteners are described by the head style, family name, thread specification, and material. Some examples follow:

$\frac{3}{4}$-10UNC-2A $\times$ 2$\frac{1}{2}$ REG FINISHED HEX BOLT, STAINLESS STEEL

HEX SCREW, M16 $\times$ 1.5 $\times$ 110

SLOTTED SET SCREW, CONE POINT, M4 $\times$ 12, BRASS

#10-32UNF-3A $\times$ 1$\frac{1}{2}$ OVAL HD MACH SCREW

1″-8UNC-2B FINISHED JAM NUT

Installation information is often added to assembly drawings. The torque, or turning force, necessary to install the fastener tightly, but without damaging it, is often specified. If the fasteners are installed with automatic machinery, the torque may be preset. If the fasteners are installed manually, a special wrench with a dial gauge built into it (a torque wrench) would be used.

8.6 OTHER MECHANICAL FASTENERS

There are other types of mechanical fasteners that are not threaded. These include rivets, pins, keys, splines, and retaining rings. There

BUTTON   TRUSS   FLAT   COUNTERSUNK   PAN

Figure 8.27 Head forms for American standard rivets.

are also special-purpose clips and latches that are usually designed for specific applications. Fastener catalogs will provide you with a wealth of information on special fasteners.

8.6.1 Rivets

Rivets function like a bolt and nut. They fasten by gripping or compression. A rivet has a smooth shaft and a large head on one end (Figure 8.27). The shaft is inserted through holes in the parts to be joined, and a new head is formed, in place, on the end without a head. This process may be accomplished with pressure only or with heat and pressure.

Have you ever seen steelworkers high on a building under construction? One worker lines up the holes in the steel shapes to be joined while another uses a bucket to catch a red hot rivet heated by a third worker. The hot rivet is inserted in the aligned holes and held in place with a "bucking bar" while the other worker forms a new head on the smooth end with an air hammer. This is hot, noisy, and dangerous work. Many new buildings are joined with high-strength bolts assembled with powered wrenches.

Rivets are still used for shop assembly of steel shapes where the workers are not walking on steel beams hundreds of feet in the air. Until recently the outer covering or "skin" of aircraft was riveted to the framing with thousands of rivets. Many small parts are assembled with rivets. Rivets are less expensive than screws, and the installation process has been automated in many cases.

8.6.2 Pins

Pins are cylinders that pass through preformed holes to join parts. When pins are used to fasten a part to a shaft, the holes will be perpendicular to the axis of the shaft. They have no threads and usually no heads, but some may have special shapes. Pins hold parts by interference, friction, or the pin may have a split end that can be deformed to lock it in place. Some pins have a hole in the end and a locking wire or another, smaller pin is inserted in this hole to keep the pin in place. Several types of pins are shown in Figure 8.28.

The taper pin has been used for many years to assemble parts to a round shaft. As its name suggests, it is not a cylinder but is slightly tapered (a frustum of a long, slim cone). To fasten parts with a taper pin, the parts are located with respect to each other and a hole drilled through both of them. The hole is then enlarged and tapered with a tool that removes only a small amount of metal—a taper reamer. Then the carefully machined taper pin is driven into the tapered hole. The parts are locked firmly in place. Today, spring pins are replacing taper pins in many applications. They do not require a tapered hole, just a drilled one. The pin itself is rolled up from a sheet of spring steel—the round rolled shape retains some "springiness," so it adapts to irregularities in the hole. The hole is made a little smaller than the pin; the pin is driven in the hole and presses tightly against the sides of the hole locking the parts together.

Groove pins are used much like spring pins. They are solid, round pins (dowel pins) with a longitudinal groove formed in them. This process deforms the pin so that it is no longer round. When the groove pin is driven into a hole the groove edges tend to grip

Figure 8.28 Some of the many varieties of pins.

the sides of the hole and the deformed shape is forced back to a round shape. These actions lock the pin in the holes.

Cotter pins are formed from a single piece of D-shaped wire. They have two parallel legs and the loop is usually opened up to form a head. The pin is inserted in a hole and the legs bent apart to keep it in place.

8.6.3 Keys

Keys are metal shapes placed in holes or slots cut longitudinally in a shaft and mating part to lock them together so that they rotate together (Figure 8.29). Keys can be square in cross section or rectangular. They may be straight or tapered. Some may have a head to facilitate removal (gib head key), and others may have rounded ends to fit a specially shaped keyseat. The keyseat is the slot or pocket cut into the shaft to receive part of the key. The other part of the key rests in a slot cut into the wheel, pulley, gear, or other device attached to the shaft. This slot is called a keyway.

One type of key that allows some freedom of movement to compensate for misalignment is the Woodruff key. It is shaped like a half moon, and the keyseat is a semicylinder. This shape allows the key to "rock" in the keyseat.

The size of the key is selected to meet the load on the shaft and the other part fastened to it. Proper selection of a key for specific design conditions is a design problem requiring some engineering analysis. Keys are seldom used that are wider than one fourth the diameter of the shaft because cutting a larger key-seat would materially weaken the shaft.

SQUARE   GIB HEAD   PRATT & WHITNEY   WOODRUFF

Figure 8.29 Four common types of keys.

If a shaft is formed with several key shapes formed on the shaft and the wheel-type part has several keyways to match, this is a splined connection (Figure 8.30). Splined shafts and holes require careful fabrication to get all the projections and cavities to match within close tolerances. Splines are used where loads are heavy, precision is needed, and where the wheel-type part must move axailly on the shaft while retaining the same radial relationships. If you have the opportunity to look inside an automobile transmission, you will see the gears riding on splined shafts.

Figure 8.30   Splined shaft with mating wheel.

8.6.4   Retaining Rings

Retaining rings hold parts on a shaft by acting as an enlargement of the shaft (Figure 8.31). Most are open rings that fit into a groove around the shaft. They are held in the groove by spring action or by squeezing them into the groove. Some are complete circles with angled spring teeth that go on a shaft easily but "bite" into the shaft if a force attempts to move them in the other direction.

Retaining rings may be stamped from flat sheets or formed from square or round wire. Some require special tools for installation and removal.

Figure 8.31   Application of external retaining ring.

8.7   *OTHER MACHINE ELEMENTS*

There are some other types of parts that are represented frequently on machine drawings. You will meet them here and will probably encounter them frequently in design courses and in your professional life. Springs are common devices for storing energy. They sometimes act as pure storage devices to collect energy in one part of a cycle and return it in another part of a cycle (valve springs in an automobile engine). They may also be used to even out movements by storing energy when moved in one direction and returning it to move in the other direction (body springs on an automobile).

A common form of spring is a helix wound from high-strength wire. Three common types are pictured in Figure 8.32, and the symbols for them are shown in Figure 8.33. They are named for the kinds of load they are designed to accept:

> **Compression**   Accept an axial load that compresses them; then they try to resume their normal length when the load is removed.

Figure 8.32 Three simple springs.

COMPRESSION   TENSION   TORSION

Figure 8.33 Symbols for three simple springs.

Tension Accept an axial load that stretches them; then they try to resume their normal length when the load is removed.

Torsion Accept a radial twisting load (clockwise or counterclockwise); then they twist in the other direction when the load is removed.

There are also flat springs (the leaf springs on auto and truck bodies are stacked, flat springs) and power springs that look like a spiral of flat wire. Energy is stored in them by winding and released through a controlled movement, as in a mechanical clock or a wind-up toy.

Bearings and bushings are used to line a hole through which a shaft passes. Electric motors always have some type of bearings so that the motor shaft can spin easily. If the shaft turns or rotates in the hole, the liner is usually referred to as a bearing. If the shaft slides back and forth (reciprocates) in the hole, the liner is called a bushing. The term *bushing* is also used to describe some bearings and liners used to restore the diameter of holes that have increased in size through wear.

Bearings and bushings are used to decrease the friction between a shaft and the part through which it passes—some materials rubbing together have less friction than others. If the bearing or bushing wears, it is easier and cheaper to replace the small part than to replace an entire machine.

Bearings may be plain, just a sleeve of material inserted in the hole; it is this type that may also be referred to as a bushing. A greater reduction in friction occurs if balls or rollers are incorporated in the bearing—rolling friction is less than sliding friction. A ball bearing is shown in Figure 8.34. The rolling elements may be balls, cylindrical rollers, or tapered rollers. Tapered rollers are used for very heavy loads and to accept thrust loads in an axial direction (car and truck axles). Rollers of very small diameter may be used in tight spaces and for light loads; they are called needle bearings.

Bushings are usually just a hollow cylinder. Sometimes they may have a flange on the end to make it easier to position or remove them (Figure 8.35). Materials for bushings and plain bearings are often made of brass or bronze. The basic material may have a lubricant such as oil, Teflon, or molybdenum disulfide impregnated in it.

Bearings and bushings are usually shown in outline form on

Figure 8.34 Ball bearing.

Figure 8.35 Flanged bushing.

the drawing and a reference given for ordering the part from a firm that makes them. Plain bearings and bushings may be fabricated for a small machine and shown with the drawings for that machine; only a bearing manufacturer would produce a ball or roller bearing because special materials and machines are required.

8.8 THERMAL PROCESSES

Threaded fasteners are generally reusable. Several processes are used for joining materials that result in a permanent bond. Some process require heating the materials. Welding involves actually melting the parts to be joined at the point of contact; brazing and soldering require melting a filler material but not the parts to be joined.

8.8.1 Welding Processes

In **welding** the parts are joined by the application of heat or pressure, often with the addition of a filler material. The parts are usually similar metals. Heat is supplied by a gas flame, electric arc, or electric resistance. The filler material is a metal of a composition similar to the parts being welded. A flux is used to improve the quality of the weld by combining with unwanted products of the heating process, such as metallic oxides. It may also generate a gas cloud that protects the heated area from oxidation. In some processes an inert gas shield (carbon dioxide or argon) is used to prevent contamination of the heated area.

Electric arc welding creates heat with an electric arc drawn between an electrode and the work to be welded (Figure 8.36). There are several variations of the process. The electrode itself may be consumable—it supplies the filler material. The electrode may be nonconsumable and a separate filler material is added in the arc. The electrode may be attached to a gas nozzle that supplies an inert gas shield. The operation may be controlled manually or automatically.

Gas welding uses a flame, commonly from burning acetylene with oxygen, to heat the parts being welded (Figure 8.37). The filler material, coated with a flux, is melted in the flame and added to the joint. Gas welding is slower than arc welding; it is often used for repair work and in small workshops. Generally it is done manually.

Resistance welding uses both heat and pressure to form a joint. The heat is generated by passing a large electric current through the work pieces while they are clamped tightly together. The weld may be continuous or intermittent (spot welding). Because the

Figure 8.36 Arc welding generates heat through electric arc between electrode and work. Courtesy of Ohio State University Engineering Publications

Figure 8.37 Gas welding heats work and melts filler rod with high-temperature gas flame. Courtesy of Ohio State University Engineering Publications

process requires high currents and clamping devices, it is usually considered a mass production process.

Other exotic processes for welding coming into greater use include laser beam welding, electron beam welding, and plasma arc welding. These processes require expensive equipment and careful control. Laser beam and plasma arc produce clean welds on thin materials; they are suitable for high-performance materials—stainless steel, titanium, and aluminum alloys. Electron beam welding can produce a deep, clean weld and is useful for welding refractory metals or joining dissimilar metals.

8.8.2 Specifying Welds on Drawings

Welding is specified on drawings by symbols, a type of shorthand standardized by the American Welding Society. The various methods for making a joint in metals are defined and the parts of the weld symbol describe the conditions of the weld (Figure 8.38).

The basic symbol is a reference line with an arrow pointing to the joint. The area below the reference line describes the *arrow* side of the weld—where the arrow touches the joint. The area above the line describes the *other* side—opposite from where the arrow touches the joint. The type of joint is described by symbols; Figure 8.39 shows several common joints and their symbols. The sizes of the welds are given around the joint symbol. If the weld is a groove weld to be ground smooth after completion, a flush symbol is added. If a part is circular and is to be welded all around,

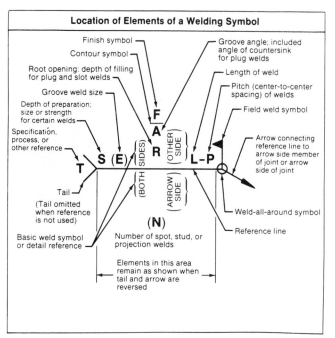

Figure 8.38 Standard weld symbol from American Welding Society standards. Courtesy of American Welding Society

FILLET	PLUG AND SLOT	GROOVE WELDS				
		SQUARE	V	BEVEL	U	J

Figure 8.39 Types of welded joints and their symbols.

Figure 8.40 Weld symbol and welds it describes.

a circle is drawn around the junction of the reference line and the arrow line; a weld to be done in the field (not in the factory) gets a solid black circle at this point. Detailed specifications for welding are called for by a reference letter in the tail of the arrow. Figure 8.40 illustrates a weld symbol and the work you would expect from this symbol.

8.8.3 Soldering and Brazing

Soldering is a process for joining metals with a low melting temperature alloy (below 840°F). The parts to be joined are not melted; the solder alloy flows over the joint and adheres to the component parts. Soldering is used for electrical assemblies and for sealing metal cans. The heat can be supplied from a flame or an iron heated by electric resistance. Most solder alloys are a mixture of lead and tin. Other alloys include tin and antimony, tin and silver, and lead and silver.

There are no standard symbols for soldered joints. Process

conditions, solder alloy, and joint design are selected based on experience and research.

Brazing requires heating the parts to more than 840°F, but the work pieces are not melted. A filler metal is used that has a melting point lower than the metal in the parts to be joined. Copper, aluminum, and silver alloys are commonly used as filler materials. Heat may be applied in a furnace, a bath of molten flux material, or by a torch, electric induction, or electric resistance. Brazing can join most all metals and is useful for joining dissimilar metals. Cast metals are difficult to weld but are readily brazed. Design of a brazing process requires knowledge of the base metals, the filler metal, the process equipment, and the service demands on the finished part.

8.9   *ADHESIVES*

Adhesives do not fasten parts mechanically, and they do not require heating the parts to be joined to high temperatures. They do form a bond that transmits forces uniformly across the adhesive joint. Adhesives of some types have been used for centuries, but engineering adhesives are a product of the last half of the twentieth century. First developed to solve difficult aerospace joining problems, they are now used in automobiles, appliances, and other common devices.

Adhesives can join thin materials and dissimilar materials (metals to nonmetals). They do not damage or mar the work part surfaces; they can seal a joint as well as bond it. Proper selection results in very high strengths. Because there are so many possible combinations of materials and adhesives, selection is difficult. Operating temperatures for the finished joint may be limited. Surface preparation may be costly for the surfaces must be clean to develop a good bond.

Some adhesives require low heat, ultraviolet light, the absence of air, or time (up to 24 hours) for curing. Several adhesives come in two parts—a resin and an activator that can only be combined at the point of use. Adhesives can be applied as liquids by brushing, spraying or dipping. Many types are available as films or tapes that can be laid between mating surface. Others come as powders, and some can be microencapsulated (in tiny beads).

One interesting application of adhesives is in combination with threaded fasteners. The threads can be coated with an adhesive, often in the form of microencapsulated beads, and when the fastener is installed, the adhesive acts as a locking agent to prevent loosening. Because not all the tiny beads are broken when the fastener is first installed, some can be reused up to four times with little loss in locking power. The adhesive can also act as a sealant to prevent fluid from escaping around the threads. One manufacturer has specified various colors of adhesives for different types of fasteners: metric fasteners are coated with one color, and inch-size fasteners with another color.

The principal types of adhesives are epoxies, urethanes, cyanoacrylates, anaerobics, acrylics, and hot melts. Selection, application, surface preparation, curing conditions, and operating environments are all important in designing an adhesive joint. The manufacturers of the adhesives all provide design information and technical service to assist in the design.

8.10 *THE FUTURE OF FASTENING AND JOINING*

Pins and threaded fasteners have been with us for many years. The taper pin, which requires careful manufacture and installation, is being replaced in some applications by spring pins. Three-part fasteners are being replaced by specially designed one-piece fasteners that perform all the functions of the three pieces for less cost. Stapling and stitching with wire are replacing separate fasteners in thin and soft materials. You will need to read, observe, and learn in order to be aware of trends and developments in fasteners.

A few areas you may want to consider as you advance in the design profession are noted. Joining of plastics can be done by mechanical fasteners, solvent bonding, welding of thermoplastics, and fusion bonding—new methods are reported each year. Welding processes are in use today using lasers, ultrasonics, and the electron beam that a generation ago were unknown. Fasteners made of materials other than steel are finding increasing applications—nylon, polyvinyl chloride, and titanium. Combinations of elements and materials such as adhesive-coated screws, nuts with integral sealing washers to seal in fluids, quick-operating fasteners, and explosive rivets will continue to be developed. If you become involved in manufacturing look for faster, more reliable ways to install fasteners or new ideas for accomplishing the fastening and joining process.

8.11 *SUMMARY*

Fasteners hold our industrial society together. Threaded
fasteners are the most common; other mechanical fasteners
such as pins, keys, rivets, and retaining rings serve specific
needs. Many types of threaded fasteners have been
standardized by the ABC agreements and the American
National Standards Institute for inch-size fasteners and by the
ISO for metric fasteners. Fasteners are seldom drawn exactly
as they look; standard symbols have been developed to
represent fasteners on drawings. This chapter also included a
brief review of other mechanical elements including springs,
bushings, and bearings.

Joining processes provide permanent bonds. Welding is a
high-technology discipline with many new processes available.
The specification of welds on drawings is standardized through
the specifications of the American Welding Society. Other
joining processes that use heat to create a bond are soldering
and brazing. Adhesives are the frontier of joining technology—
the design of an adhesive-bonded joint requires much
information and careful design. Fastening and joining
technology is moving rapidly—you will need to be aware of
advances in this area to do a good job of design.

PROBLEMS

These problems may be done on your computer or manually as specified by your instructor.

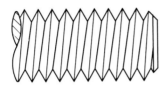

Problem 8.1
Copy this figure as a freehand or computer sketch. On your sketch label these features:

Major diameter
Minor diameter
Pitch
Depth of thread
Thread angle

Problem 8.2
Sketch the profile of these threads with the noted pitches:

Sharp V	Pitch $\frac{1}{4}$
Unified	Pitch 0.50
Square	Pitch 20 mm
Knuckle	Pitch 1 in.

Problem 8.3
A threaded shaft with a specification of 0.50-20UNF-3A DOUBLE is turned five revolutions in a mating nut. How far does it advance?

Problem 8.4
Copy and complete the following table:
Which of the threads per inch that you entered has the smallest pitch? Which has the largest pitch?

MAJOR DIAMETER	COARSE	SERIES FINE	EXTRA FINE
#6	UNC−32	UNF−?	——
?	UNC−24	UNF−32	——
.375	UNC−?	UNF−24	UNEF−32
.50	?	?	?
?	UNC−10	?	?
1.00	?	?	UNEF−20

Problem 8.5
Name a tapered thread (noncylindrical). Why is it tapered?

Problem 8.6
Write specifications for the following threads:

   (a) $\frac{1}{2}$ major diameter, fine external threads with precise tolerances
   (b) 0.75 major diameter, coarse internal threads with very loose tolerances
   (c) 12 mm major diameter with coarse threads

Problem 8.7
Name the illustrated fasteners:

Problem 8.8
Pick up two large hex screws and examine the heads. One looks like (*a*) and the other looks like (*b*). What do you know about each?

Problem 8.9
Schematic thread symbols are used on assembly drawings; simplified symbols are usually used on detail drawings. Copy and complete the following assembly and detail drawings.

ASSEMBLY   DETAIL

Problem 8.10
Write the specifications for the following fasteners:

   (a) 1 in. regular semifinished hex-head bolt, $3\frac{1}{2}$ long with a coarse thread

(b) Metric cap screw with a 24-mm fine thread
(c) 0.50 finished jam nut made of Monel metal and having
precisely made fine threads

Problem 8.11
Complete the assembly drawing four times the size shown in your
text. Add a socket head set screw at point A and a Woodruff key at
point B.

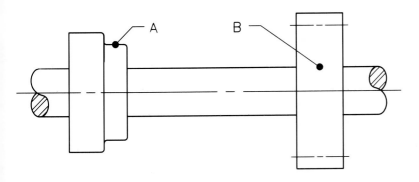

Problem 8.12
Prepare the weld symbol for welding the end of a steel pipe to a plate
using a $\frac{1}{4}$ fillet weld and welding all around.

Problem 8.13
Name three fastening processes that require heat and a filler material.

Problem 8.14
Indicate that part x is to be welded to part y with an intermittent fillet
weld on both sides. A $\frac{3}{8}$ fillet, $1\frac{1}{2}$ long, spaced at 3 intervals is required.

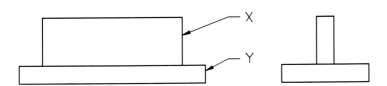

Problem 8.15
You need to fasten two plastic spheres together as shown. List at least
three ways to accomplish this.

TERMS YOU WILL SEE IN THIS CHAPTER

BILL OF MATERIAL   List of all the parts necessary to make up an assembly and some details about these parts.

PICTORIAL ASSEMBLY   Three-dimensional drawing showing how parts fit together.

EXPLODED VIEWS   Three-dimensional drawings showing the parts separated from each other but along predetermined axes to show how parts are assembled together.

SECTIONED PICTORIAL DRAWINGS   Similar to exterior assemblies except some portion of the assembly has been "removed" so the viewer can see the interior features of the assembly.

ZONES   Coordinate grid used to identify areas of a drawing.

WIRE FRAME DRAWINGS   Traditional pictorial drawings generated from a computer data base. Hidden elements may be included or deleted.

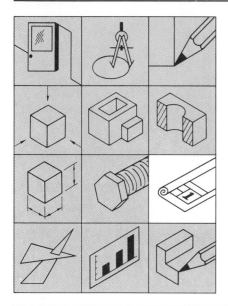

9

PRODUCTION DRAWINGS

9.1 *INTRODUCTION*

Production drawings are used to transmit and communicate information for the production of objects and assemblies (Figure 9.1). Production drawings are classified in two major categories.

Detail drawings are drawings of single parts. They may be drawn one part per sheet or there may be several parts detailed on a large sheet. These detailed drawings include such additional information as dimensions and notes that relate to material, finish, weight, or standard tolerances (Figures 9.2 and 9.3).

Assembly drawings show how parts fit together or are functionally related. Dimensions are often found on assembly drawings. However, the dimensions refer to relationships among the parts, not to the size of the individual objects. Assembly drawings often include a listing of all parts necessary

Figure 9.1   Flow of production drawings.

Figure 9.2 Single detail drawing. Courtesy of Jones and Vining, Incorporated

to make up the total assembly. This list is called a **bill of material** (Figure 9.4). Large products requiring many parts may require several sets of subassembly drawings in order for the product to be manufactured and assembled.

9.2 APPLICATIONS OF PRODUCTION DRAWINGS

Production drawings are used in the design process. The first step usually includes sketches (freehand or computer generated) that describe the concept in general and provide some size informa-

Figure 9.3 Several detail drawings on one sheet. Computer generated image. Courtesy of American Electric Power Service Corp.

tion. Purchased parts to be used may be noted on sketches. From the sketches, designers prepare a layout drawing that shows the relationships of the parts and defines key dimensions. After the layout is approved by the responsible engineer, designers begin

BILL OF MATERIAL				
PART NO	NAME	MATL	REQD	NOTES
1	BASE	CI	1	
2	SHAFT	1030STL	1	
3	GEAR	3120STL	1	
4	DOUBLE V-PULLEY	3110ALUM	1	
5	SLEEVE	1020STL	1	
6	BEARING RETAINER	1020STL	1	
7	BEARINGS	—	2	NEW DEPARTURE #3207,
				LIGHT SERIES
8	GREASE SEAL	—	1	PERFECT OIL SEAL
				#235120
9	WOODRUFF KEY	—	1	#1008
10	PRATT & WHITNEY KEY	—	1	#B
11	HEX SLOTTED NUT	—	1	1-14NF-2B SEMI-FIN REG
12	REG HEX JAM NUT	—	1	1-14NF-2B SEMI-FIN
13	COTTER PIN	—	1	$\frac{1}{4}$ – SAE
14	HEX HD CAP SCR	—	4	$\frac{1}{2}$-13-UNC-2A×2LG.SEMI-FIN

Figure 9.4 Small assembly drawing with bill of material. Courtesy of OSU Dept. of Engineering Graphics

the detail drawings of parts that will meet the functional needs of the device and fit the key dimensions.

The principal reason for making assembly drawings is to show how pieces fit together. An assembly drawing is made by putting the detailed parts together on paper or on a computer screen to learn if the parts fit properly into the assembly. This is an important stage in the design process because the parts must fit together to be able to function properly. They may be used in the shop to guide the assembly process. Drawings of this type are also used for maintenance purposes. They are a great tool for the service repair person since they help the service person identify parts that need repair or replacement. Assembly drawings help identify the individual items within the assembly by part number to speed up the process of locating a replacement.

Production drawings affect many people. Once these drawings are prepared, different professional people will refer to them during the manufacturing and assembly process. The components and assemblies will be analyzed for strength, producibility, and cost. Also, you need to remember that the person who designed the product or the person who made the drawings will not be present to answer questions. For this reason, standard graphic and written descriptions must be given on the drawings to ensure that all parties who see the drawings will interpret the information in the same way.

Pictorial assembly and **exploded views** are often used in technical manuals and sales literature. These types of drawings are three dimensional and make it easier for a person not familiar with a device to understand the technical information.

9.3 *TYPES OF PRODUCTION DRAWINGS*

There are several types of drawings that can be considered as production drawings. You have read about details, assemblies, and pictorial drawings. More information about each type is given in the following paragraphs.

9.3.1 Orthographic Drawings

Orthographic drawings are very commonly used as production drawings. These are the easiest to dimension and are easy to interpret for the person trained to read technical drawings. You learned about orthographic drawings in Chapter 4. By now, you should be proficient in making and reading them.

An orthographic detail drawing may contain a single part or several parts. If threads are included on a detail drawing, they are usually shown with simplified symbols and totally described in the thread specification. If assembly drawings are sectioned, the section line patterns for each part will be the same, but adjacent parts will have a pattern of a different direction or spacing to clearly identify each part, as shown in Figure 9.4. You learned about this in Chapter 6. If threaded parts that have been detailed or threaded fasteners are included on an assembly drawing, they are usually shown with schematic symbols. The threads on detailed parts will have been described on the detail drawings; threaded fasteners will be described in the bill of material.

A tabular drawing is a drawing of a part that includes dimensions identified by letter. The viewer refers to an adjoining table to read the actual dimensions. You first learned about these drawings in Chapter 7 (Figure 9.5). Standard drawings are drawings of parts an organization uses frequently. They are drawn once, and copies are included in any set of production drawings where they are needed (Figure 9.6).

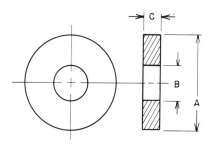

NOMINAL	A	B	C
.250	.625	.281	.065
.375	.812	.406	.065
.500	1.062	.531	.095
.625	1.312	.656	.095
.750	1.469	.812	.134
.875	1.750	.938	.134
1.000	2.000	1.062	.134
1.250	2.500	1.375	.165
1.375	2.750	1.500	.165

Figure 9.5 Tabular detail drawing.

Figure 9.6 Standard detail drawing of often-used part. Computer generated image. Courtesy of Ford Motor Company

Figure 9.7 Exterior assembly shows outside of product in easy-to-read pictorial form. Computer generated image.

9.3.2 Pictorial Drawings and Diagrams

Exterior assembly drawings show several parts assembled together. These drawings show only the surfaces and edges that are visible to the eye and exclude hidden surfaces and edges. Pictorial representations are easy for anyone to visualize. These drawings may or may not be dimensioned (Figure 9.7).

Sectioned drawings and diagrams are similar to exterior assemblies since they show many parts assembled together (Figure 9.8). The difference is that some portion of the assembly has been "removed" so the viewer can see certain interior features of the assembly. These drawings usually employ section lines to indicate the surfaces actually cut by the cutting plane. They may be full sections, half sections, or broken-out sections, as described in Chapter 6 (Figure 9.9).

Exploded pictorial drawings represent several parts assembled according to the axes of their assembly. However, the parts are not shown assembled but are moved apart along the principal axes of the product. This type of drawing typically is not dimensioned but is very helpful for assembly purposes on the production

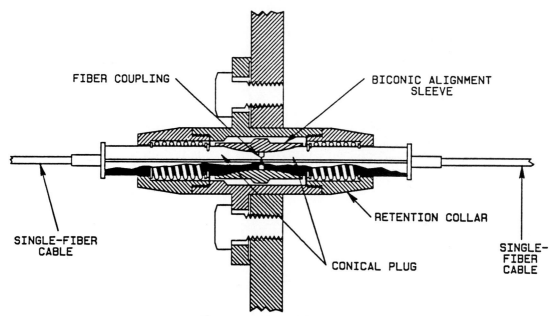

Figure 9.8 Assembly section: Cross hatching is same for any one part but different for adjacent parts. Computer generated image. Courtesy of AT&T

Figure 9.9 Sectional pictorial assembly helps in understanding inner workings of product. Computer generated image. Courtesy of AT&T

Figure 9.10 Exploded assembly shows how parts relate along principal axes of product and its components. Computer generated image.

line. The worker can easily visualize the order and arrangement of assembly of parts that make up an assembly (Figure 9.10).

An installation assembly shows how one assembly attaches to another assembly. These drawings may be later used for maintenance purposes when disassembly is required (Figure 9.11).

9.3.3 Drawings for Other Disciplines

The drawings described as required for production in the preceding paragraphs apply primarily to the manufacture of mechanical devices (appliances, machinery, and vehicles). Some engineering disciplines are not involved in the production of mechanical devices: Civil engineers design highways, bridges, and waterways; chemical engineers create process systems; architects design buildings; and electrical engineers often deal only with circuit design and electrical components (although these components may end up in a mechanical assembly). Each of these disciplines may have a slightly different system for naming various types of drawings or may need other types of drawings for production.

Architects' drawings are commonly assembly drawings. Architects make plan-and-elevation drawings of buildings (Figure 9.12). These are really assemblies of the parts that make up the structure. They also prepare section drawings where they open up a wall to show how the components fit together inside the wall: These section drawings may not be crosshatched (Figure 9.12).

Figure 9.11 Installation assembly tells how components fit assembly. Computer generated image. Courtesy of Ford Motor Company

Architects also make detail drawings of special parts that need to be built to fit the structure.

Electrical designers may prepare schematic wiring plans showing where wires are run, leaving the exact locations and distances to the field electricians. On the other hand, small electronic components, such as circuit boards, may be detailed very carefully to scales of 100:1 or larger to be certain that elements of the circuit are the right size and in the correct relationship (Figure 9.13).

Civil and structural engineers often design assemblies such as steel trusses and leave it to the fabricator of the structure to prepare the detail drawings of the individual pieces of steel that go into the assembly. This is often done because different fabri-

Figure 9.12 A building drawing shows plan and elevation views and may show section views. Courtesy of Timothy C. Whiting

cators have different equipment for production. Each organization will design to take advantage of its own facilities.

Chemical engineers often make schematic assemblies of a process concept, including the major components, piping, and instruments for process control. These are called *process and instrumentation diagrams* (Figure 9.14). Chemical engineers, nuclear engineers, and others assembling large quantities of pipe and large components often have three-dimensional assemblies made. The finished job is built to scale of miniature components and wires

Figure 9.13 Printed circuit drawing for electronic component was produced with computer graphics and printed. Computer generated image.

Figure 9.14 Chemical engineers prepare process and instrumentation diagrams such as this drawing. Courtesy of Timothy C. Whiting

Figure 9.15 Three-dimensional model of part of chemical plant.
Courtesy of O.S.U. Engineering Publications

or rods for the piping. Sometimes this model becomes the principal assembly and the construction workers obtain their information by "reading" the model (Figure 9.15).

9.4 CALLOUTS

Callouts are identifiers placed at the end of a leader line that direct the eye of the viewer from the part to the identifying note. The callouts identify a part by name, number, or letter and number. These help the viewer relate that particular part to a list of parts or details, as in a bill of material. Figure 9.16 shows compounds identified by number.

Callouts are placed around the perimeter of the assembly and are usually identified in sequence by letter or number. These identifiers may be placed in clockwise order or may be assigned according to the importance or size of the parts.

SECTION 'B-B'

Figure 9.16 Callouts on body of drawing referenced to bill of material. Computer generated image. Courtesy of American Electric Power Service Corp.

9.5 *METRIC IDENTIFICATION*

When the dimensions or notes of a production drawing are given in the metric system, the word *metric* should appear within a box in an open area of the drawing or near the title block.

If metric dimensions are given on the drawing, the standard unit of measure for small-parts production is the millimeter. The dimensions given in millimeters need not be followed with mm since this is understood if the note "metric" was drawn within the box (Figure 9.17). Conversely, if your organization makes both metric and inch drawings, you should identify the English system drawings with the "inch" in a box near the title block.

Figure 9.17 Metric drawing dimensioned in millimeters and identified in title block. Computer generated image. Courtesy of Ford Motor Company

9.6 PAPER SIZES, FOLDS, AND BORDERS

There are three basic sizes from which standard drawing papers are derived. They are 9 × 12 and $8\frac{1}{2}$ × 11 in. in the English system and 210 × 297 mm in the metric system.

The most common sizes in the English measure are the multiples of $8\frac{1}{2}$ × 11. The A size sheet can be copied in an ordinary office copy machine using standard $8\frac{1}{2}$ × 11 paper. Standard B, C, D, and E sheets are multiples of the $8\frac{1}{2}$ × 11.

An A size sheet will fit into standard file folders without covering the raised identification tab. The larger size sheets of the 9 × 12 multiples will not fit into standard office file folders and will become torn, or "dog eared," because the edges of the paper are exposed to wear. These and larger size sheets are usually filed in shallow, flat file drawers.

There is a standard method for folding the larger letter size

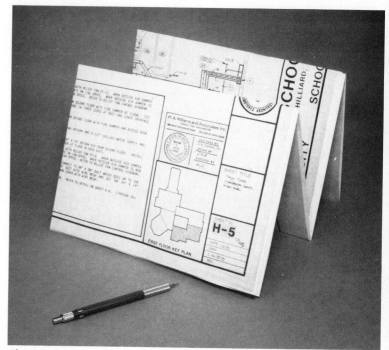

Figure 9.18 Copies of large drawings are folded with "accordion" folds so drawing number is visible. Courtesy of McDonald-Cassel & Bassett, Inc.

sheets. This procedure provides for the exposure of the identifying information (such as the drawing number) on the drawing in a clearly visible location on the outside of the folded sheet (Figure 9.18). Some drawing sheets have fold marks printed along the edge of the sheet. Industry standards are specific about how larger sheets should be folded. The accordion style fold makes it easy to locate particular identifying information and allows the sheet to be easily opened for reading, similar to reading a road map.

Border lines may be drawn or printed on standard size drawing sheets. These improve the appearance of the sheet for presentation purposes. Also, the borders are printed a short distance from the edge of the paper to prevent the drawing from extending to the edge of the paper. Since well-used drawings become damaged on the edges, the border space provides a safety area to prevent the loss of important information.

Letters or numbers are often printed to identify **zones.** If the lines between zones were extended across the working area of the paper in a coordinate grid, then identification of various areas of the drawing can be made more quickly and accurately. You

Figure 9.19 Zones help locate areas on drawing quickly. Computer generated image. Courtesy of AT&T

can relate to the zones on a road map when visualizing the grid on a drawing. The zoned grid is a particularly important aid when talking on the telephone. The speaker can help orient the listener to a specific area of the drawing quickly if proper use of the zone identifiers is made (Figure 9.19).

9.7 TITLE BLOCKS

The title block is usually found at the bottom of the drawing. The title block is an area within a box that contains a concentrated amount of information. Although title blocks vary according to

DATE	NO.	DESCRIPTION	APPD.
		REVISIONS	

8

9

DWG. NO.

ARCH.		ELEC.		MECH.		STR.	

SCALE•			ENGINEERING DIVISION	
DR•		APPROVED BY		
CH•			DESIGN DIVISION	
SQ. LDR•				
DATE•				

AMERICAN ELECTRIC POWER SERVICE CORP.

O [.]

Figure 9.20 Typical title block. Computer generated image. Courtesy of American Electric Power Service Corp.

the information required by the industry, some items are standard (Figure 9.20). Items usually included within title blocks are as follows:

Drawing number   Identifying number of the drawing. The numerals are the largest numerals in the lettering block.

Sheet number   Sheet ____ of ____ . This notation means the drawing sheet number followed by the total number of sheets in the entire set of plans. It is very important to know the total number of drawings in order to establish that you have a complete set of drawings.

Title   Name of the part or assembly. This identifier usually consists of larger letters than other items in the lettering block.

Company name   Name of the company for whom the drawing was created.

Company logo   For quick identification, the company logo or other simplified graphic may appear near the company name.

Scale   Scale at which the drawing was made. This shows the relationship of the size of the drawing to the size of the actual part or assembly.

Date   Date the drawing was created. This date may be very important for establishing the date the work was created for patent or other protection purposes.

Tolerances   Standard tolerances for the company or industry. They may be entered in decimals, fractions, millimeters, or degrees. These tolerances usually state a range of acceptable limits that can be applied to dimensions on the drawing. Tolerances are given as fractional, decimal (to a given number of places), or angular for angles and bends. This is described in detail in Chapter 7.

Drafter — Name or initials of the person who created the drawing. With this entry, questions or comments related to that drawing can be addressed to the person who created the drawing.

Checker — Name or initials of the person who checked the drawing for drafting accuracy and compliance with standards.

Supervisor — Name or initials of the supervisor who approved the drawing. The supervisor is the person ultimately responsible for the quality and accuracy of the drawing.

Optional items that might appear in the lettering block could include such things as the following:

Weight — This may be very important for shipping reasons. This will allow someone to calculate the gross weight of a single part or the weight of a total shipment as well as the total weight of an assembly.

Material — If only a single part is detailed on the drawing, then it is appropriate to identify the material used to make the part. However, if there are details of several parts on the same sheet, especially if the parts are made of different materials, then the material should be listed under the part being detailed, not in the lettering block. When this condition arises, the term *as noted* should be lettered in the lettering block box reserved for identifying the material.

Number required — Number of parts to be produced.

Surface — Surface roughness, texture, or other such characteristics as may be required.

Hardness

Hardness of the material may be listed. This value is given for a standard hardness test such as a Rockwell hardness test.

Heat treatment

If the part is to be heat treated, there may be an entry to note the type or degree of heat treatment required.

Sheet size

Seldom will it be necessary to identify the size of the drawing sheet, but given within some title blocks is a place to check which corresponds to the standard letter size sheets A–E.

9.8 *REVISIONS*

All revisions made to an original drawing must be recorded. For this purpose a revision chart or schedule is shown on the drawing (Figure 9.21). The date is important for legal or patent purposes and to inform any reader of the time of the change. A brief de-

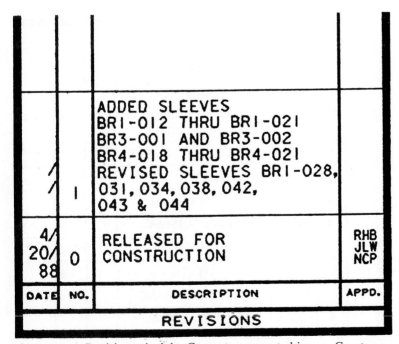

DATE	NO.	DESCRIPTION	APPD.
	1	ADDED SLEEVES BRI-012 THRU BRI-021 BR3-001 AND BR3-002 BR4-018 THRU BR4-021 REVISED SLEEVES BRI-028, 031, 034, 038, 042, 043 & 044	
4/20/88	0	RELEASED FOR CONSTRUCTION	RHB JLW NCP

REVISIONS

Figure 9.21 Revision schedule. Computer generated image. Courtesy of American Electric Power Service Corp.

scription of the change is given as well as the zone, if on a large drawing. The changes must also be made on the specific parts or specifications affected by the change. A change letter or other code is noted beside the change to identify it.

Since information concerning revisions is relatively standard, it is common to list the revisions near the title block. If the revision schedule starts at the title block, notice that revision number 1 is the closest to the title block. This allows for additional revisions to be added to the "open-ended" schedule.

9.9 PARTS LIST

A bill of material may be listed along the side of an assembly drawing or on a separate sheet. This list typically includes all items necessary to make an assembly. This includes standard items such as bolts, nuts, and washers and other purchased parts. The listing of these standard items eliminates the necessity for detailing many standing parts (Figure 9.22).

The parts list typically includes but is not limited to the part number, part name, number required, and a place to enter sizes or other details.

BILL OF MATERIAL

NO. REQ'D.		MARK	DESCRIPTION	LENGTH		WEIGHT
ITEMS SHIPPED	SHOP ASSEMBLY			FT.	IN.	
1		4801	8'-0 HIGH FRAME			
	4	4801A	WELDED ASSEMBLY			
	4	a	$\frac{1}{2}$x4	0	4	
	4	b	L $2\frac{1}{2}$x$2\frac{1}{2}$x$\frac{3}{16}$	7	11	
	4	4801B	WELDED ASSEMBLY			
	4	c	L $2\frac{1}{2}$x$2\frac{1}{2}$x$\frac{1}{4}$	1	$3\frac{1}{2}$	
	20	4801C	L $1\frac{3}{4}$x$1\frac{3}{4}$x$\frac{3}{16}$	1	$10\frac{3}{8}$	
1		4802	10'-0 HIGH FRAME			
	4	4802A	WELDED ASSEMBLY			

Figure 9.22 Typical bill of material (parts list). Computer generated image. Courtesy of American Electric Power Service Corp.

Parts lists, revisions, and other notes are primarily words rather than lines and symbols. There may be pages of description defining how various operations will be performed, how testing will be done, packaging and shipping requirements, and other legal requirements. These descriptions are called specifications. Because of the complexity of specifications and the time needed to prepare them, word-processing programs are commonly used on the computer to help generate specifications. Also, additions and deletions can be made to the master set of specifications very quickly using word-processing programs. They are discussed further in Section 9.14.

The specifications will normally be produced on $8\frac{1}{2} \times 11$ paper. Notes, bills of material, and revision schedules appear on the face of the drawings, which may be much larger than $8\frac{1}{2} \times 11$. Some organizations use wide-carriage printers that allow the information to be printed directly on the drawing. In other cases the information is printed on adhesive-coated transparent sheets that can be attached to the drawing.

9.10 GENERAL NOTES

Some general notes may be appropriate. These should be neatly lettered and placed near but outside the lettering block. Occasionally, these notes are lettered at an angle to attract the viewer's attention. Also, the size of the lettering is generally larger than other lettering on the drawing. They might include such notes as:

All dimensions in millimeters

All undimensioned radii = _____

FAO (finish all over)

Not approved

For estimate only

For quotation purposes only

Not to scale

Sometimes notes are placed for security purposes. This is particularly appropriate on government drawings. Examples of such security notes might include:

Top secret

Security clearance required

Restricted drawing

"In house" drawing only

Confidential document

Do not copy

9.11 *COMPUTER-GENERATED PRODUCTION DRAWINGS*

There are many advantages to making production drawings on the computer using available graphics packages. For example, if the graphics package has the capacity to generate a data base, then after some views have been entered, others can be generated without having to draw them individually. The points can simply be generated from the data base.

Various types of views may be generated from the data base. For instance, once the points that define a building have been entered, orthographic views and perspective pictorials can be generated from this data base. Perspective pictorials can be generated after noting such variables as the distance the eye is from the object, the height of the eye above ground level, and whether a one-, two-, or three-point type of perspective drawing is desired (Figures 9.23 and 9.24). The same techniques apply to mechanical parts: A pictorial drawing can be generated from the data base created by preparing the orthographic drawings.

One Point

Two Point

Three Point

MEGA^{CADD}

Figure 9.23 Computer-plotted perspective generated from three-dimensional data base. Computer generated image. Courtesy of MEGA-CADD

Lower Level

Upper Level

MEGA^{CADD}

Figure 9.24 Elevation views from same data base. Computer generated image. Courtesy of MEGA-CADD

The same data base can be used in auxiliary programs that do stress analysis (finite-element analysis programs). A few large manufacturing organizations also analyze the part data base for producibility and cost to manufacture. One manufacturer has publicized a program that performs economic optimization—changing material thicknesses and contours to give a part with maximum performance using minimum material.

Wire frame drawings are the traditional pictorial generated from the computer data base (Figure 9.25). These drawings indicate the edges, points, and surfaces that make up the object. You generally see the lines for all the surfaces, both front and back with no regard for hidden lines. These drawings may be further refined by solid modeling. However, some programs allow wire frame diagrams to be viewed showing only visible surfaces and edges (Figure 9.26).

Solid modeling programs use the data base information to produce a diagram that displays the object as a three-dimensional pictorial clearly depicting the visible surfaces of the object (Figure 9.27). Hidden surfaces are seldom pictured. Solid modeling is very effective when displaying industrial parts and assemblies. The object can be examined from many points of view, and clearances and relationships among adjacent parts can be checked.

The architectural application of solid modeling is illustrated by generating the visible surfaces of a building from the data base (Figure 9.28). Relationships with adjacent structures as viewed from different points can also be evaluated.

Figure 9.25 Ordinary wire frame drawing from computer data base. Computer generated image.

Figure 9.26 Wire frame drawing with hidden lines deleted. Computer generated image.

Figure 9.27   Solid model from computer data base. Computer generated image. Courtesy of O.S.U. Engineering Publications

Zoomed portions of detail drawings allows the CADD operator to produce large images of detailed areas of production drawings. This can be very effective when dimensions, notes, or details must be added to a complicated portion of the drawing. Zoomed images are also appropriate for showing enlarged views of complex sectioned details (Figure 9.29).

Layered drawings were discussed in Chapter 4. They have many advantages and can easily be obtained from the data base (especially if you plan ahead and assign information to different layers). For instance, plans for a chemical processing plant might be developed with drawings with limited information for certain

MEGA^CADD

Figure 9.28   Solid model for building from architectural data base. Computer generated image. Courtesy of MEGA-CADD

Figure 9.29 Complex drawing magnified by zooming. Computer generated image.

Figure 9.30 Layered drawing: Electrical plan only. Computer generated image. Courtesy of Consulting Engineers, Mechanical/Electrical/Energy Consultants, Inc.

contractors. A floor plan with only the electrical details added would be appropriate for the electrical contractors (Figure 9.30). A plan with the heating and air-conditioning ducts would be appropriate for the heating, ventilation, and air-conditioning contractors (Figure 9.31). Another plan with the location of the equip-

EF-1 (TYP.)

EF-1 (TYP.)

Figure 9.31 HVAC plan only; different layer. Computer generated image. Courtesy of Consulting Engineers, Mechanical/Electrical/ Energy Consultants, Inc.

ment superimposed on the floor plan would be appropriate for the general contractor installing the equipment (Figure 9.32). Once the data base has been entered, it is easy to develop drawings with only the required information for particular applications.

Large companies that must provide many sets of drawings for the construction of facilities can save time and money by producing only the plans necessary for each of the bidders. In the past, entire sets of drawings would be provided to all parties. This was very expensive and time consuming. However, once a computer data base has been entered, then it is a relatively simple process to rearrange the data in a form to produce only the needed information for specific bidders.

The use of the computer to manipulate data bases can also be useful in the transmittal of information over the telephone lines. Information for drawings as well as written material can be transmitted from one location to another by using a telephone modem (Figure 9.33).

The same data base used to produce drawings can be used to drive computer-aided manufacturing (CAM) machines such as milling machines, lathes, or profilers. There are various acronyms for this type of integration. It has been referred to as CAD/CAM; because of the progress achieved in getting all types of manufacturing facilities to communicate electronically. It has also been called computer-integrated manufacturing (CIM) (Figure 9.34).

Figure 9.32 Equipment plan; only equipment shows on this layer. Computer generated image. Courtesy of Consulting Engineers, Mechanical/Electrical/Energy Consultants, Inc.

Figure 9.33 Computer work station with MODEM for telephone transmission of data. Computer generated image. Courtesy of AT&T

Figure 9.34   Machine tool with integrated numerical control. Courtesy of O.S.U. Engineering Publications

9.12 *DRAWING MEDIA*

In our contemporary industrial world, two media are used almost exclusively for making production drawings. They are vellum and film. Vellum is a translucent paper that may be used to make large copies. It is semitransparent, which allows for tracing; this can save the drafter a considerable amount of time.

Film is nearly transparent and is very dimensionally stable. This means the drawings will not expand or contract due to changes in temperature and humidity. It also ages very slowly (deteriorates slowly over long periods of time). This can be important for archival storage (many years) of master drawings. Another advantage of film is that it is so tough that it may be run through a copy machine thousands of times with little wear. If vellum is used for a large number of copies, the wear and tear on the vellum will degrade the quality of the copies. Film masters are dimensionally stable.

This stability makes film quite attractive for those industries making drawings of electronic circuitry that will be reduced photographically for the making of printed circuits. Because of the accuracy required, the complexity of the drawings, and the need for excellent line quality, these drawings can best be produced using a computer-driven plotter. For working purposes, some computer-generated prints can be printed on opaque paper.

Sepias are intermediate copies. They are brown line drawings on a semitranslucent paper or film base. You may draw on a sepia to alter an original, use it as a print, or make copies from it.

9.13 *REPRODUCTION OF DRAWINGS*

Original drawings never leave the engineering office. Only copies of the originals are sent to other professionals or the production or construction site. Because these original drawings must be used to make many copies, they must be very clear and have legible lettering. Generally speaking, if the original drawing is difficult to read, the copy will be worse.

Small drawings, details, or specifications made on $8\frac{1}{2} \times 11$-in. opaque paper may be copied on a common office copy machine. Drawings produced on vellum or film may also be copied on these office copy machines, but a backing sheet of white paper may have to be placed behind the vellum or film before the copy is made. Some of these machines will copy larger sheets, but most business copiers are limited to small sheets (Figure 9.35).

Copy machines are now readily available for copying larger sheets. These machines can easily handle D and E size master drawings (Figure 9.36).

Large drawing sheets have commonly been reproduced by the diazo process. This is a two-stage process involving exposure to light and then exposure to ammonia fumes. In the first stage, the vellum or film is placed on a sheet of copy paper that has been specially treated with diazo salts. When the two sheets are inserted into the diazo machine, the ultraviolet light passes through the master and ''burns off'' the salt crystals on the emulsion surface of the copy paper. The pencil or ink lines prevent the light from affecting the salts, so the salts remain under the areas covered by the lines.

In the second stage, the exposed copy paper is subjected to ammonia fumes. The areas exposed to the light will not be affected by the fumes, but the remaining salt crystals that were under the lines of the master during the exposure stage will turn to a deep blue (or other color if special paper is ordered) (Figure 9.37).

Contemporary reproduction methods include many products and methods that add greater flexibility for the user. Clear film

Figure 9.35 Standard office copier accepts originals $8\frac{1}{2} \times 11$ in. and may take originals up to 11×17 in. Courtesy of Xerox Corp.

appliques are often applied to master drawings where repetitious shapes must appear on the drawing. This avoids the time-consuming process of redrawing common shapes. Appliques are made on clear film and have a clear, light adhesive on the back.

Large architectural, engineering, and design firms now use overlay reproduction on complicated sets of drawings of major projects. The master drawings are actually drawn on separate layers of film. The film layers are aligned through a pin registration system. Each sheet is accurately punched with a series of up to seven holes that are spaced to fit a standard pin registration bar.

Once the main registration drawing has been completed (say, the floor plan wall layout), other professionals can each begin on a different overlay (electrical, plumbing, dimensions, etc.). This helps reduce drawing time. This process also makes each layer much less complicated, and thus changes can be made easily to any layer without disturbing the information that is now on another layer.

These layers can be easily copied on clear film with the lines on each layer being of a different color. These are used for design and checking purposes. When the set is complete, they will be sent to the printer, and the final copies will be printed in color by offset printing. One might first think this is the most expensive

Figure 9.36 Copier used to copy large-size drawings. Courtesy of Xerox Corp.

Figure 9.37 Diazo process machine for reducing engineering drawings. Courtesy of Bruning, Div. Am. Intl., Inc.

route to follow, but this process really saves a great amount of time and money on large projects. For instance, those bidding or contracting for a particular segment of a project (say, the electrical portion) only need a limited number of drawings. This process results in more accurate and quicker reading of the set of drawings. This overlay process also aids in the bidding process by reducing time in cost estimation.

9.14 *FLOW OF INFORMATION*

You need to understand the role of production drawings in the manufacturing process. Many people's activities are affected by the information on the drawings, so they must communicate a very clear message. This message takes into account graphic, numeric, and written information. The flow of engineering information from the designer to the final product that is ready for the user may take several paths. Two possible paths within an organization are shown in Figure 9.38. Many variations of these paths are possible. Sometimes this information travels on paper and sometimes it may travel electronically. A delay in the flow because of misunderstanding, missing information, or unclear line work costs time and money.

9.15 *ACCOMPANYING WRITTEN SPECIFICATIONS*

If all the needed information cannot be placed on the detail or assembly drawings, then this additional information is supplied in written form. General standards for testing and quality control, types of finish, and packaging requirements are written in specifications rather than lettered directly on the drawing (Figure 9.39).

Assembly drawings related to the assembly at the job site (large machinery, conveyors, chemical process equipment, etc.) will require many pages of specifications to describe the needs for packaging, shipping, protection of parts, storage conditions, and procedures for assembling the parts together at the job site. Testing and checking procedures are also included in the written specifications. The specifications for a large-machine installation may be the size of several large city telephone books.

9.16 *SUMMARY*

Production drawings carry information from the engineering office to the sites where it will be used to plan, buy, produce, and assemble the components and final form of useful devices. These drawings may also be used for maintenance and repair. The information may be transmitted on paper or electronically.

The two most common types of production drawings are detail drawings of individual parts and assembly drawings, which show how parts are installed relative to each other. Drawings may carry information in orthographic or pictorial form; they may be produced manually or with a computer. Computer-generated drawings create a data base that may be

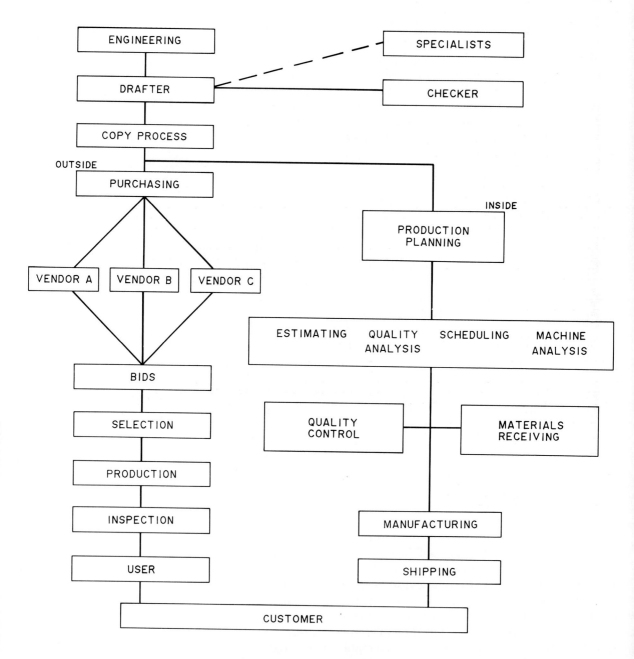

Figure 9.38 Flow of engineering information and production drawings may take several paths.

SLEEVE SCHEDULE										
STEAM GENERATOR BLDG. -ZONE BRI										
				DIMENSION		LOCATION				
SLEEVE NO.	ELEVATION	SIZE	TYPE	A	B	DRAWING NO.	COORD.	COL. NO.	LINE NO.	MATERIAL
BRI-001	135'-0"	6"	I	13"	4"	5230-01	(B8)	G/2	3" FW-050-B1	C.S.
BRI-002	135'-0"	18"	I	13"	4"	5230-02	(F5)	G/8	10" FW-131-B1	C.S.
BRI-003	229'-9"	14"	I	9"	2-1/2"	5270-01	(B3)	H/2	8" DC-015-P5	C.S.
BRI-004	292'-0"	14"	I	12-1/2"	2-1/2"	5270-01	(L3)	H/2	8" DC-015-P5	C.S.
BRI-005	214'-0"	14"	I	9"	2-1/2"	5270-03	(C3)	G/10	8" DC-028-P5	C.S.
BRI-006	229'-9"	14"	I	9"	2-1/2"	5270-03	(C3)	G/10	8" DC-028-P5	C.S.
BRI-007	262'-9"	14"	I	12-1/2"	2-1/2"	5270-03	(H3)	G/10	8" DC-028-P5	C.S.
BRI-008	135'-0"	14"	I	17"	4"	6270-06	(E8)	H/10	10" RD-055-F3	C.S.
BRI-009	217'-1-1/2"	10"	I	9"	2-1/2"	6270-04	(G6)	H/4	6" RD-092-F3	C.S.
BRI-010	135'-0"	14"	I	13"	4"	5280-37	(G3)	H/2	10" FP-432-H2	C.S.
BRI-011	262'-9"	14"	I	12-1/2"	2-1/2"	5270-03	(G2)	H/9	8" DC-032-P5	C.S.
BRI-012	292'-0"	8"	I	12-1/2"	2-1/2"	5208-26	(M5)	G/1	2" HS-814-F1	C.S.
BRI-013	292'-0"	8"	I	12-1/2"	2-1/2"	5208-26	(M5)	G/1	1-1/2" HS-848-F1	C.S.
BRI-014	309'-6"	8"	I	12-1/2"	2-1/2"	5208-26	(M8)	G/1	1-1/2" HS-947-F1	C.S.
BRI-015	309'-6"	8"	I	12-1/2"	2-1/2"	5208-26	(M8)	G/1	1" HS-975-F1	C.S.
BRI-016	262'-9"	8"	I	12-1/2"	2-1/2"	5208-26	(F8)	G/10	2" HS-623-F1	C.S.
BRI-017	262'-9"	8"	I	12-1/2"	2-1/2"	5208-26	(F8)	G/10	1-1/2" HS-645-F1	C.S.
BRI-018	279'-6"	8"	I	12-1/2"	2-1/2"	5208-26	(H8)	G/10	1-1/2" HS-978-F1	C.S.
BRI-019	279'-6"	8"	I	12-1/2"	2-1/2"	5208-26	(H8)	G/10	1" HS-979-F1	C.S.
BRI-020	135'-0"	1-1/2"	I	13"	4"	6202-08	(F3)	H/2	1/2" DW-018-M1	C.S.
BRI-021	229'-9"	8"	I	9"	2-1/2"	33401	(C5)	H/1	————	C.S.
'-"	'-"	"				-	( )	/	" - -	
'-"	'-"	"				-	( )	/	" - -	
BRI-024	144'-3"	10"	2	12"	————	6270-04	(D2)	H/10	6" RD-061-F3	C.S.
BRI-025	229'-9"	10"	I	9"	2-1/2"	6270-04	(G6)	H/4	6" RD-092-F3	C.S.
BRI-026	135'-0"	8"	I	13"	4"	5280-37	(G5)	H/6	6" FP-424-H2	C.S.
BRI-027	229'-9"	14"	I	9"	2-1/2"	5270-01	(B1)	G/1	8" DC-025-P5	C.S.
BRI-028	229'-9"	10"	I	9"	2-1/2"	33320	(B3)	G/4	————	C.S.
BRI-029	135'-0"	14"	I	13"	4"	5208-13	(J8)	H/4	6" HS-501-F1	C.S.
BRI-030	135'-0"	12"	I	13"	4"	5208-13	(J8)	H/4	4" HS-737-F1	C.S.
BRI-031	135'-0"	48"x81-1/2"	5206-04	13"	4"	5206-04	(L3)	H/8	36" FT-300-D6	C.S.
BRI-032	135'-0"	32"	I	13"	4"	5203-06	(L6)	G/6	18" AS-118-G1	C.S.
BRI-033	135'-0"	6"	I	13"	4"	6202-08	(K3)	H/2	3" VENT	C.S.
BRI-034	135'-0"	4"	I	13"	4"	6202-08	(K3)	H/2	2"	C.S.
BRI-035	135'-0"	14"	I	13"	4"	5280-05	(D6)	G/8	10" FP-223-H2	C.S.
'-"	'-"	"				-	( )	/	" - -	
'-"	'-"	"				-	( )	/	" - -	
BRI-038	262'-9"	5"	I	12-1/2"	2-1/2"	5260-27	(G7)	G/9	2-1/2" DA-570-F2	C.S.
BRI-039	279'-6"	5"	I	12-1/2"	2-1/2"	5260-27	(K7)	G/9	2-1/2" DA-570-F2	C.S.
BRI-040	135'-0"	6"	I	13"	4"	6202-08	(F3)	H/2	4" DW-013-M1	C.S.
BRI-041	135'-0"	4"	I	13"	4"	6202-08	(F4)	H/2	1-1/2" DW-015-M1	C.S.
BRI-042	149'-0"	6"	I	9"	2-1/2"	33309	(C2)	H/1	————	C.S.
BRI-043	135'-0"	6"	I	13"	4"	5270-13	(N8)	G/1	4" SD-576-F3	C.S.
BRI-044	149'-0"	6"	I	9"	2-1/2"	5270-13	(N3)	G/1	4" SD-576-F3	C.S.
'-"	'-"	"				-	( )	/	" - -	
'-"	'-"	"				-	( )	/	" - -	
'-"	'-"	"				-	( )	/	" - -	
'-"	'-"	"				-	( )	/	" - -	
'-"	'-"	"				-	( )	/	" - -	

Figure 9.39 Specifications include engineering information too lengthy to include on drawings. Computer generated image. Courtesy of American Electric Power Service Corp.

useful for producing other types of drawings, engineering analysis, or direct transmission to computer-aided manufacturing.

The notes, bills of material, title blocks, and accompanying specifications are important parts of the total set of production drawings. The total package needs to be clear, accurate, and correct in order to avoid delays, which cost time and money.

PROBLEMS

Problems 9.1–9.5

Problems for this chapter consist of several assembly problems. Any individual part from these assemblies may be drawn and detailed as a single unit, or may be drawn in isometric as a pictorial view.

The instructor may elect to assign the assemblies to be drawn as exploded views, assemblies, sectioned assemblies, or as assembled pictorials (which may or may not be rendered).

Problem 9.1

Problem 9.2

CHAIN RIVET
EXTRACTOR

CROWN BOTH ENDS R 1.00

$\phi \frac{.1900}{.1903}$ .13 DEEP

.75
.38

1.50
.50
.75

R .19
.19
.31-24 UNF-2B

.53
.31
.75
.47

.13
.09
.19
.09

① BASE

CROWN BOTH
ENDS R .19

.1906
.1904

1.75
1.50

ϕ .13

ϕ .31
.13

② HANDLE
SHAFT

CROWN BOTH
ENDS R .19

2.00

.1906
.1904

③ SCREW
HANDLE

.1600
.1605 .50 DEEP

.31-24 UNF-2A

.06×45°

1.56

.56
.25
.03×45°
$\phi \frac{.1900}{.1903}$

④ SCREW

ϕ .31
.50 DEEP

$\phi \frac{.1300}{.1303}$

.75 SPHERE

⑦ HANDLE BALL

.09
.03×45°

.88
.44

.1609
.1606

⑤ SCREW
POINT

CROWN BOTH
ENDS R .38

.75

.1306
.1304

⑥ PIN

114

22

51

64

114

13-13 UNC-2B

57

∅ 11, 3 HOLES

6

73

64

38

57

35

13

10

① BASE

⑤ M10×1.5
HEXHEAD CAPSCREW
2 REQ'D

④ M12×1.75
HEXHEAD CAPSCREW

∅ 11, 2 HOLES

51

44

22

76

13

16

⑥ 10 PLAIN WASHER
2 REQ'D

R 24

R 16

22

32

25

13

102

SYMMETRICAL ABOUT ℄

② CLAMP BRACKET

⑦ 12 EXTERNAL TOOTH
LOCK WASHER

79

32

22

R 24

102

70

51

76

51

13

16

19

44

57

R 16

∅ 14

22

44

25

13

70

10-16 UNC-2B, 2 HOLES

③ SWING ARM

Problem 9.2

4.50

.88

2.00

2.50

4.50

2.25

.50-13 UNC-2B

φ .44, 3 HOLES

.25

2.88

2.50

1.50

2.25

1.38

.50

.38

① BASE

φ .44, 2 HOLES

2.00

1.75

.88

3.00

.50

.63

R .94   R .63

.88

1.25

1.00

.50

4.00

SYMMETRICAL ABOUT ₵

② CLAMP BRACKET

⑤ .38-16 UNC-2A,
   HEXHEAD CAPSCREW
   2 REQ'D

④ .50-13 UNC-2A,
   HEXHEAD CAPSCREW

⑥ .38 PLAIN WASHER
   2 REQ'D

⑦ .50 EXTERNAL TOOTH
   LOCK WASHER

⑤

②

④

⑥

③

①

3.13

1.25

.88

R .94

4.00

2.75

2.00

2.00

3.00

.50 .63 .75

R .63

1.75

2.25

φ .56

.88

1.75

1.00

.50

2.75

.38-16 UNC-2B, 2 HOLES

③ SWING ARM

Problem 9.3

Problem 9.3

R .50
4 PLACES

Ø 1.25
Ø 2.38 C'BORE
.31 DEEP

1.76
.88
3.00

1.50

.88
3.88
4.88

Ø .38
4 HOLES

.62

1.19

.50
1.00
1.44
2.88
.50

① BASE

⑦ ④ ⑥ ② ① ⑤ ③

Ø 1.75

Ø 1.75
.19
1.07
Ø 1.25

Ø 1.50
45°x.13

.75
.94
2.38

Ø .25

Ø .19

45°x.06

.38
Ø .88

② SHAFT

Ø 2.38
.88

⑤ CASING

Ø .88

⑥ BUSHING

.50 TYP

Ø 4.00

Ø .88
Ø 1.50 C'BORE
.38 DEEP

Ø 1.25

Ø 2.38
.25

③ DISK

1.50
Ø .25

④ PIN

Ø .38
.31
.81
1.25
.94
Ø .25
Ø 1.25
Ø 2.38

⑦ WHEEL

Problem 9.4

① BASE

SYMMETRICAL
ABOUT ℄

② FLOOR SEAT

③ ADJUSTER NUT

④ COLLAR

⑤ SEAT SUPPORT

⑥ SHAFT SUPPORT

Problem 9.4

ø .38, 4 HOLES   ø .88, .75 DP

4.00
2.00
2.00
4.00
R .13 TYP

2.75
1.38
1.38
2.75
R .25

ø 1.13
1.00
.69
.25
.13 TYP
2.50

① BASE

SYMMETRICAL
ABOUT ₵

1.06 TYP
.13 TYP
R .06 TYP
1.06
45°
1.19 1.50
.31
ø 1.00
ø 1.25

② FLOOR SEAT

②
④
⑥
⑤
③
①

R .06
ø .56 THRU HOLE

.88-9 UNC-2A×2.5
.13×1.88

.88-9 UNC-2B
R .06 TYP
1.25
1.25
.13 TYP
.63 .31
R .38

③ ADJUSTER NUT

ø .88
ø 1.25
ø 1.00
.50
.13
.19
.09 H-SPACE TYP

④ COLLAR

ø .88
ø 1.00
2.00

⑤ SEAT SUPPORT

.13×45°
.81 .94
.06×45°
ø .88
ø 1.13

⑥ SHAFT SUPPORT

Problem 9.5

Problem 9.5

HARDWARE SCHEDULE		
PART	NO. REQ'D	DESCRIPTION
(A)	4	#10-24 UNC-3A x .50 LG, CUPPOINT, HEADLESS, SETSCREW
(B)	4	#10-24 UNC-3A x .50 LG, RHMACH SCREW
(C)	4	#10 N WASHER
(D)	2	#5-40 UNC-3A x .13 LG, CUPPOINT, HEADLESS, SETSCREW
(E)	6	#5-40 UNC-3A x .25 LG, CONEPOINT, HEADLESS, SETSCREW
(F)	2	#5-40 UNC-3A x .38 LG, CUPPOINT, HEADLESS, SETSCREW
(G)	3	#10-24 UNC-3B HEXNUT
ALL HARDWARE-STEEL		

BILL OF MATERIALS			
PART	NO. REQ'D	MAT'L	DESCRIPTION
(1)	1	CI	BASE
(2)	1	CI	INTERMEDIATE SLIDE
(3)	1	CI	JAW
(4)	1	CI	SLIDE BASE
(5)	2	3140 STL	JAW SPACER
(6)	2	1040 STL	PIN
(7)	2	1040, STL	COLLAR
(8)	2	BRZ	COLLAR BRACE
(9)	3	1020 STL	HANDLE
(10)	3	1020 STL	HAND WHEEL
(11)	2	BRASS	PRESSURE SPACER
(12)	1	1020 STL	ALIGNMENT ROD
(13)	1	1040 STL	SPINDLE A
(14)	1	1040 STL	SPINDLE B
(15)	1	1040 STL	SPINDLE C

3.88 — .06 — .34 — 1.12 — 1.12 — .88

ø .19, DRILL TO DIMPLE, 3 DIMPLES

(11) PRESSURE SPACER

#5-40 UNC-3B — .06 — .38 — 1.25 — .56 — .25 TYP — 1.38

.25

R .44 — 1.75 — .50 — #10-24 UNC-3B, 2 HOLES — 45° — .44 — .75 .81 — .44 — .31 — .50 — R .44 — ø .38 — ø .38 .88 DP

(3) JAW

#10-24 UNC-3B, 2 HOLES — 1.00 — .25 — 1.75 — .50 — 2.75

(5) JAW SPACER

Problem 9.5

SYMMETRICAL ABOUT ₵

BASE

Problem 9.5

#5-40 UNC-3B

.44

15°

2.75

1.38

.26

.06

4.81

6.00

.81

.25

.88

.25

.13

VIEW A

VIEW B

.38-12 ACME-3B

R .44

1.38

.50

1.19

VIEW A   ⌀ .38

SYMETRICAL
ABOUT ℄

(4)   SLIDE BASE

1.38

.50

.25

#10-24 UNC-3B,
2 HOLES

.88

2.25

1.06

.31

.31

.56   .75

.50

.25

1.00

⌀ .38

#10-24 UNC-3B
2 HOLES .50 DEEP

VIEW B

Problem 9.5

Problem 9.5

5.13

.50

.19

3.25

.38

.19

.38

.19

ø .38

ø .19

ø .38

.251 / .252 SQ

.19-24 UNC-3A-.25 LG

.38-12 ACME-3A-3.25 LG

(13) SPINDLE A

7.88

6.00

1.25

.38

ø .38

.251 / .252 SQ

.38-12 ACME-3A-6.00 LG

.19-24 UNC-3A-.25 LG

(14) SPINDLE B

10.88

9.00

1.25

.38

ø .38

.251 / .252 SQ

.38-12 ACME-3A-9.00 LG

.19-24 UNC-3A-.25 LG

(15) SPINDLE C

6.00

R 2.00

ø .38

(12) ALIGNMENT ROD

2.00

1.38

.38

ø .75

.5003 / .5005

45° x .06

.38-12 ACME-3B

(6) PIN

2.25

.63

5°

ø .38

ø .19

R .31

(9) HANDLE

TERMS YOU WILL SEE IN
THIS CHAPTER

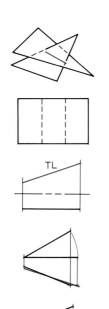

LINE OF INTERSECTION   Points that are common between two planes that intersect.

PATTERN DEVELOPMENT   Process by which a solid object can be made from sheet stock.

TRUE LENGTH LINE   View of a line in which it is parallel to the picture plane.

ROTATION   Method of solving descriptive geometry problems in which all the points of the object revolve about an axis.

PARALLEL LINES   Two or more lines that are nonintersecting where all the points of one line are separated from the other line(s) by the same distance.

PERPENDICULAR LINES   Two or more lines that cross each other at 90°.

SKEW LINES   Two or more lines that are nonintersecting and are not parallel.

10

DESCRIPTIVE GEOMETRY

10.1 *INTRODUCTION*

We live in a three-dimensional world and must solve three-dimensional problems. Engineers must think in three dimensions and be able to visualize their three-dimensional problems (Figure 10.1). For many years engineers and scientists sought ways to describe these three-dimensional

Figure 10.1 Engineers must be able to visualize three-dimensional shapes.

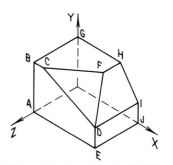

A (0, 0, 1.25)
B (0, 1.25, 1.25)
C (0.25, 1.25, 1.25)
D (1.75, 0.5, 1.25)
E (1.75, 0, 1.25)
F (1.25, 1.25, 0.5)
G (0, 1.25, 0)
H (1.25, 1.25, 0)
I (1.75, 0.5, 0)
J (1.75, 0, 0)

Figure 10.2 (*a*) Pictorial of cut block. (*b*) Wire frame pictorial drawing showing axis system and vertice coordinates.

designs and problems without having to build models. They wanted to be able to use pencils or pens and paper to draw representations of their problems. Although orthographic projection and pictorials as presented in Chapters 4 and 5 did much to provide the information required, they did not help solve the problems of finding true distances, true angles, and true areas.

Descriptive geometry was developed by Gaspard Monge for solving design problems for military fortifications while he was a student in France. His graphical solutions to these complex problems were deemed so valuable that they were considered to be a defense secret for the French. He published his work in a book entitled *Descriptive Geometry* in 1795. Monge found that any view could provide two of the three dimensions and thus two adjacent views could provide all three dimensions.

Descriptive geometry is a key part of an engineering student's graphic and problem-solving tool kit. It allows you to represent three-dimensional problems in two dimensions. Studying descriptive geometry will also improve your ability to visualize. Today, two dimensions refers to a drawing either on paper or on a CRT screen.

Fundamental concepts will be introduced using a cut block. Figure 10.2 shows pictorial and wire frame views of the block. The wire frame view also shows the axis system to be used and the coordinates of the block's vertices based on the axis system. This block contains principal, inclined, and oblique planes. These planes are labeled and highlighted in Figure 10.3. This figure also shows the line of sight, or normal vector, for each of the designated planes. The block will be used to present the physical location in space of the planes and lines for the basic problem-solving concepts.

The flat pattern for the block has been printed on heavier paper on the inside back cover of this text. The three-dimensional coordinate information for this block, shown in Figure 10.2, can be used to create a data file for CADD software that has three-dimensional capability.

This chapter uses three methods to solve three-dimensional problems: (1) the direct method using reference planes, (2) the glass box with fold lines, and (3) rotation or revolution. The direct method and fold line method work well using manual drawing techniques. Normally, rotation using manual methods is very difficult; however, a computer does rotation quickly and is better suited for rotation or revolution methods.

Each demonstration problem for the fundamental line and plane problems is solved using the three methods where possible. When these three methods are presented, all three problem solutions will be shown on the same or adjacent pages so that you can compare them. Later in the chapter, problems will be solved using the most commonly used method(s).

Figure 10.4 shows the pictorial and orthographic views of the cut block. Figure 10.5 shows the block with principal surfaces highlighted and their normal vectors as they appear in the three

Figure 10.3 Pictorial showing (*a*) principal planes highlighted; (*b*) inclined plane highlighted; (*c*) oblique plane highlighted.

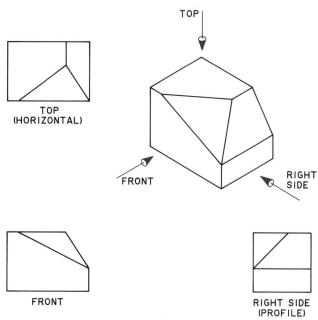

TOP
(HORIZONTAL)

FRONT

RIGHT SIDE
(PROFILE)

Figure 10.4 Pictorial and orthographic views of cut block.

views. Normal vectors are the vectors that are perpendicular to the surfaces. Figure 10.6 shows the pictorial view of the block with the inclined surface highlighted and the inclined surface's normal vector appearing as a point. Figure 10.7 shows the three orthographic views with the inclined surface highlighted and the arrows representing the line of sight. The transparent reference planes are added in Figures 10.8 and 10.9 to show the

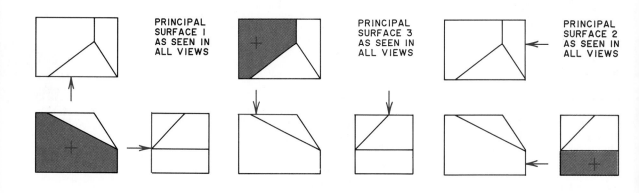

PRINCIPAL SURFACE I AS SEEN IN ALL VIEWS

PRINCIPAL SURFACE 3 AS SEEN IN ALL VIEWS

PRINCIPAL SURFACE 2 AS SEEN IN ALL VIEWS

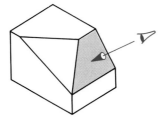

Figure 10.6 Pictorial with inclined plane highlighted.

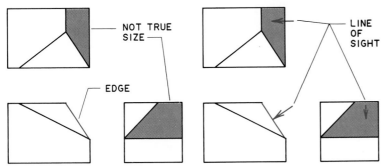

NOT TRUE SIZE

LINE OF SIGHT

EDGE

Figure 10.7 (*a*) Three views with inclined plane highlighted. (*b*) Line of sight or normal vector added in each view.

relative location of the block and the reference planes. The latter of these illustration sets shows which planes would be visible in individual views. Figure 10.10 shows how combinations of reference planes are then used to make up a glass box that surrounds the object. The glass box is unfolded to describe the three-dimensional object using two-dimensional views. The two-dimensional views are called principal views, and the particular planes are called principal planes. Refer to Chapter 4 for more information on principal planes. Also look ahead to Figure 10.26.

Rotations will always be done around the X, Y, and Z axes where the X and Y axes are in the plane of the paper or CRT screen for the front view and the Z axis is coming out of the paper or the CRT screen toward you. The principal views of the cut block are created by rotating the block about the axes. Lines that are true length and planes that are true size will appear in another color or shade of gray in the figures.

Your first step is to cut out the pattern on the inside back cover of this book and assemble the model of the cut block.

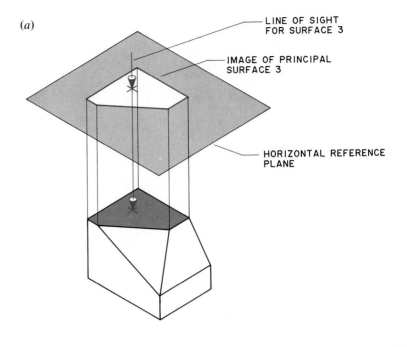

(a)

LINE OF SIGHT
FOR SURFACE 3

IMAGE OF PRINCIPAL
SURFACE 3

HORIZONTAL REFERENCE
PLANE

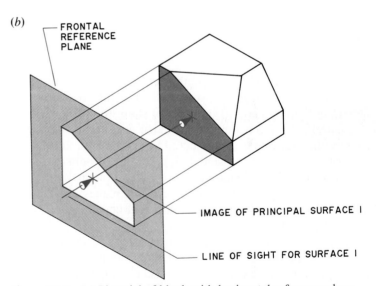

(b)

FRONTAL
REFERENCE
PLANE

IMAGE OF PRINCIPAL SURFACE I

LINE OF SIGHT FOR SURFACE I

Figure 10.8 (a) Pictorial of block with horizontal reference plane added. Principal plane highlighted. (b) Pictorial of block with frontal reference plane added. Principal plane highlighted. (c) Pictorial of block with profile reference plane added. Principal plane highlighted.

(c)

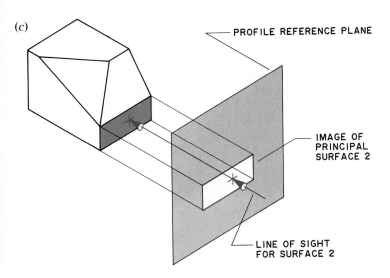

PROFILE REFERENCE PLANE

IMAGE OF
PRINCIPAL
SURFACE 2

LINE OF SIGHT
FOR SURFACE 2

Once assembled, rotate the finished model so that it matches the examples shown in Figures 10.2–10.10 so that you can see how each face looks in each of the principal views. The faces have been numbered, and they will be referred to by their numbers. You should note whether the faces are visible or hidden and whether they are true size, foreshortened (smaller than true size), or an edge.

If you are using three-dimensional software on your computer, load your current CADD package. You will need to create the files for the cut block using the data shown in Figure 10.2. As you study the material in this chapter, it is helpful to have both the physical model and the computer available for understanding the solutions being presented. The manipulation practice will increase your visualization skills, which are vital in communicating your ideas.

10.2 *APPLICATIONS OF DESCRIPTIVE GEOMETRY*

Houses, other buildings, chairs, cars, and tables are objects that we use that have combinations of planes and lines that intersect. The questions are: Where do they intersect? How long is the **line**

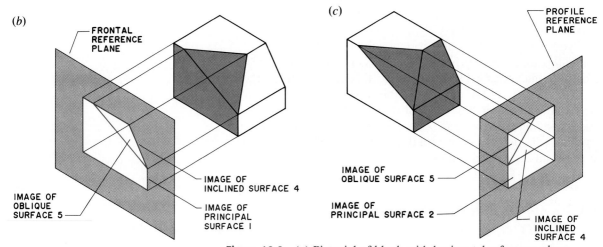

Figure 10.9 (*a*) Pictorial of block with horizontal reference plane added. All visible surfaces highlighted. (*b*) Pictorial of block with frontal reference plane added. All visible surfaces highlighted. (*c*) Pictorial of block with profile reference plane added. All visible surfaces highlighted.

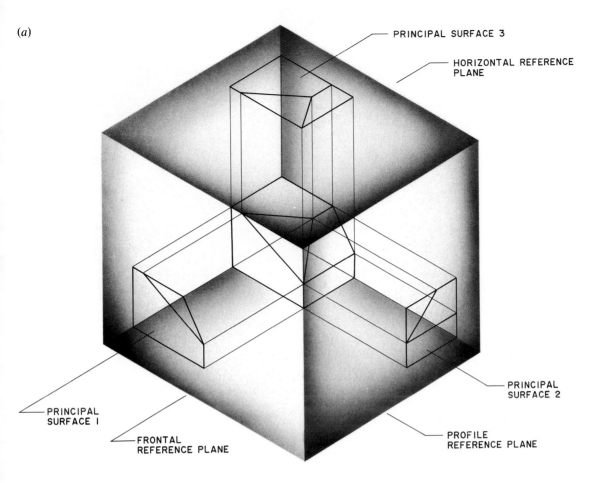

PRINCIPAL SURFACE 3

HORIZONTAL REFERENCE PLANE

PRINCIPAL SURFACE 2

PRINCIPAL SURFACE I

FRONTAL REFERENCE PLANE

PROFILE REFERENCE PLANE

(b)

PRINCIPAL VIEWS OF SOLID

TOP VIEW CAPTURED BY HORIZONTAL REFERENCE PLANE

SIDE VIEW CAPTURED BY PROFILE REFERENCE PLANE

FRONT VIEW CAPTURED BY FRONTAL REFERENCE PLANE

Figure 10.10 (a) Pictorial of block with all three reference planes added to form glass box. (b) Glass box unfolded to provide three orthographic views.

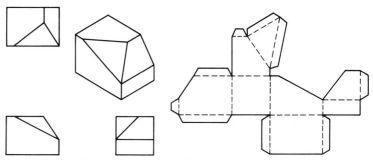

Figure 10.11 Tool box.

Figure 10.12 Pictorial, orthographic views, and pattern development for cut block.

of intersection? If the line of intersection is created by two planes of a toolbox, we need to know how many spot welds, how many rivets, or how much weld bead is required to join the two surfaces. Figure 10.11 shows a photo of a spot-welded tool box.

Another aspect of complex objects such as cartons is how to create the three-dimensional object from a flat sheet. This requires a pattern. In the **pattern development** process, the object is drawn first as a three-dimensional object, usually a pictorial sketch and then as three or more orthographic views; then the development is done by laying off true length lines. Figure 10.12 shows the orthographic views, the pictorial, and the flat pattern for the example object. Developments are important in areas other than packaging: Applications are in heating and air-conditioning duct work, aircraft design, building construction, luggage, and clothing, to name a few.

Another important application of descriptive geometry lies in the area of mapping. Mapping includes subdivision layouts, contour maps, and construction plans for roads, water mains, sewers, and dams. Directions on maps are important. Section 10.8 will define bearings and azimuths, which are two methods used in descriptive geometry to define directions. Contour lines on a map define the location of constant elevations. See Figure 10.13. Descriptive geometry methods can be used to transform these contour lines into cross sections through the earth to show better detail. Determining slopes and scaling are important problem-solving methods when applying descriptive geometry to mapping. Three-dimensional visualization is needed in creating multilayer printed circuit boards and plans for chemical processing plants.

10.2.1   Problem-solving Skills

The only way to solve a complicated problem is to reduce the complicated problem to a series of simple problems. You then

Figure 10.13   Contour map showing north arrow. Courtesy of USGS

determine which problem to solve first and proceed to solve the simple problems until the complicated problem is solved. In most engineering problems there are a variety of correct solutions. For example, count the number of different makes of cars, bicycles, chairs, television sets, and so on; all of the products in each category solve the same problem.

In descriptive geometry you will be given a statement of what you are required to find and enough information to solve the problem. This text provides a set of rules and techniques to get from the given information to the solution. There will usually be two possible solutions that will provide the desired information. Thus, descriptive geometry provides experience in basic engineering problem-solving skills and helps you learn how to attack any problem. However, in the real world you may find that not all of the information is given and you may have to make assumptions.

10.2.2 Developing Visualization Skills

Working with a series of views of real and abstract objects (individual lines and planes in space) can develop visualization skills. Looking at individual lines and planes and being able to attach them to an object or a set of axes for location in space can be invaluable in solving real-world problems.

One of the fastest growing areas in the computer-aided engineering field is that of geometric modeling. Geometric or solid modeling is done by working with primitive shapes and requires the ability to visualize. Primitive shapes are chosen as the building blocks to create more complicated objects. This is done by preschool children on a daily basis as they allow their imaginations to think of castles, roads, houses, buildings, airplanes, and other solid objects. With building blocks, all that can be done is to add one block to another and the blocks are of constant size. Figure 10.14 shows some of the primitive shapes used to create models of real objects.

The engineer working with solid modeling has some advantages

Figure 10.14 Primitive shapes for solid modeling.

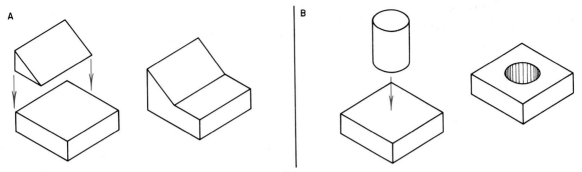

Figure 10.15 (*a*) Adding two blocks to form new solid. (*b*) Differencing block and cylinder to produce hole.

over the preschool child. Computer-based modeling systems provide the ability to select a primitive and resize it. A set of primitives can be chosen and modified for use by the engineer. Then the engineer can add them together to create a more complicated object or an assembly of complicated objects. However, the engineer can also subtract (difference) objects. For example, the difference between a rectangular solid and a cylindrical solid can be a hole in the rectangular solid. Figures 10.15*a,b* show the adding and differencing operations. Other mathematical operations are also possible, but these simple examples give an idea of the power of such tools.

In the coming years, solid modeling will expand to cover much of what we now think of as descriptive geometry. In the meantime, a good understanding of three dimensions through descriptive geometry will pave the way for efficient learning and use of solid modeling systems.

10.3 *OPERATIONS*

10.3.1 The Direct Method

The direct method as used in solving descriptive geometry problems is a manual method. This means that problem solutions are more easily found using a pencil, paper, and drawing tools rather than using a computer. The basic concept of the direct method is that an infinite reference plane placed anywhere in a given view will appear true size in that view and as an edge in any adjacent view. Figure 10.16 shows the front view of the cut block with the reference plane installed and several adjacent views of the block that show the reference plane as an edge. You may find useful a

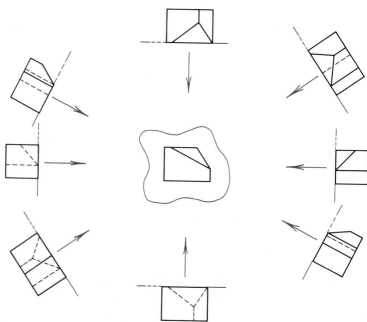

Figure 10.16 Figure showing front view with reference plane installed and several adjacent views.

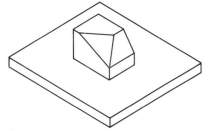

Figure 10.17 Drawing of cut block sitting on desk.

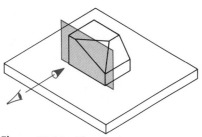

Figure 10.18 Plastic sheet used as frontal reference plane.

sheet of clear plastic so that you can model the reference plane with the cut block.

Set your model on the table in front of you. See Figure 10.17. As you sit at a desk, you can see several of the planes that make up the object. You are seeing a pictorial view of the object. Orient the block so that the sides are lined up with the desk and plane 1 is facing you, plane 2 is to the right, and plane 3 is on top. Drop your eye to desk level and look at the block. You are seeing the front view of the object. If you have a clear plastic sheet, hold it so that it is between you and the object and is vertical and parallel to the front of the object. See Figure 10.18. If you were to trace the outline and edges of the object's planes on the clear plastic sheet, you would have the front view of the object and plane 1 is seen true size. The clear plastic sheet represents the picture plane or reference plane.

Without moving the plastic sheet or the object, crouch down by the side of the desk and look at the object again. See Figure 10.19. This is the side view and plane 2 is seen true size. Notice that the plastic sheet has now become an edge when you look at the side view. Therefore, the reference plane for one orthographic view becomes an edge for the adjacent orthographic views. Now stand up and look directly down on the object while continuing to hold the plastic sheet in the location where it was when viewing the front view. See Figure 10.20. This is the top view and plane 3 appears true size. Again, note that the reference plane for the front view appears as an edge when viewing the object from the top.

Auxiliary views are orthographic views. This was discussed in Chapter 4. They are not principal views. There are an infinite number of auxiliary views possible for any given view; however, only six principal views are available as described in Chapter 4. In this next series of steps you will be positioning yourself so that you can see auxiliary views. Continue to hold the reference plane so that it is a frontal reference plane and do not move the object. Now, slowly move from your position when viewing the object from the top and crouch beside the desk again. Figure 10.21 shows various positions of a viewer and some of the possible lines of sight from the top to the right side views. In this process, note that the reference plane continues to appear as an edge in all of the positions. Any position between the top view and the side view is considered to be an auxiliary view.

When you moved from the top view to the side view, you moved

Figure 10.19   Looking at cut block to see plane 2 true size and reference plane as edge.

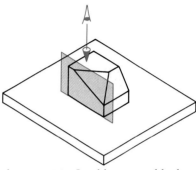

Figure 10.20   Looking at cut block to plane 3 true size and reference plane as edge.

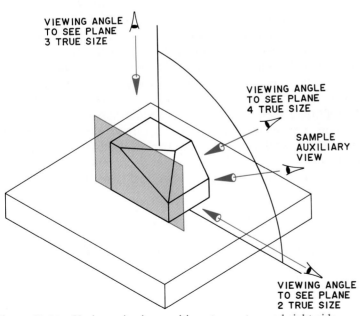

Figure 10.21   Various viewing positions to see top and right side views and some of auxiliary views between top and right side views.

through a position where the inclined plane, plane 4, appeared true size. This occurred when your line of sight was perpendicular to plane 4. See Figure 10.21.

Finding the true size of surfaces is one use of descriptive geometry. The process provides us with ways to find the true size and shape of inclined and oblique planes. When we do this, we also get **true-length lines** and true-size angles.

If you had desired to draw any of the auxiliary views on paper, you could have had another reference plane to hold between you and the object. See Figure 10.22. On that reference plane you could trace the object as you see it. What is the relationship between the two reference planes? They are perpendicular. It is important to point out that you saw the frontal reference plane in the same position relative to both the top and side views as well as to any of the auxiliary views in between. This allows us to measure from the frontal reference plane in one view to get the depth of a point (distance from the reference plane) and be able to transfer it to another view where we can see the frontal reference plane as an edge, as shown in Figure 10.23. This illustrates an important concept. In any situation where we see the reference plane true size in one view, it appears as an edge in *any* of the adjacent views. Hold the reference plane parallel to surface 3 in the top view and then look at the edge of the plane in the front and right side views. See Figure 10.24.

The direct method uses a systematic method of drawing and labeling reference planes. There are two approaches. One uses a standard black pencil for drawing the representations and labels. The other uses colored pencils. Both approaches require sketching a rough-cut plane around the view where the reference plane appears true size. This rough-cut plane is called the "cloud" view. The reference plane is installed in the adjacent views as an edge or a line. Labels are attached to both the cloud view and edge views of the reference plane. Examples of these labels are RP1 or RP2 or, in general, RPn. The other approach is to use colored pencils to draw the reference planes, and each plane is drawn in

REFERENCE PLANE I
(FRONTAL)

REFERENCE PLANE 2
(AUXILIARY VIEW FOR
PLANE 4)

Figure 10.22 Block with frontal and auxiliary reference planes.

Figure 10.23 (*a*) Frontal reference plane used for transferring distances. Reference plane, RP1. (*b*) Frontal reference plane used for transferring distances. Reference plane colored.

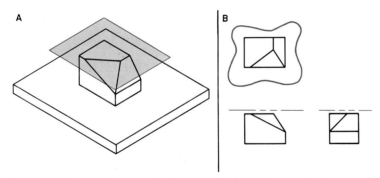

Figure 10.24 (*a*) Horizontal reference plane (HRP) installed on block. (*b*) Orthographic views showing HRP.

a different color. Then the adjacent views have a line drawn in the same color representing the edge view of the reference plane. The colored planes can also be labeled, but the labels may be superfluous unless copies are to be made of the drawing. See Figures 10.23*a,b*.

Measurements for construction are then made either away from the given view in both adjacent views or toward the given view in both adjacent views. When a reference plane is attached to a principal plane, it is often designated the horizontal, frontal, or profile reference plane (HRP, FRP, or PRP) rather than RPn.

10.3.2 The Glass Box: Fold Line Method

Another way of looking at the reference plane and objects is the fold line method. Leave the reference planes located in the position for each of the three principal views. If three more planes were added, a box would be created around the object. Refer again to Figures 10.10a,b. Look through each of the three reference planes and draw what you see inside the box. You have drawn three of the principal views. This can be a very convenient approach for beginners because it fixes the reference planes relative to each other. If we were to put hinges on the box between the frontal and top and the frontal and right sides, we could unfold the box and we would have the orientation of three of the six principal views. This glass box approach for the six principal planes is also presented in Chapter 4.

In addition, Chapter 4 presents the concept of the auxiliary view. With the three principal views in place, it is possible to add a fourth plane that is parallel to plane 4. The hinges would be placed between the frontal plane and the seventh, or auxiliary, plane. When the box is unfolded, it is easy to see the location of the auxiliary view relative to the principal views. Figure 10.25a shows the complete glass box constructed for six principal views and one auxiliary view. Figure 10.25b shows the box open to show the relative positions of the views.

You need to know at this point that while creating the glass box for a simple figure with an inclined surface is relatively easy, it can be much more difficult to think of the glass box for oblique planes such as surface 5.

In the glass box construction, the edge views of the reference

A

B

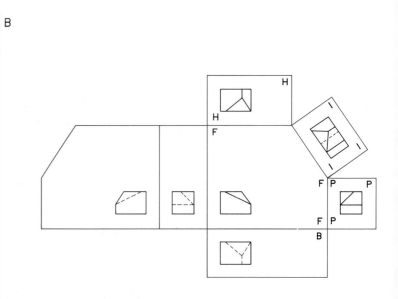

Figure 10.25 (a) Auxiliary plane added to glass box. (b) Seven planes opened to show relative position.

planes are called fold lines because this is where the box is folded or hinged. When working on paper, fold lines, the edge views of the reference planes, are labeled by the planes that they are between. For example, the fold line between the front and top views is labeled F/T or F/H and the fold line between the front and side, or profile, views is labeled F/P.

At this point in the discussion, it is important to point out that the principal views are known by more than one name. The top view is sometimes called the plan or horizontal view. The front view is sometimes called an elevation, and the right and left side views can be called profile views or elevations. These names are used interchangeably in this text, but the use of a particular name, such as "plan view," for example, is used by civil engineers to designate the top view.

The fold lines have traditionally been labeled F/T or F/H between the top, or horizontal, and front views. The fold line between the front and right side, or right profile, view is labeled the F/R, or F/P, line. It is not important which letters are used, but they should be used consistently.

The designation for the fold lines between auxiliary and principal views normally uses a letter and a number. For example, H/1 indicates that the fold line is between the horizontal and first auxiliary view. When the fold line is between two successive auxiliary views, it might be designated by 1/2 or 2/3.

In Figure 10.25b, the glass box with seven planes is unfolded so that you can see the reference planes and fold lines with the proper labeling. Note that the fold line method uses a 1 for the number representing the first auxiliary views. Subsequent views are labeled by 2, 3, and so on.

10.3.3 Rotation

When working with objects, it is important to know the frame of reference within which you are working. *Frame of reference* means the set of three axes to define an origin and three mutually perpendicular axes in space. In this case we shall discuss **rotation,** and it is important to know what is rotating and what is considered to be the frame of reference. We consider that the frame of reference stays fixed in space and the objects move (rotate or translate).

When you rotate something, there must be an axis about which the rotation takes place. There are three dimensions and therefore we have three axes. For this text, as was discussed in Chapter 5, we use a set of axes where the X axis goes to the right, the Y axis is vertical, and the Z axis points toward the person looking at the object.

For the previous discussion on reference planes and glass boxes,

Figure 10.26 CRT with axes.

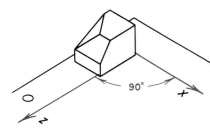

Figure 10.27 Axes shown on sheet of paper.

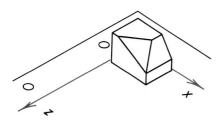

Figure 10.28 Cut block placed on paper.

Figure 10.29 Block rotated 90° clockwise about *Y* axis.

Figure 10.30 Viewing plane added to Figure 10.29.

we placed the cut block on the desk so that the top surface of the block was parallel to the desk surface and the front surface was perpendicular to the desk surface and was facing you. This would mean that the *X* axis points to the right, the *Y* axis points toward the ceiling, and the *Z* axis points toward you. It may be easier to think of this orientation as a CRT sitting on a desk with the *Z* axis coming out of the CRT. Figure 10.26 is a copy of Figure 5.32 showing the CRT and axes.

When we start this rotation process, the width of the object is along the *X* axis, the height is along the *Y* axis, and the depth is along the *Z* axis. This is called the "home" position. In order to preserve the frame of reference, take out a clean sheet of paper and draw the *X* and *Z* axes on it. See Figure 10.27. Put the sheet on your desk with the axes aligned as shown in the figure. This will be your frame of reference and will not move. Set your object on the paper so that it is in the home position as shown in Figure 10.28.

First, we rotate the object around the *Y* axis. When we start, the object is oriented to give the object's front view to the person seated at the desk. If we rotate the object 90° clockwise with the axis through the back left corner, then we see the right side view. See Figure 10.29. Note that in this case, you are rotating the object with the reference plane fixed between you and the object and perpendicular to the *Z* axis of the frame of reference. Figure 10.30 shows the viewing plane added to Figure 10.29. Now return the object to the home position.

As shown in Figure 10.31*a*, rotate the object about the *Y* axis 45° clockwise. This allows you to see both the front and right side although both are foreshortened. We want to rotate about the *X* axis approximately 35° counterclockwise. If the desk were made of sand, we could push the front edge of the object down into the sand and see a pictorial view. The desk is not made out of sand. Therefore, you will have to raise the back of the object so that the base of the object forms an approximate 35° angle with the desk top. You have now positioned the object as an isometric view, as shown in Figure 10.31b. Once you have done this, return the object to the home position.

The purpose of these rotations is not to find a way to display an isometric pictorial view of the object. The purpose is to prepare you to find the true size of planes or the true length of lines using rotation as a method.

Two rotations are normally sufficient to find the answer to most problems. The following two operations will display the plane 4 true size. Note that plane 4 makes an angle of 45° with the horizontal. First rotate about the Y axis 90° clockwise as shown in Figure 10.32a. Next rotate about the X axis approximately 35° counterclockwise, as shown in Figure 10.32b. This set of rotations shows you, the viewer, the true size of plane 4 just as drawing the auxiliary view showed the true size of plane 4 in Figure 10.25b.

You now have seen the three methods for solving descriptive geometry problems. Each has its applications and conventions. As you use them, you will develop a set of tools that gives you confidence to solve problems. The choice of the direct method or the fold line method is a matter of personal preference. Both can be used to solve the same type of problems. Rotation, on the other hand, is required in development problems. Rotation is also a very natural approach for computer displays, and you should be comfortable with the concepts of rotation.

10.4 *LINE RELATIONSHIPS*

In engineering drawings, figures are composed of collections of lines, planes and curved surfaces. In order to work effectively with engineering drawings, you will need to know how lines appear. In this section we look at lines as part of solid objects or lines as individual items.

10.4.1 Principal Lines

In Chapter 4, you learned about principal lines. These are lines that appear true length in one of the principal views and are referred to as frontal, horizontal, or profile lines. Figure 10.33a shows some principal lines highlighted in the three given views of the block. Figure 10.33b shows these same lines in space. Note that, first, the true-length view is the longest of the given views of the lines and, second, the line has to be parallel to the reference plane for the view in which it appears true length. This means that it appears as a line parallel to the edge view of the reference plane or the fold line in the adjacent views. As you look at a principal line, your line of sight is perpendicular to it.

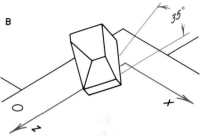

Figure 10.31 (a) Rotation of block 45° clockwise about Y axis. (b) Second rotation of block approximately 35° counterclockwise about X axis.

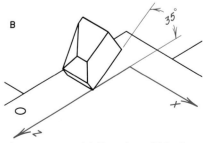

Figure 10.32 (a) Rotation of block 90° clockwise about Y axis. (b) Second rotation of block approximately 35° counterclockwise about X axis.

A

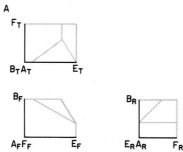

Figure 10.33a Principal lines highlighted in three views of block.

B

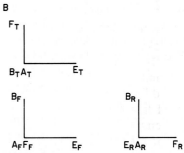

Figure 10.33b Principal lines highlighted in Figure 10.33a shown in three views without block.

A

B

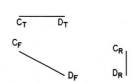

Figure 10.34 (*a*) Inclined line highlighted on block. (*b*) Inclined line shown in three views without block.

A

B

Figure 10.35 (*a*) Oblique line highlighted on block. (*b*) Oblique line shown in three views without block.

10.4.2 Inclined Lines

Inclined lines are lines that appear true length in one of the three principal views and foreshortened in the other two principal views. Figure 10.34 shows three views of one inclined line as part of the block and the line in space. Note that the foreshortened views of inclined lines are parallel to the edge view of the reference plane or fold line between the views. When the fold line or the line representing the edge view of the reference plane is parallel to a view of a line, the line will appear true length in the adjacent view.

10.4.3 Oblique Lines

The term *oblique* refers to lines that are not parallel to any of the principal planes. Figure 10.35 shows three views of an oblique line as part of the block and in space. This presents the problem of determining the length of that line. Why do we need the true length of lines? In engineering applications, a line can represent the wire bracing used on a utility pole to keep it vertical, the structural component that stiffens a framework, the pipeline that leads from a pump in a refinery to the top of a storage tank, or the overflow pipe on an earthen dam. Figures 10.36–10.38 show some of these applications.

Descriptive geometry provides two solutions for the problem

Figure 10.36 Utility pole with wire bracing.

Figure 10.37 Building structural components. Courtesy of Ohio State University Engineering Publications

Figure 10.38 Radio telescope. Courtesy of Ohio State University Engineering Publications

of finding the true length of a line. One way is using auxiliary views. This means that the viewer's line of sight is perpendicular to the given view of the line but from a direction other then a normal view of the principal plane. Figures 10.39 and 10.40 show the auxiliary view for a line as part of a three-dimensional figure and the line isolated in space, respectively. Figure 10.41 shows the steps required for construction of the true length of the line using the direct method. Remember that when an auxiliary view is being projected, there are three views involved: the auxiliary

FOLD LINE

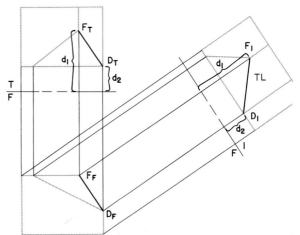

Figure 10.39 Auxiliary view to get true length of line on block.

FOLD LINE

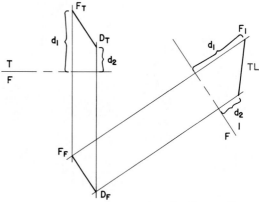

Figure 10.40 Auxiliary view to get true length of same line not attached to block.

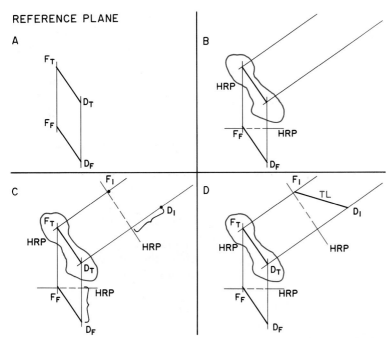

REFERENCE PLANE

Figure 10.41 Steps to solve for true-length line using direct method.

view that is being constructed and two given adjacent views. The projection is done from one of the given views, and measurements for the construction must be made in the other given view. Figure 10.40c shows where the measurements are taken in the front view and where they are positioned in the auxiliary view.

The measurement to point *D* in the front view is made *away* from the top view and the location of point *D* in the auxiliary view is made *away* from the top view as well. Figure 10.42 shows the same construction using the fold line method and nomenclature.

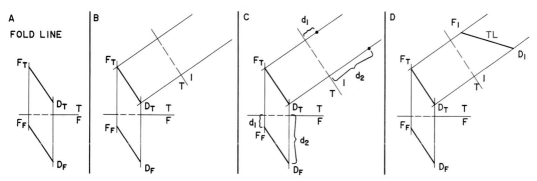

Figure 10.42 Steps to solve for true-length line using fold line method.

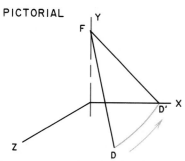

Figure 10.43 Pictorial showing rotation of oblique line to find true length.

Figure 10.44 Steps of rotation method to solve for true length of oblique line.

It is also possible to find the true length of the line by rotating the line until it is parallel to one of the principal planes. This approach requires two basic operations: a rotation in one view and a translation in an adjacent view. Translation is a movement of one end of a line to new coordinates with that end at the same distance from the reference plane. In this case, you choose one of two views for the rotation and a point (one end of the line) in this view to locate the axis of rotation.

The axis is installed as a point in this view and as a true-length line in the adjacent view. In the case shown in Figures 10.41 and 10.42 the front and top views were provided. Figure 10.43 shows the rotation in a pictorial. Figure 10.44*a* shows the given information. We shall choose the top view for rotation and choose the end *F* as the pivot point. The axis of rotation is installed through *F* in both views. In the top view, the axis of rotation appears as a point and in the front view it appears as a center line. In Figure 10.44*b*, *D* is rotated in the top view until *FD* is parallel to the frontal reference plane. A projector is then drawn from *D* in the top view perpendicular to the reference plane in the front view. Then *D* in the front view is translated parallel to the reference plane until it crosses the projector. This locates *D* in the front view. When *F* is connected to *D* in the front view, the problem has been solved, and *FD* in the front view is the true length, as seen in Figure 10.44*c*. Compare the true length found by rotation with the true length found by projection. These true lengths are identical.

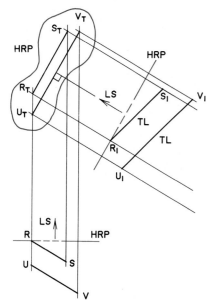

Figure 10.45 Parallel lines in top, front, and auxiliary views. Line of sight 90° to top view gives auxiliary view that shows true length of lines.

10.4.4 Parallel Lines

Parallel lines appear parallel in all views. Therefore, if lines do not appear parallel in one view, they are not parallel. Figure 10.45

shows two lines *RS* and *UV* that are parallel in the front and top views. When an auxiliary view is drawn to determine the true length of one of the lines, the lines still appear parallel and both appear true length. Figure 10.46 shows another auxiliary view where the line of sight is drawn at an arbitrary angle. This new view shows the two lines as being parallel. Therefore, if two lines are parallel, they are parallel in any view.

The distance between the lines may also be important. If the two lines represent center lines of two pipes, it is possible for the two pipes or the insulation around the two pipes to interfere. The distance between parallel lines can be determined with two auxiliary views. Figure 10.47 shows the steps to find the distance between two parallel lines. The first auxiliary view shows both lines true length. This second auxiliary view is drawn with the line of sight parallel to the true-length lines. This makes them appear to be points, and the space between points is the true-length distance. This procedure is similar for both the first method or the fold line method.

Whenever a second auxiliary view is required and you are using the rotation method, a second rotation is required. While this can be confusing, it is a little easier if the given views are in black, the intermediate step (first rotation) is done in one color, and the

Figure 10.46 Parallel lines in top, front, and auxiliary views. Line of sight at arbitrary angle to top view gives auxiliary view that shows lines are still parallel but not true length.

Figure 10.47 Distance between parallel lines.

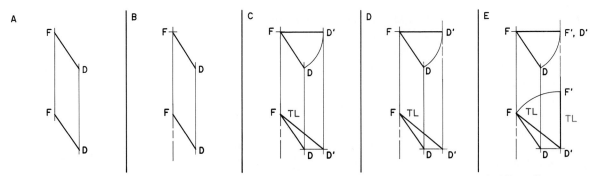

Figure 10.48 Double rotation to get point view of oblique line.

second rotation is done in another color. Where this cannot be done, an alternative is to label each point *carefully* so that first rotations are primes and second rotations are double primes (*A'* and *A''*). Figure 10.48 shows the double rotation to find the point view of a line. More than two rotations are difficult to show with a two-dimensional drawing on paper; however, a computer system that allows rotation does not retain previous images and can continue to rotate without confusion as long as you define the axis and the angle of rotation.

10.4.5 Perpendicular Lines

Figure 10.49 shows **perpendicular line** relationships. Two lines representing the edges of a box have been highlighted with color in the front and right side views. It is easy to see that they are perpendicular when looking at the front view. What makes you sure of this fact? Both lines appear true length in the front view. When both lines are true length and the angle between them is 90°, they are perpendicular. Is it necessary for both lines to appear true length to be able to determine perpendicularity? Do the lines have to intersect in order to be perpendicular? When the relationship between lines is unknown, you must find a view where *one* of the lines appears true length. If the two lines then appear to be perpendicular, they in fact are perpendicular.

Figure 10.49 Pictorial, front, and right side views of block showing two lines highlighted that are perpendicular.

Figure 10.50*a* shows two **skew lines** (lines that do not intersect in space) and an auxiliary view with one of the lines true length. Figure 10.50*b* shows the second auxiliary view. It also shows the shortest perpendicular distance from one line to the other. The second auxiliary view is drawn with the line of sight parallel to

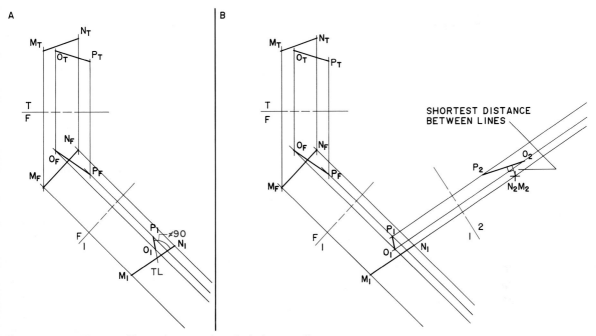

Figure 10.50 Two auxiliary views used to find shortest distance between skew lines.

the true-length line. A perpendicular line is drawn from the point view of one line to the other line. Do lines have to be perpendicular to be able to determine the shortest distance between them? If you think about this question, the answer is no. Skew lines are lines that are neither perpendicular nor parallel to one another; however, there is a shortest distance line between two skew lines that can be determined.

10.4.6 Angular Relationships

Angular relationships apply to lines whether or not they intersect or whether or not they are in the same plane. To find the angular relationship between lines, the lines have to be true length in the same view. This requires a three-step procedure for the general case.

Figure 10.51 shows the steps to find the angle, or apparent angle, between two lines AB and CD. The first step is to find the true length of one line CD and project the other line AB into the

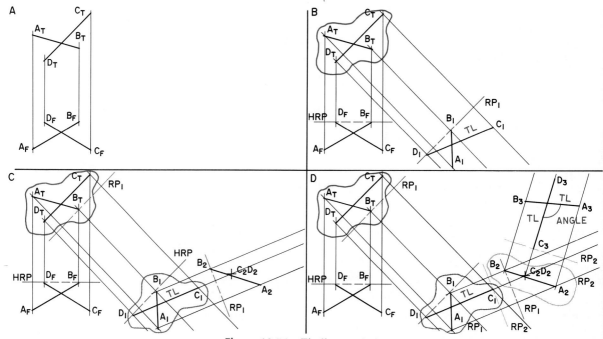

Figure 10.51 Finding angle between two skew lines using three auxiliary views.

Figure 10.52 Definition of vector.

same view. The second step is to find a point view of line *CD* and again project *AB* into this view. The third step requires a view taken off the point view of *CD* such that the line of sight is perpendicular to line *AB*. This view will show line *AB* true length. It will also show *CD* true length because any view projected from the point view of a line will show its true length.

Lines that intersect form planes. Planes are discussed later in this chapter.

10.4.7 Applications

There are a variety of applications for line relationships. Some were mentioned in the preceding. One of the most important is the application to graphical vectors.

A vector has a magnitude and a direction. The magnitude is determined by the length of the line and the direction is determined by the arrowhead on one end of the line, as shown in Figure 10.52. Two-dimensional vectors require only one view because they are all true length. See Figure 10.53. Three-dimensional vectors require at least two views, as shown in Figure 10.54. The techniques discussed here, single and multiple auxiliary views or single or multiple rotations, can be used to solve vector problems. Vector problems are discussed in Chapter 11.

10.5 *PLANE AND LINE RELATIONSHIPS*

10.5.1 Principal Planes

The cut block that you have been studying has six surfaces that are principal planes. They are numbered 1–3 and 6–8. It also has two surfaces, 4 and 5, that are inclined and oblique, respectively. We shall study these surfaces individually, with some lines that can be represented by your pencil, and in pairs to learn the relationships necessary to solve problems. See Figure 10.55.

The principal planes are those that are parallel to the surfaces of the glass box. Pick up the cut block and hold it in the home position so that you can see plane 1 true size. Then study each of the other planes in turn. Rotate the block so that you can see planes 2, 3, and 6–8. In this process start with the block in the home position and make the X or Y rotations in order to see each of the principal planes. Make some notes for yourself either mentally or on paper as to whether the rotations were clockwise or counterclockwise and the number of degrees required. In each of these cases also note that when you see one principal plane true size, you will see the adjacent principal planes as edges. Keep in mind that each plane's normal vector, the unique vector perpendicular to the plane's surface, will appear as a point when the plane is viewed true size.

Your computer software can be used for this exercise as well.

10.5.2 Inclined Planes

Repeat the studying process outlined in the preceding with plane 4, the inclined plane. First, how did this plane appear when you were studying each of the principal planes? When you can see plane 1 true size, 4 appears as an edge. See Figure 10.56. However, when you rotate the block 90° clockwise about the Y axis to see plane 2 true size, you see 4 foreshortened. See Figure 10.57. In order to see 4 true size, you also have to rotate the block about the X axis counterclockwise by an angle of arctan $\frac{2}{3}$ or approximately 34°. See Figure 10.58.

Figure 10.53 Two-dimensional vectors. Two views.

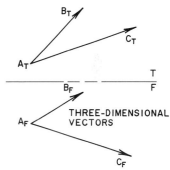

Figure 10.54 Three-dimensional vectors. Two views.

Figure 10.55 Block with surfaces numbered and vertices identified with letters.

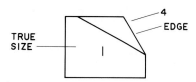

Figure 10.56 Block oriented to show plane 1 true size.

Figure 10.57 A 90° *Y*-axis rotation to show plane 2 true size.

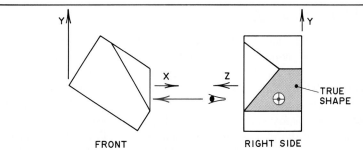

Figure 10.58 Arctan $\frac{2}{3}$ or ~34° *Z*-axis rotation to show plane 4 true size in right side view.

Figure 10.59 Block on desk in home position.

Figure 10.60 Pencil perpendicular to plane 4.

As an alternative to rotating the block, set it on the desk in the home position. See Figure 10.59. Get up from your chair, walk around to the right side of the desk, and stand in a position so that the line of sight from your eye is perpendicular to plane 4. Place a pencil on 4 so that it appears as a point. See Figure 10.60. Hold it in that position and sit down again. Make your line of sight perpendicular to plane 1. See Figure 10.61. What angle does the pencil form with the surface of the desk? You must use this angle to project a view of the cut block that will show plane 4 true size.

How does this projection process take place? There are some principles involved that must be reviewed. First, any orthographic view of an object only shows two dimensions. In the case of the front view, these are the height and the width. The depth is missing. You know from previous discussions in this text that you can find depth dimensions in the top and right side views. Therefore, if you are going to project from one view, the front in this case, you must have an adjacent view, the top view, either side or profile view, or the bottom view. You also need a reference from which to measure the depths for each feature of the block.

In the previous sections, we have talked about using a clear plastic sheet as a picture plane, and in the preceding section, we

Figure 10.61 Viewer's line of sight perpendicular to plane 1; pencil perpendicular to plane 4.

Figure 10.62 Pictorial of block with reference plane installed.

used it as a reference plane. Here again, we use the plastic sheet as a reference plane. Put the sheet against plane 1 on the cut block. See Figure 10.62. The distance from that reference plane to any individual point on the block remains constant. We use this property to solve the problem of an auxiliary view that gives the true shape of the inclined plane 4.

Study Figure 10.63a, which shows the front and top views of the block. Figure 10.63b shows the reference plane attached to the front view as a cloud and to the top view where it appears as an edge. Figure 10.63c shows the **edge view** of the reference plane drawn perpendicular to the line of sight. The front view locates the starting points for the projection lines, which are drawn parallel to the line of sight in Figure 10.63c. Measurements are made in the top view, as shown in Figure 10.63d, and transferred to the auxiliary view. The completed auxiliary view is shown in Figure 10.63d.

Figure 10.63 Orthographic and auxiliary views of block with frontal reference plane.

Compare the auxiliary view in Figure 10.63d with the rotated view of the block shown in Figure 10.58. They should be the same view because the line of sight of the viewer is parallel to the normal vector of plane 4. Therefore, you have two ways to solve for the true size of an inclined plane assuming that you work with a three-dimensional CADD system that allows you to manipulate a cut block, have an exact model of the block, or are willing to use the reference plane method to solve the problem.

There are two more paper methods to solve the same problem.

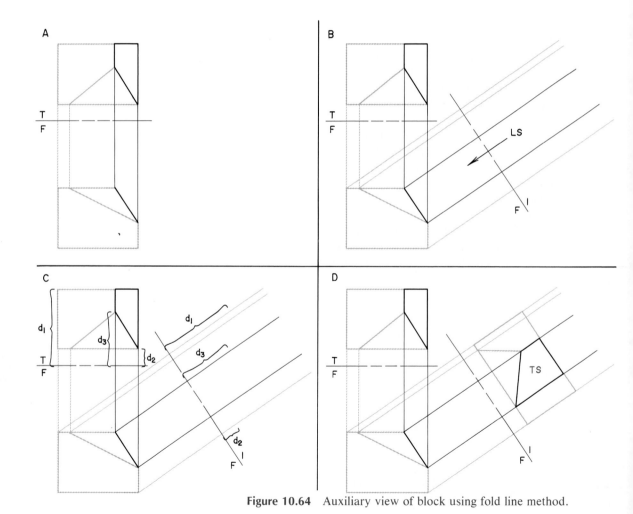

Figure 10.64 Auxiliary view of block using fold line method.

Figure 10.65a Pictorial showing cut block inside glass box.

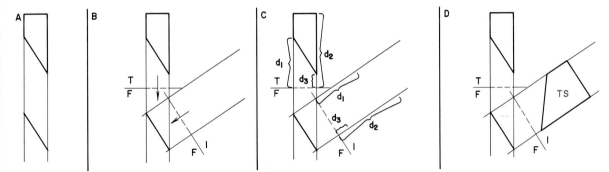

Figure 10.65b Auxiliary view of plane 4 using fold line method.

Figure 10.64 shows the fold line method of solving the problem. Figure 10.65a also shows the cut block inside the glass box for reference. Note that it is very similar, again, to the direct method. The exceptions are the labeling of the fold lines instead of the edge views of the reference plane. It is more difficult to think of the fold lines as being attached to the object as we did with the reference plane. Figures 10.65 and 10.66 show both of these solutions with just the plane rather than the plane as part of the block. Figures 10.65*a* and 10.66*a* are pictorials that provide a reference for the two methods shown in Figures 10.64–10.66.

The final method is to solve the problem by rotation on paper. The steps are the same as they were in the revolution of the block by hand or rotating on the computer. However, your work has to be done more neatly, and you have to keep track of labels because the first and second rotations are drawn directly over the given views of the object.

Two adjacent views with one showing the inclined plane as an edge are sufficient to solve for the true size of the plane. Figure 10.67*a* shows top, front, and right side views as the given information. However, only the front and right side views are necessary to solve the problem. In this case, a single rotation about the Z axis of arctan $\frac{2}{3}$ degrees or 33.7 degrees is sufficient to solve the problem.

If you are restricted to displaying the true size of the plane in the front view, as with some computer software, then three views are required. This method takes longer and requires a 90° rotation about the Y axis first and a 33.7 degree counterclockwise (arctan $\frac{2}{3}$) rotation about the X axis second. Figure 10.67*b* shows the top, front, and side views during the rotation processes.

The available tools will dictate how you solve your problems. The goal here has been to get you to think about the possible methods that are available and to strengthen your visualization skills.

10.5.3 Oblique Planes

Finding the true size of oblique planes requires the same basic steps and information. First, you have to find a view that shows the edge view of the oblique plane and then a view that shows the normal vector of the plane parallel to your line of sight.

Figure 10.66*a* Pictorial showing the reference plane.

Figure 10.66*b* Auxiliary view of plane 4 using direct method.

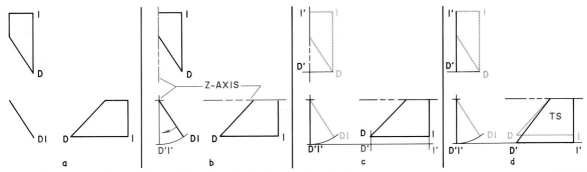

Figure 10.67a Auxiliary view of plane 4 with single rotation.

Figure 10.67b Auxiliary view of plane 4 with double rotation.

In this case, the direct method will be shown first. The fold line method will be shown next. The third method will be to rotate the block with the correct angle of rotation about the Y and X axes to display the true size of plane 5. The last will be to perform the same operations on a computer.

The key to many descriptive geometry problems involving planes is to find a true-length line in the plane. In this case we have one true-length line in each of the given views. See Figure 10.68*a*. Lines *CD* and *CF* are true length in the front and top views, respectively. This means that either view can be used to start the construction. When there is no true-length line in the plane, you must construct one.

In this case we choose the top view to start the construction. The first auxiliary view will be drawn to the right and above the top view. See Figure 10.68*b*. Put in all the projection lines and the reference plane between the front and top views and touching the front view so that measurements can be made in the front view. Show the cloud view of the reference plane in the top view. Put in the projectors for the auxiliary view parallel to line *CF* in the top view and install the reference plane far enough away to clear any of the given views. Label this reference plane HRP in all views. See Figure 10.68*b*. Make measurements in the front view, transfer them to the auxiliary view, and connect the points with visible lines. See Figure 10.68*c*, which shows the completed first auxiliary view. Figure 10.68*d* shows the projectors from the first auxiliary view, which are drawn perpendicular to the edge view of plane 5.

A second reference plane, RP1, is drawn in between the first auxiliary view and the top view so that it touches the top view. The cloud view of this reference plane is attached to the first auxiliary view, and the edge view of RP1 for the second auxiliary view is drawn perpendicular to the projection lines. See Figure 10.68*d*. Measurements are made in the top view from RP1 to the specific points on the top view and laid off from RP1 in the second auxiliary view. Figure 10.68*d* shows the finished views with plane 5 labeled true size. Figures 10.69*a–d* show the fold line solution to the same problem.

The key to solving the problem by revolution is the same as for solving the problem by projection. You must find the true-length line. In this case, the same true-length line will be used. See Figure 10.70a for the given views. Note that in this case we need three views, the top, the front, and the right side views and that all of the points on the block have been labeled with letters. The first step here is to rotate the object in the top view clockwise about the Y axis until the true-length line *CF* is aligned with the X axis. You can do this with the model and with the drawing. When the top view has been rotated, you must also show the location of all the points in the front and right side views by labeling them with their letter and a prime. In this case, Figure 10.70b shows the views in the block's new position as dark lines and the original position as gray lines. The right side view shows plane 5 as an edge. The next step in the solution is to rotate the block in the right side view counterclockwise about the X axis

Figure 10.68 True size of plane 5 by direct method.

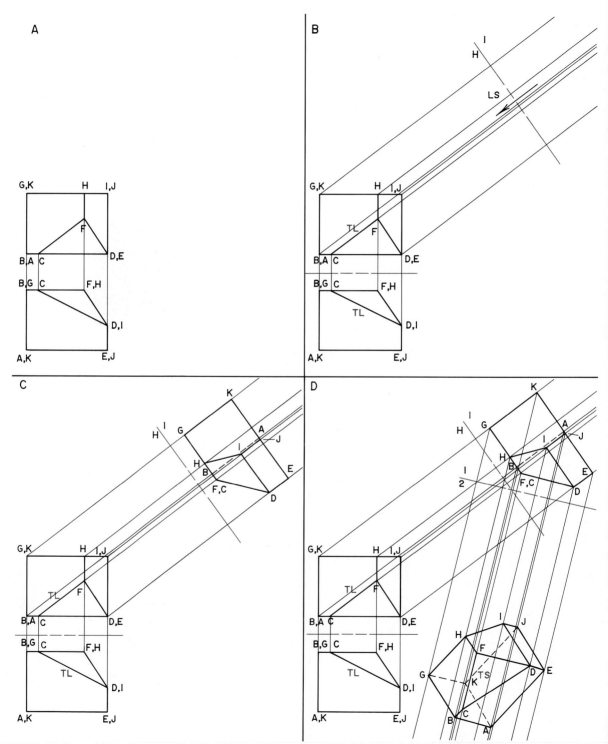

Figure 10.69 True size of plane determined by fold line method.

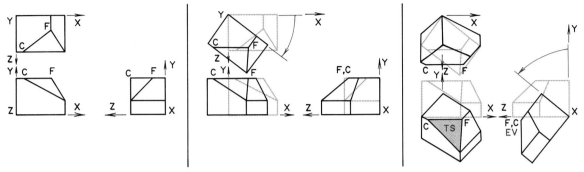

Figure 10.70 True size of plane 5 by rotation.

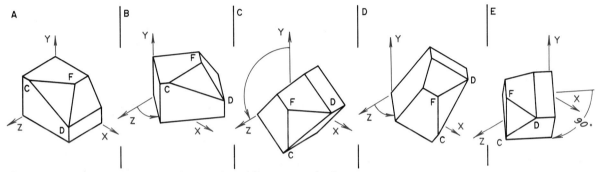

Figure 10.71 True size of plane by rotation. Alternate method.

until the edge view is perpendicular to the Z axis. Figure 10.70c shows this rotation with the resultant views where the front view shows plane 5 true size.

This particular series of steps was chosen so that the same or similar operations could be done on a computer when all three views are shown simultaneously. In this example, as in the manual example, the front view provides the finished view.

Extra steps will probably be required if you are working on a computer system that displays only one view at a time and allows only X and Y rotations. The steps might be as follows. You are looking at the front view or down the Z axis. First, rotate the block counterclockwise about the Y axis until you can see plane 5 as an edge. Then rotate 90° counterclockwise about the X axis so line CF appears true length and parallel to the Y axis. The third step is a counterclockwise rotation about Y so that the plane appears as an edge perpendicular to the X axis. Fourth, rotate about the Y axis 90° clockwise so that the plane appears true size. Figure 10.71 shows these steps.

When you have only one view available at any one time and can rotate about all three axes, the required steps are as follows: Rotate about Y counterclockwise until line CF appears as a point

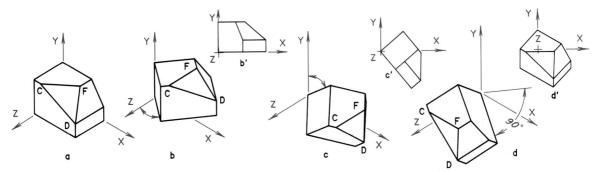

Figure 10.72 True size of plane by rotation. Alternate method.

and plane 5 appears as an edge. Rotate the block about Z clockwise until plane 5 is perpendicular to the X axis. Finally, rotate 90° clockwise about Y again, and you will see the true size of the plane in the front view. See Figure 10.72.

One more axis of rotation provides you with the ability to do one less rotation.

If your system also provides a normal to each plane, you can reduce the rotations to two. First, rotate the block about the Y axis until the plane 5 normal is parallel to the Y axis and then rotate about the X axis until the normal vector appears as a point in the front view. When the normal appears as a point in the front view, you see the plane true size. See Figure 10.73a–c.

A review of the procedures points out the following facts. When you are solving problems involving planes, no matter which method you use, you will need to know how to find the true length of a line in a plane. The next step is the **point view** of the line, which produces the edge view of the plane. The last step is either viewing

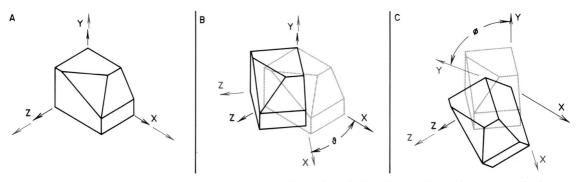

Figure 10.73 True size of plane by rotation. Alternate method.

the plane so that you see the true size or rotating the block so that you see the true size. Assuming that you can either construct or have been given the direction normal, the plane will appear true size when the normal vector appears as a point.

10.5.4 Lines Parallel to Planes

A line is parallel to a plane when it is parallel to any line in the plane. If plane 4 is used as an example, we can see that lines *EJ*, *BG*, and *AK* are parallel to the plane because each is parallel to both the lower and upper edges of the plane. When looking down the *Z* axis, you see the edge view of plane 4, and you see lines *EJ* and *BG* as points. With a pencil, draw a line parallel to line *CF* on plane 3 on your block. See Figure 10.74. Observe what happens to that line and to plane 5 as the block is rotated or when you walk around the block.

If you wish to determine whether a given line is parallel to a given plane, draw a line on the plane in one view that is parallel to the line in that view. Project that line to the other view and check to see whether it is still parallel to the constructed line. If it is not, then the line is not parallel to the plane, as shown in Figure 10.75.

10.5.5 Lines Perpendicular to Planes

In the cut block there are a series of lines that are perpendicular to planes. Line *BG* is perpendicular to plane 1. Rotate the model and study the relationships between *BG* and plane 1. When *BG* is a point, you can see the plane true size. When *BG* is true length, you can see plane 1 as an edge. See Figure 10.76.

How do you determine whether a line is perpendicular to a plane? Think about the last two statements in the last paragraph.

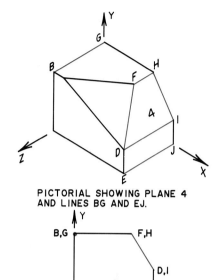

PICTORIAL SHOWING PLANE 4 AND LINES BG AND EJ.

VIEW DOWN Z AXIS SHOWING PLANE 4 AS AN EDGE AND LINES BG AND EJ AS POINTS

Figure 10.74a Pictorial and front views of block highlighting plane 4 and lines *EV* and *BG*.

  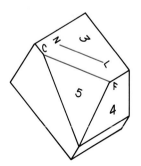

Figure 10.74b Line *NL* parallel to plane 5 and line *CF* in plane 5.

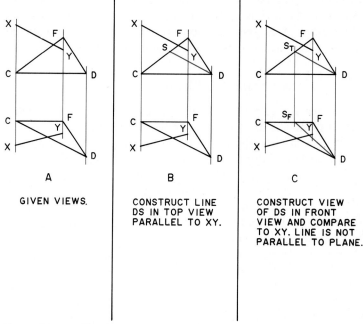

Figure 10.75 Checking whether line is parallel to plane.

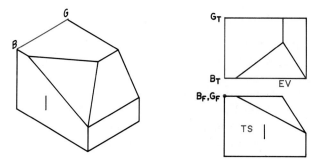

Figure 10.76 Lines perpendicular to planes.

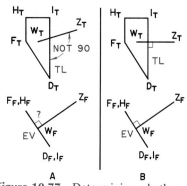

Figure 10.77 Determining whether line is perpendicular to plane.

If you know that the line is true length and the plane is an edge, then you have a line perpendicular to the plane. If the plane is an edge and the line appears perpendicular to the plane, do you necessarily have perpendicularity? Study Figures 10.77a,b. In Figure 10.77a, the line appears perpendicular to the edge view of the plane in the frontal view. However, the top view shows that it is not perpendicular because the line *WZ* is not perpendicular to *DI* and *FH*, which are true length in the top view. The frontal view in Figure 10.77b appears just the same as the frontal view in Figure 10.77a, but the top view shows line *WZ* is perpendicular to the true-length lines.

On your model, points P and Q are located with crosses on planes 2 and 5. Push a straightened paper clip or toothpick through these points. See Figure 10.78, a drawing of the cut block. In the drawing, the line ST is shown in color. Is the line perpendicular to either plane? Does it appear to be perpendicular to either plane from any view?

10.5.6 Angles between Lines and Planes

What angle does line ST make with plane 5? How do you go about measuring the angle? There are projection methods to accomplish this task. One is to find the plane's edge view, then the true-size view, and then the edge view of the plane in a direction that shows the line true length. Figures 10.79a–d show the steps using the direct method. Figures 10.80a–d show the same steps using the fold line method. Figures 10.81a–d show the rotation steps required on paper. The first step is to rotate plane 5 counterclockwise until it appears as an edge in the front view. The rotation is determined by making sure that line CF is parallel to the Z axis in the top view. Rotate the block clockwise around the Z axis until the edge view appears perpendicular to the X axis in the front view. The plane should now appear true size in the right side view. Rotate the block clockwise about the X axis until the line ST appears parallel to the Y axis in the right side view and the line will appear true length in the front view. Plane 1 still appears as an edge and ST is true length. In each of these cases the last step in the solution process is to measure the angle.

When you are holding the sample block in your hands, it is much easier to find the true angle because you are able to rotate the block about any axis.

10.5.7 Lines Intersecting Planes

One of the key solutions for descriptive geometry problems is finding where a line intersects a plane. There are two methods. If you find the edge view of a plane, it is easy to determine the intersection point on the line. Once you have found the intersection point on the line, that point can be transferred into the other views to locate the point on the plane. Figure 10.82 shows the given views and the auxiliary view using the direct method with the intersection point located in each view. Either the fold line method or rotation also can be used to solve this problem.

You need to know what part of the line is visible in each view. In the auxiliary view the entire line is visible. It can be determined with which part of the line is visible in the top view. The part of

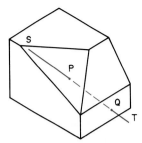

Figure 10.78 Checking perpendicularity with block and colored line ST through PQ.

A

B

C

D
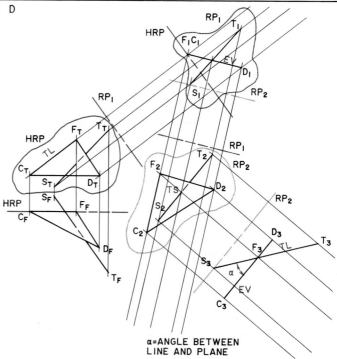

α=ANGLE BETWEEN
LINE AND PLANE

Figure 10.79 Direct method to determine angle between line and plane.

Figure 10.80 Fold line method to determine angle between line and plane.

Figure 10.81 Rotation method to determine angle between line and plane.

the line "above" the plane (closest to the top view) in the auxiliary view is visible in the top and the part of the line "below" the edge view of the plane is hidden.

The remaining problem is to determine visibility in the front view. Where is the line visible and where is the plane visible? You can determine this by checking the apparent intersections between line *ST* and the edges of the plane in the front view. Trace projectors from these points up to the top view. Which does each projector hit first, the line or the plane? If they hit the line first, as the projector on the left does in this case, then this portion of line *ST* is visible in the front view. This means that the line will not be visible on the other side of the point of intersection. To check this, use the projector on the right side. This time your projector hits the plane first, and therefore the plane is visible on the other side of the point of intersection and the line must be represented by a dashed (hidden) line. If you could not see the visibility between the top view and the auxiliary view, this method would also find the visibility for that case. See Figure 10.82 for the steps for determining visibility.

There is another method that can be used to find the intersection point between a line *ST* and plane 5, or *CDF*: the cutting plane method. This method allows you to find the solution for this problem using only the two given views. Therefore, it saves time. More important is the fact that this concept can be used effectively in all intersection problems, as will be shown later in this chapter. The steps are shown in Figure 10.83. The key point is that the edge view of an imaginary cutting plane is constructed that contains the given line in one view where the edge view of the cutting plane is shown in color or labeled CPn. In the example problem,

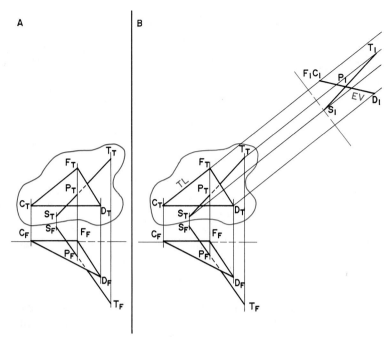

Figure 10.82 Intersection point between line and plane using direct method and edge view of plane.

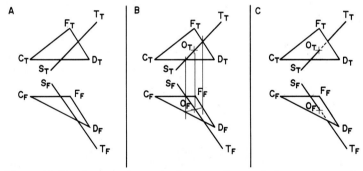

Figure 10.83 Intersection point between line and plane using cutting plane method.

the edge view of the cutting plane is constructed in the top view. The points where this cutting plane crosses the edges of the plane are marked and labeled *L* and *M*. These points are projected to the front view and connected by a line called the trace. The point of intersection, *O*, is where line *LM* in the front view crosses *ST*. Point *O* is projected back to the top view.

The remaining problem is to determine visibility. The method described for determining the visibility in the front view must be used twice: once for the front view and once for the top view. The procedure was described for the front view. For visibility in the top view, you must start in the top view with the apparent intersections and look to the front view to find whether you see the line *ST* or the plane 5 first.

10.5.8 Planes Parallel to Planes

At times you have enough information to define two planes and yet do not know whether they are parallel. You have to find the edge view of one plane and see if the other one also appears as an edge *and* is parallel. Figure 10.84a shows two planes that are not parallel while Figure 10.84b show two planes that are parallel.

10.5.9 Dihedral Angles

Another important problem is finding the angle between planes. This is termed the *dihedral angle*. The key to this particular prob-

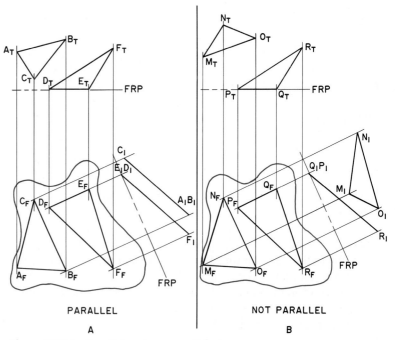

Figure 10.84 Parallel and nonparallel planes.

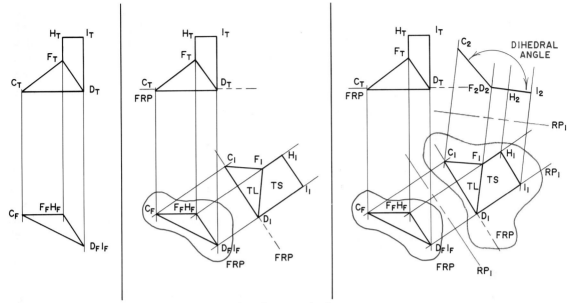

Figure 10.85 Determining dihedral angle using direct method.

lem is identifying the line of intersection between planes and then getting the true length and the point views of that line. When the point view has been constructed, both planes will appear as edges and the angle can be measured.

In the example block, the oblique and inclined planes, planes 5 and 4, respectively, have a line of intersection that is labeled *FD*. Figure 10.85 shows the steps to get the true angle between the planes. In this case only the planes have been shown for the reference plane solution. Because the direct and fold line solutions are so similar, the fold line solution has been omitted.

To solve the problem by rotation, you start with the front view by rotating clockwise around the Z axis until *FD* appears parallel to the Y axis. This will make *F'D'* true length in the right side view. The next rotation occurs in the right side view by rotating the planes around the X axis until line *F"D"* again appears parallel to the Y axis. The top view now shows *F"D"* as a point and the angle between the planes can be measured. Figure 10.86 shows the required steps.

If you have only Y and X rotations available and the front view, then rotate the block clockwise about Y until *F'D'* appears parallel to the Y axis. Now rotate the block counterclockwise about the X axis until you see *F"D"* as a point.

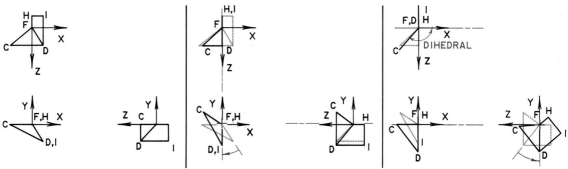

Figure 10.86 Determining dihedral angle by rotation.

10.5.10 Planar Intersections

When the line of intersection between two planes is not known, another projection method becomes important. In this case we think of passing two cutting planes, labeled ZW and XY, through both of the given planes. If need be, return to the section on intersection of lines and planes for a review of cutting planes. Figure 10.87a shows the given planes. Figure 10.87b shows the cutting planes being installed in the top view. The traces of the cutting planes are shown in the front view in Figure 10.87c, which

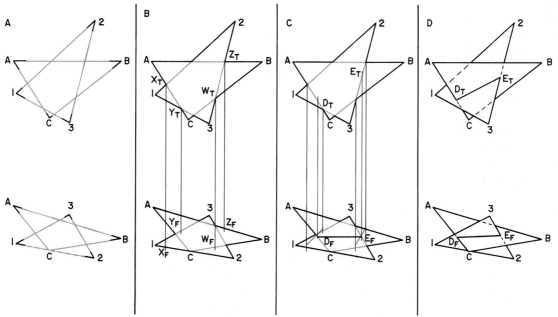

Figure 10.87 Finding line of intersection between two planes using cutting planes.

also shows the line of intersection, *DE*. The line of intersection, *DE*, is transferred to the top view in Figure 10.87*d*. Figure 10.87*d* also shows the visibility of both planes. Visibility must be checked as before.

The rotation solution requires the revolution of the planes as a pair to get one of the planes perpendicular to one of the axes, for example, the Y axis, and both planes as edge views. In order to do this, we first must construct a true-length line in the top view of one of the planes (the black plane or plane 1 was chosen). Rotate the planes about the Y axis until the true-length line is parallel to either the X axis or the Z axis. In this case the planes are rotated clockwise so that the true-length line is parallel to the X axis. We now have the edge view of plane 1 in the right side view. Next, rotate the planes clockwise about the X axis until the edge view of plane 1 is coincident with the XZ plane. This produces the true size of the plane in the top view.

Now construct a true-length line in plane 2 (gray) in the top view. Rotate the planes until the new true-length line is parallel to either the X or Z axis. For this example, the planes are rotated clockwise about the Y axis so that the second true-length line is parallel to the Z axis. The front view shows the edge view of both planes and the point view of the line of intersection. If the intersection does not occur, extend the edge views to find the point view of the line of intersection so that the dihedral angle can be measured. Any adjacent view will show the line of intersection true length. See Figure 10.88.

10.5.11 Plane–Solid Intersections

When planes and solids intersect and the intersection between the two is sought, it becomes a matter of applying the principles of line–plane intersections. We shall deal with two different plane–solid intersections. The first is the intersection of a plane with a prism and the second is the intersection of a plane with a pyramid. In order to make it easy to see the principles, the prism will have a triangular cross section and the pyramid will have three sides and a base (tetrahedron).

If the plane is an edge in the front or any of the given views and we can see the sides of the prism as edges in the other view (top in this case), then the solution is easy. See Figure 10.89. The edge view of the plane appears as a line, and the places where the line cuts across the edges of the prism mark the intersections on the surface. The line of intersection is simply marked as a line in the front view and assumes the cross-sectional shape of the prism in the top view.

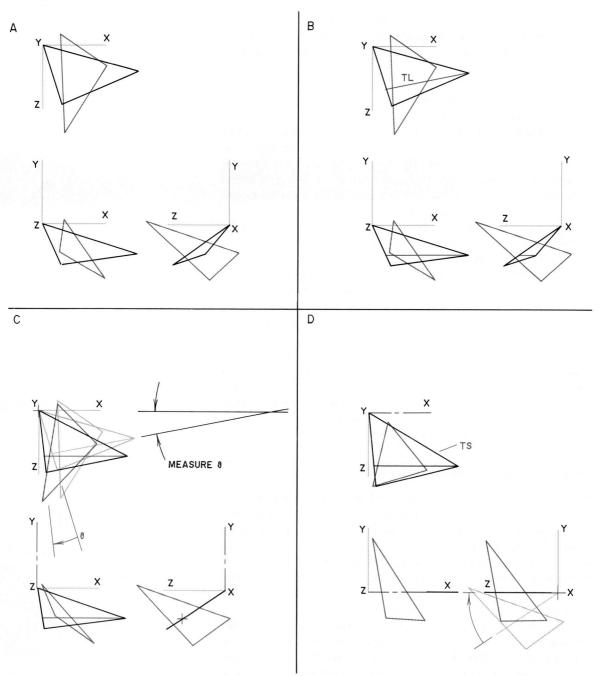

A

B

C

MEASURE θ

θ

D

TS

Figure 10.88 Finding line of intersection by rotation. Computer generated image.

E

TL

F

ϑ

G

LINE OF
INTERSECTION

H

Figure 10.89 Plane intersecting three-sided prism. Front view shows plane as edge.

If the plane does not appear as an edge in any of the given views, then there are two possible solutions. The first is to construct an edge view of the plane. See Figure 10.90. Label each of the intersection points where the prism goes through the plane. Do this in both the top and auxiliary views. Then project the intersection points to the front view by measuring from the reference plane or fold line in the auxiliary view and transferring those measurements to the front view.

Once the points that form the line of intersection have been located in the front view, it is time to determine the visibility. Here, with the help of the auxiliary views, it is relatively easy to see which part of the prism is above and which part is below the plane.

The same problem is presented in Figure 10.91. This time we use the cutting plane method. Starting in the top view, three cutting planes are installed so that they contain the edges of the prism. This creates three traces on the plane. The three traces are then located in the front view, and the points where the traces cross the constructed edges of the prism locate the line of intersection. However, without the edge view provided by the aux-

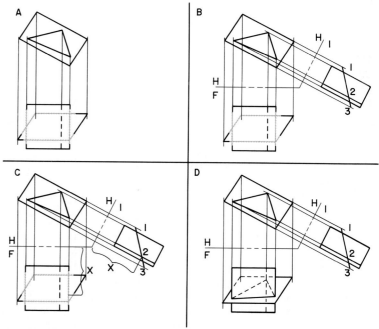

Figure 10.90 Plane intersecting three-sided prism. Top view shows end view of prism. Solve by constructing edge view as auxiliary from top view.

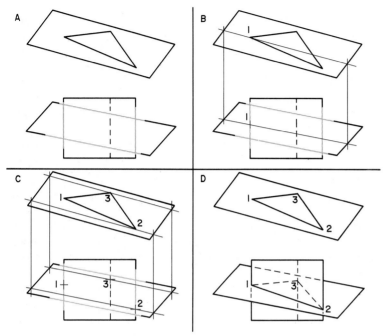

Figure 10.91 Plane intersecting three-sided prism. Top view shows end of prism. Problem solved by cutting plane method.

iliary view, it is a little more difficult to determine whether the plane or the prism is visible. You will have to observe in both views whether the traces are located above or below in the front view and are in front of the prism, touch, intersect, or are behind the prism in the top view. You can always check visibility by constructing the edge view. Another method for checking your problems is to take your hexagonal pencil and sketch the plane on a piece of paper. Then push the pencil through the constructed plane and compare your model with your drawings.

Although there are modeling programs that will determine the intersections of two- and three-dimensional objects, many CADD programs do not yet provide the capability. Therefore, the methods described in the preceding paragraphs will have to be used. If the prism happens to have its axis aligned with one of the display axes on the system, then many systems will allow you to rotate the figure until the plane appears as an edge. Do not forget to construct a true-length line in the view where the rotation is to take place. This is the key to solving these problems because it can be used to estimate or determine the required angle of

Figure 10.92 Three-sided prism intersected by plane, solved by rotation. Computer generated image.

rotation. Figure 10.92 shows a three-sided prism intersected by a plane. The CADD system allows you to see only one view at a time, and this figure shows the given top and front view, the construction of the true-length line, and the rotation so that the edge view of the plane can be seen. Once the edge view is seen in the front view, the points of intersection can be marked on the edges of the prism and the figure can be rotated back to the given position. Now the points can be connected with lines to form the line of intersection.

The intersection of a plane and a pyramid is treated in much the same way as the intersection of a plane and a prism. The techniques for determining the intersection using a cutting plane are slightly different.

When given two adjacent views of a plane intersecting a prism where the line of intersection must be determined, it is always possible to construct an auxiliary view that shows the edge view of the plane. The intersection of the edges of the pyramid with the edge view of the plane are easy to determine. These points

Figure 10.93 Three-sided pyramid intersected by plane. Solved by direct method using an edge view of the plane.

are then located in the front view and the line of intersection can be drawn. See Figure 10.93.

You can solve this problem using cutting planes in the two given views. In this case, the key is to install the cutting planes through the vertex of the pyramid coincident with the corners of the pyramid. See Figure 10.94. The traces from the cutting planes are then drawn in the front view where they provide the intersection points. These points can be connected, creating the line of intersection.

10.5.12 Solid–Solid Intersections

The problem to be dealt with in this section on planar figures is the intersection of two solids. Here, the key is to remember the elements already covered: line–plane, plane–plane, and plane–solid intersections. Solving for the intersection of two solids is nothing more than breaking down the solids into manageable pieces. As a demonstration we solve for the intersection of a triangular prism with a three-sided pyramid. See Figure 10.95.

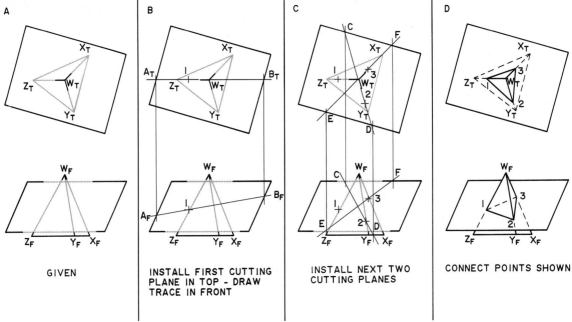

Figure 10.94 Three-sided pyramid intersected by plane. Solved with cutting planes.

A — GIVEN

B — INSTALL FIRST CUTTING PLANE IN TOP - DRAW TRACE IN FRONT

C — INSTALL NEXT TWO CUTTING PLANES

D — CONNECT POINTS SHOWN

A — GIVEN INFORMATION

B — LOCATE POINTS BY LINES INTERSECTING EDGE VIEWS OF PLANES

C — LOCATE POINTS 5 AND 6 BY INSTALLING CUTTING PLANE AND TRACES

D — DRAWING COMPLETE SHOWING LINE OF INTERSECTION AND VISIBILITY

Figure 10.95 Three-sided pyramid intersected by three-sided prism. Front view shows end view of prism so that sides of prism are edge views.

In this case, two of the pyramid edges intersect the upper and lower surfaces of the prism. Because these surfaces appear as edges, it is easy to mark the four intersection points in the front view and project them to the top view. The final two intersection points must be located by passing a cutting plane from the vertex of the pyramid through the point view of the line forming the prism edge and extending through the edge view of the pyramid base. This cutting plane creates two traces, one on the front surface of the pyramid and one on the rear surface of the pyramid. The intersection points are located where the traces intersect the line forming the edge of the prism in the top view. The six points are connected with lines to give the line of intersection. Note that part of the intersection is visible on the upper surface of the prism and part is hidden on the lower surface in the top view.

In situations where the end view of a prism is not provided, it is easiest to solve the problem by constructing an end view of the prism. The methods shown here work for the intersections of prism with prism and pyramid with pyramid as well.

The key points in this section are the relationships between lines and planes. By knowing how to determine angles between lines and planes or planes and planes, we can see the three-dimensional relationships that are important in design. You should now know when lines are perpendicular or parallel to each other and when a line is perpendicular or parallel to a plane. You should also know how to solve for the intersections between lines and planes through solids intersecting with other solids. All are solved by breaking an apparently difficult problem into a series of simple problems.

10.6 *CURVED SURFACES*

The previous sections have dealt with lines and planes. You also need to learn about curved surfaces. Curved surfaces include cylinders, cones, spheres, and warped surfaces (Figure 10.96). The material presented here deals mostly with right-circular cylinders and cones. *Right circular* means that the cross section of the cylinder or cone taken perpendicular to the cylinder or cone axis appears to be a circle.

Although cylinders and cones have been shown in preceding chapters, it is important to review this material again and look at the possible orientations for these primitive shapes. A series of views are shown that display the cylinder and cone in a variety of positions. The series begins with the axis vertical so that the top view appears circular. The subsequent views show the primitives rotated about the base. See Figures 10.97 and 10.98. Note that the front view did not change its shape while the top view

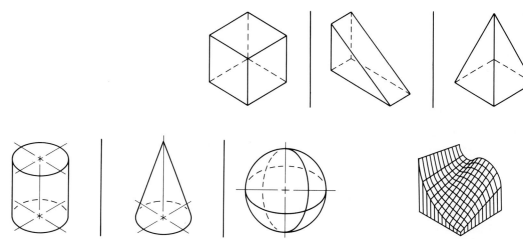

Figure 10.96 (*a*) Planar primitives, (*b*) curved surface primitives, (*c*) warped surface. Computer generated image.

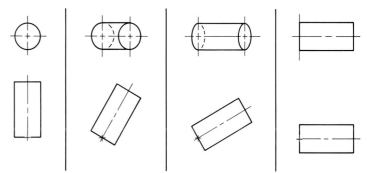

Figure 10.97 Successive views of rotated cylinder. Computer generated image.

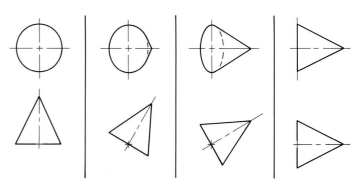

Figure 10.98 Successive views of rotated cone. Computer generated image.

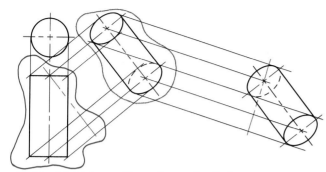

Figure 10.99 Multiple auxiliary views of cylinder.

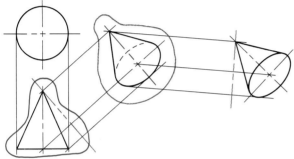

Figure 10.100 Multiple auxiliary views of cone.

shows the circular portion of the object first becoming an ellipse and then a line. When the bases of the primitive shapes are seen as lines in the top view, the top view and front views become identical.

This series of top views could have been drawn by using a series of auxiliary views (changing lines of sight), as shown in Figures 10.99 and 10.100.

The construction of the ellipses can be accomplished in several ways. First, we must define the relationship between the line of sight and the edge view of the plane. See Figure 10.101. The angle

Figure 10.101 Edge and top views of circle with line of sight.

Figure 10.102 Ellipse templates.

Figure 10.103 Ellipse partially drawn with template.

measured is used to specify the proper ellipse template. Templates are available in 5° increments. See Figure 10.102. If the measured angle is a multiple of 5°, the proper template can be chosen and the ellipse completed. Practically, if the angle is other than in 5° increments, the closest angle is used. Figure 10.103 shows how to use an ellipse template to draw a partial ellipse.

A second method is known as the trammel method. This method allows construction to proceed if both the major and minor ellipse diameters are known. Figure 10.104 shows this construction method. It is important here to note and remember that the major diameter of the ellipse is always perpendicular to the axis of the cone or cylinder.

A third method is always possible when a true-size view of the circular base can be drawn. Auxiliary views constructed with either a reference plane or rotation can provide the true-size view of the circular base. In this case both the circular view and the elliptical view of the base are projected from the edge view. The measurements are made in the circular view and transferred to the elliptical view. See Figure 10.105. Note that when using the

Figure 10.104 Trammel construction.

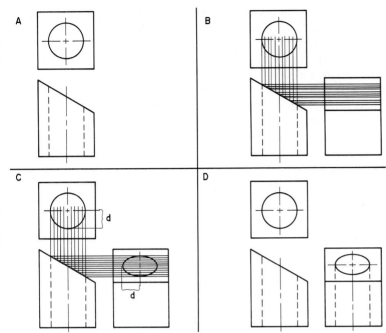

Figure 10.105 Ellipse construction. Edge–true size method.

reference plane, it is convenient to locate the plane along the axis so that only one set of measurements needs to be made for all four quarters of the circle.

Although it seems obvious from studying a spherical shape, it is important to remember that we always show the sphere as a circle on paper. We use this fact as we learn about the intersections between various lines, planes, and primitive figures.

You will need to know how to find the intersections between lines, planes, and solid primitives. These principles can be applied to complicated shapes. The two techniques for line–plane intersections are useful here. The first was to find the edge view of a plane and the second was to use the cutting plane method. If the problem is simple, the cutting plane is convenient because it requires no extra views. The methods described here apply both to manual and CADD drawings where the CADD package does not have the ability to determine intersections.

10.6.1 Line–Cylinder Intersection

We start with the equivalent of the edge view of the plane. This is the end view of a cylinder, which gives the true cross section of the cylinder. If we do not have the end view of the cylinder (point view of the cylinder axis), it is possible to obtain it by getting the true length of the axis and then the point view of the axis. This can be done by auxiliary views with reference plane–fold line techniques or by rotation. Once the end view is constructed, the intersection point(s) are visible and can be projected back to the given views. See Figure 10.106 for an example using one of the auxiliary view methods. This provides a very sure approach to solving the problem but is not necessarily the most efficient.

The second method requires an edge view of the base through the axis and an adjacent view with this base shown as a circle or an ellipse. If an appropriate base is not available, you can choose the type of cross section needed and install a cutting plane to produce that section. In this case, install a cutting plane through

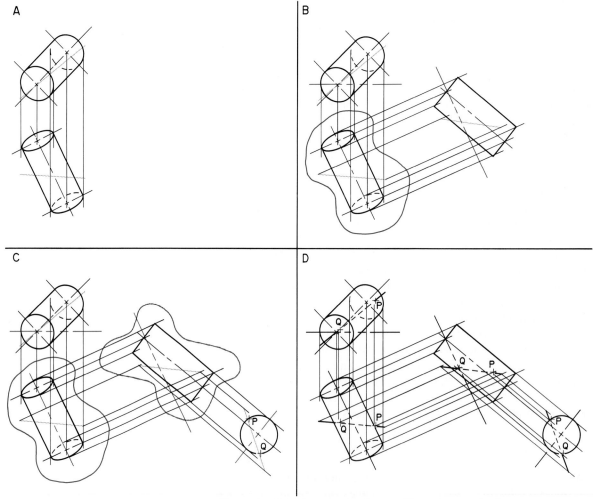

Figure 10.106 Intersection by auxiliary view method.

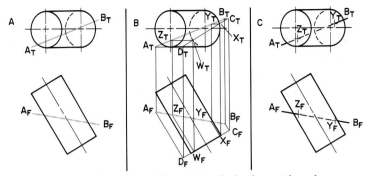

Figure 10.107 Intersection of line and cylinder by cutting plane method.

the cylinder so that it is parallel to the cylinder axis and contains the line. The cutting plane is defined by colored construction lines drawn through points on the intersecting line and parallel to the cylinder axis. (Two parallel lines define the cutting plane.) This is done in both of the given views. In the view that shows the edge of the cylinder cross section, a colored line is drawn that extends the cross section until it intersects the two construction lines. These points are marked in this view and projected into the adjacent view and connected. This line should cross the elliptical view of the cross section at two points. (If it does not, the line does not intersect the cylinder.) These are two points on the trace of the cutting plane and the feet of elements on the surface of the cylinder. Element lines are drawn from these two points parallel to the cylinder axis, and these lines are on the surface of the cylinder. The points at which these elements cross the intersecting line are the intersection points between the line and the cylinder. They are on the surface of the cylinder *and* on the line. These points can be projected to any views. See Figure 10.107.

The intersection of a line and a cylinder can be solved by rotation. The key to the solution is to get a point view of the axis. In Figure 10.108, the axis cannot be seen as a point in either of the given views. The first step is to rotate the cylinder axis about the Y axis until it appears parallel to the X axis. The cylinder axis now appears true length in the front view. The second step is to rotate the cylinder axis about the Z axis in the front view until it coincides with (or is parallel to) the Y axis. The cylinder now appears as a circle in the top view, and the intersection points can be determined as they were in the auxiliary view method.

10.6.2 Line–Cone Intersection

The cutting plane method for cones requires an edge view of a cross section or base through the axis and an adjacent view with

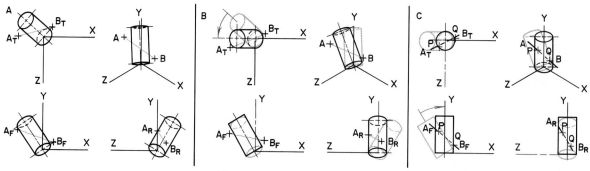

Figure 10.108 Intersection by rotation method. Computer generated image.

this cross section shown as a circle or an ellipse. In this case, the cutting plane is drawn through the cone so that it goes through the tip of the cone axis and contains the line *DE*. The cutting plane is defined by colored construction lines drawn through points on the intersecting line (the points can be the ends of the line) and the tip of the cone. Two intersecting lines define the cutting plane. This is done in both of the given views.

In the view that shows the edge of the cone cross section, a colored line is drawn that extends the cross section until it inter-

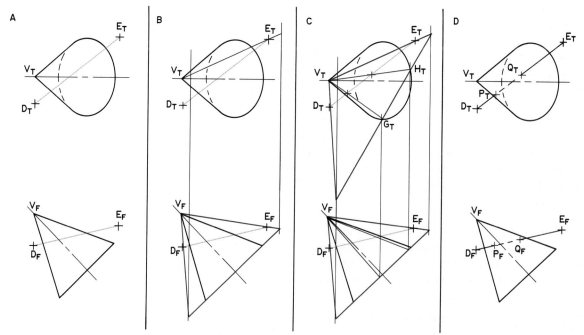

Figure 10.109 Intersection of line and cone by cutting plane method.

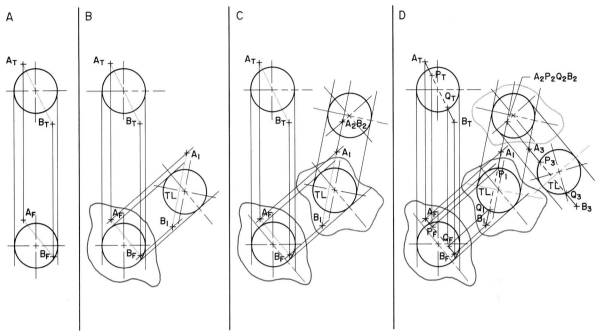

Figure 10.110 Intersection of line and sphere by combination cutting plane and auxiliary views.

sects the two construction lines. This line is in the cutting plane. These points are marked in this view and projected into the adjacent view and connected. This colored line should cross the elliptical view of the cross section at two points. These points are the feet of elements on the cone and are on the trace of the cutting plane. Element lines are drawn from these two points to the tip of the cone. These lines are traces of the cutting plane on the surface of the cone. The points at which these elements lines cross the intersecting line are the intersection points between the line and the cone. These points can be projected to any views. See Figure 10.109.

10.6.3 Line–Sphere Intersection

When a line intersects a sphere, it is possible to find the intersection by constructing a cutting plane that contains the line and then finding the edge view and true-size view of the plane. This provides a true-size view of the portion of the sphere and the cutting plane that coincide. Once this view is obtained, it is possible to see the intersection and project the intersection points to the other views. See Figure 10.110.

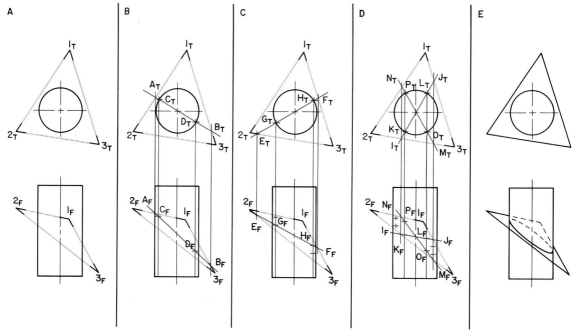

Figure 10.111 Intersection of plane and cylinder.

10.6.4 Plane–Cylinder Intersections

Finding the intersection of a plane with a cylinder requires the components of the intersection methods described in the preceding. The intersection will be a circle when the cylinder is right circular and the plane is perpendicular to the cylinder axis. It can also be an ellipse if the plane is not perpendicular to the axis. The method for solving this problem utilizes cutting planes and a point view of the cylinder axis. The example is shown where the point view of the cylinder is provided. If there is no point view of the cylinder axis, then auxiliary views or rotation must be used to get a point view of the cylinder axis.

The cutting planes are installed in the top view through the cylinder axis. They are evenly spaced to provide a sufficient number of intersection points on the cylinder to produce a smooth line of intersection. The cutting planes are extended to the edges of the plane in the top view. The cutting planes produce traces on the plane and elements on the cylinder in the front view. The traces intersect the elements to provide intersection points that are connected to show the line of intersection. When doing this

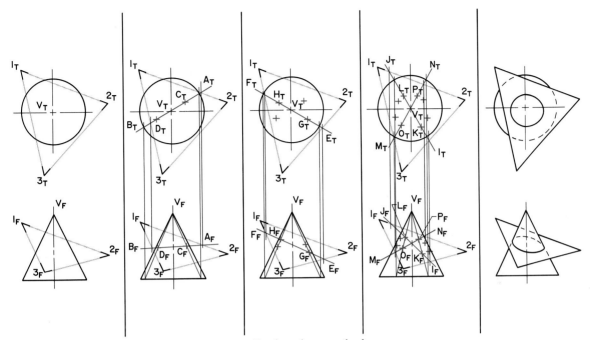

Figure 10.112 Intersection of plane and cone. Cutting plane method.

construction, you should use either numbers to label all points or a series of colored lines to represent the edge views and traces of the cutting planes. See Figure 10.111.

10.6.5 Plane–Cone Intersection

When solving for the intersection of a plane and a cone, it is important to have a point view of the cone axis. A series of cutting planes can then be constructed through the vertex of the cone so that they intersect both the plane and the base of the cone. The cutting planes produce traces on the plane and elements on the surface of the cone. The intersection points can be transferred to the other views and connected to form the line of intersection. In Figure 10.112, the cutting planes were installed in the top view and the resulting traces and elements constructed in the front view. The intersection of the traces and elements produce intersection points on the line of intersection. These points are transferred back to the top view to produce the line of intersection in the top view. When the point view of the cone axis is not provided, it must be constructed using either auxiliary views or rotation.

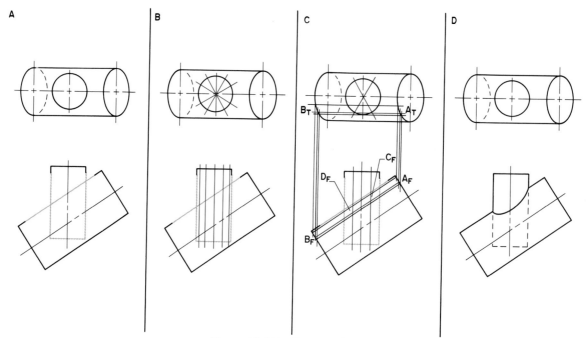

Figure 10.113 Intersection of cylinder with cylinder. Cutting plane method.

10.6.6 Cylinder–Cylinder Intersection

Cylinder-to-cylinder intersections have complex lines of intersection. A combination of auxiliary views and cutting planes is used to find the line of intersection. The example shown in Figure 10.113 shows the cylinders oriented so that the axis of one appears as a point in one view. The other cylinder is oriented so that the axis appears true length in the other view. This allows a series of cutting planes to be constructed in the top view. When working with several cutting planes, it is easier to keep track if the planes are colored or numbered. The traces of the cutting planes in the front view are found by projection from the top view. When the traces are plotted on each of the cylinders, the intersection points are determined by the locations where the traces cross. Once the points have been plotted, they are connected using an irregular curve. On the computer, they might be connected using a spline. It is easier to solve the problem if one cutting plane is constructed, the traces projected, and the points of intersection found before starting on the next one.

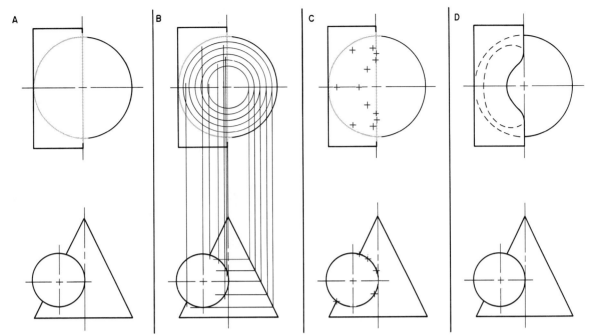

Figure 10.114 Intersection of cylinder with cone. Special case with axes perpendicular.

In real problems of intersecting cylinders you may not have the convenience of the point view of one axis and the true-length view of the other axis. This will require the construction of sufficient views so that both the point view of one axis and the true-length view of the other axis are in adjacent views.

10.6.7 Cylinder–Right Circular Cone Intersection

The key for determining any solid–cone intersection is to have a point view of the cone's axis in one view, in this case the top view, and a point view of the cylinder's axis in an adjacent view. This allows cutting planes to be constructed perpendicular to the cone axis in the front view and parallel to the cylinder axis. Then in the top view with the cone's axis as a point, the traces of the cutting plane on the cone are circles while the traces on the cylinder are lines parallel to the cylinder axis. This approach works well for situations when the cylinder axis is 90° from the cone axis. This is a special case and is Figure 10.114. The general case where the axes are not perpendicular is described next.

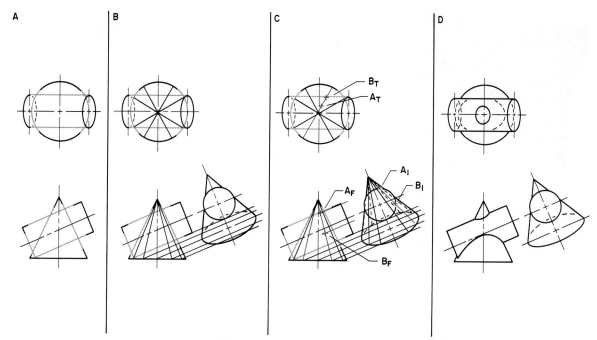

Figure 10.115 Intersection of cylinder with cone. Combination of auxiliary view and cutting plane method.

When the cylinder and cone axes are not perpendicular, an auxiliary view must be constructed that gives the point view of the cylinder axis. Cutting planes are constructed as edges in the auxiliary view through the vertex of the cone. The traces are then constructed in both the front and top views on both the cone and the cylinder. The intersection points are located and then connected either with the irregular curve by hand or with an irregular curve (in some cases a spline) function if you are working on a computer. See Figure 10.115.

10.7 DEVELOPMENTS

10.7.1 Principles and Examples

Engineers are faced with many problems that require knowing the true size of all surfaces of an object. The word *development* defines a process that takes two or three given views and creates the flat pattern or patterns for all of the necessary parts. There are relatively simple development problems associated with con-

Figure 10.117 Two cardboard boxes, one assembled and one laid flat.

Figure 10.116 Tool box.

tainers such as tool boxes, fiberboard milk cartons, and frozen juice cans. There are also some relatively difficult problems that are involved in bulk materials handling that require transition pieces between square or rectangular shapes and circular or elliptical shapes.

Figures 10.116–10.121 show some of the examples outlined in the preceding. However, all of the items have several things in common. First, each began as a flat piece of a raw material called flat stock. Second, to cut the flat stock, you must know how to find the true shape of each of the surfaces that define the item that is to be made from the flat stock. Third, the only way to solve the assembly problem is to know how to efficiently cut the extra material that is used to join the surfaces together (tabs). Joining processes include tape, screws, or rivets (mechanical processes); soldering, brazing, or welding (thermal processes); and adhesives (chemical processes). See Chapter 8 for further information on fastening and joining processes.

10.7.2 Right Solids

The simple cardboard box shown in Figure 10.118 can be developed from front and top views of the box. In this case the box opens at the top and bottom. We would like to connect as many

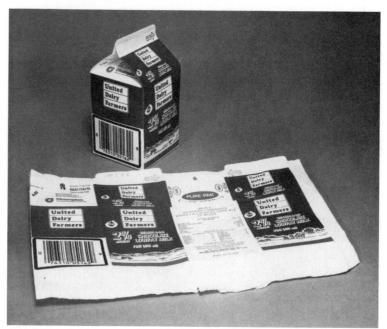

Figure 10.118 Two milk cartons, one assembled and one torn open and laid flat.

Figure 10.119 Funnel, conical to conical transition.

of the equal-height surfaces as possible. In this case those would be the front, right side, left side, and back surfaces. See Figure 10.122.

If a string is wrapped all the way around the box, marked, and

Figure 10.120 Transition piece, square to circular.

Figure 10.121 HVAC ductwork section.

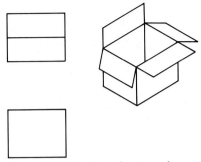

Figure 10.122 Top, front, and pictorial of box.

then pulled out straight, it provides the length of material needed for these surfaces, as shown in Figure 10.123. This is called the stretch-out line, and it is drawn next to the front view. For construction purposes, this line begins close to the front view and is drawn to the right or left of the front view an indeterminant length. It is a good idea to label the points in the top and front views so that you can keep track of the panels. When doing this, imagine that you are standing inside the box and looking at the side panels. Turn clockwise (to the right) and label the corners as you turn. The length of this line can be determined from the given views by taking dividers and measuring the perimeter in the top view. As the length of each side is measured, it is marked and labeled on the stretch-out line. The height of each panel or side can be projected from the front view. Figure 10.124 shows the steps in the development process for the box. In this particular example, all of the sides are the same height, and it does not make any

Figure 10.123 Box with string.

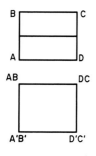

Figure 10.124 (*a*) Top and front views of box with stretch-out line and points labeled on given views. (*b*) Top and front views showing *A* and *B* marked and measuring the distance from *C* to *D*. (*c*) All points marked on stretch-out and first projection. (*d*) Side panels complete.

difference where you start and where you finish. This is not true for many other problems.

A complete box also requires a top and a bottom. For most packing boxes, the top is formed by having a flap on each panel above and below so that the box can be assembled with tape.

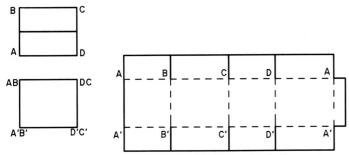

Figure 10.125 Figure 10.124 pattern with flaps added. Computer generated image.

Another flap is required at the beginning or end of the set of side panels so that tape or glue can be used to hold the side panels together to form the body of the box. See Figure 10.125.

Once the pattern has been drawn on paper, it is transferred to the flat stock so that it can be cut out. Most patterns are constructed as inside patterns so that the markings on the stock do not show once the material is assembled. In order to conserve material and avoid scrap, multiple patterns are arranged on a large sheet of the stock material so that there is minimum scrap after the patterns are cut. See Figure 10.126.

The steps to solve a second example with a rectangular cross section are shown in what follows. In this case, both the heights of the corners and sides of the base are of unequal length. Notice that the greatest heights are used for folding the material while the shortest height is used for the location of the seam. This saves money because a smaller number of rivets or spot welds are required or less adhesive is required. This also promotes the integrity of the finished package. Notice also that the attachment

Figure 10.126 Flat stock shown with more than one pattern nested. Computer generated image.

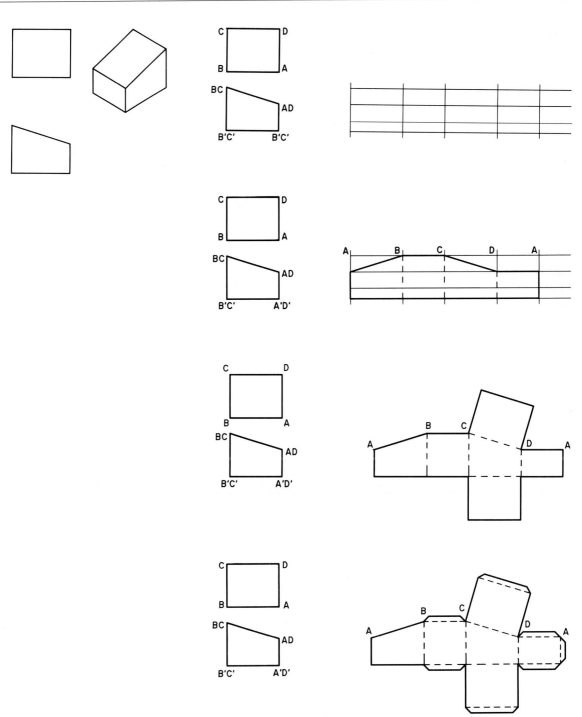

Figure 10.127 (*a*) Given views and pictorial of truncated, rectangular prism. (*b*) Stretch-out line and height projection. (*c*) Finished sides. (*d*) Top and bottom lids added. (*e*) Flaps added for gluing or riveting. Computer generated image.

for the "lid" is on one of the long sides, and the "flaps" to create the seams are on the two short sides and only one long side. See Figure 10.127.

10.7.3 Right Cylinders

Normally the objects with circular or elliptical cross section are drawn so that the top view shows the cross section. The front view shows the "rectangular" or longitudinal view of the object. The stretch-out line is drawn to the right or left of the front view. The height, either constant or variable, can be projected from the front view to the pattern view. See Figure 10.128.

When working with circular cross sections, there are two approaches that work reasonably well. In both cases, the circle is subdivided into a number of equal arcs and the subdivisions are labeled. In the first case, the circumference of the circle is calculated and then is divided by the number of arcs to give the arc length. The arc length is measured on the stretch-out line once,

Figure 10.128 (a) Top, front, and pictorial views of cylinder. (b) Top and front views with stretch-out line. Calculations shown for arc length. Figure shows scale laying off arc length on stretch-out line. (c) Construction for projection of height and finished pattern. Computer generated image.

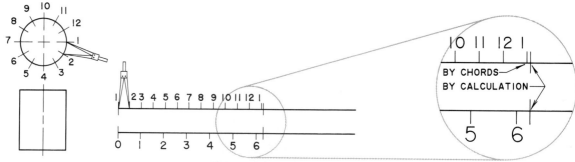

Figure 10.129 Figure 10.128 repeated but showing chord of arc being measured with dividers. Computer generated image.

Figure 10.130 Two HVAC cylindrical tubes intersecting.

and the dividers are used to mark the remaining arc lengths. Each mark is labeled. The heights for the ends of each vertical section are projected from the front view and the flat pattern is completed by providing a tab at one end. If the height is constant and the resulting pattern rectangular, the construction is relatively simple. The second approach to creating the proper length on a stretch-out line assumes that relatively small arcs and the cords for the small arcs are close enough to the same size that a calculation does not have to be made. The first approach works all of the time. Figure 10.129 shows a comparison.

When a tube with a circular cross section is to be joined with another circular cross section tube or to a flat surface but where the plane of the joint is at an angle to the axis of the cylinder, the height will vary so that the pattern will be curved at the top. Figure 10.130 shows two intersecting tubes. Figure 10.131 shows a tube that has been cut at an angle. The top view appears as it did before, a circle. The same processes take place except that labeling is very critical and the height projections must be done from the vertical line in the front view to the matching vertical line in the pattern. Here, again, it is important for the drafters to assume that they are standing inside the object in the top view and turning to their right (clockwise) to see the successive panels in the correct order.

Once each of the heights have been marked, an irregular curve can be used to create the correct shape of the top of the pattern. See Figure 10.132. Where the pattern is symmetrical, you can speed up your work by marking your irregular curve and flipping it over so that you can draw the same curved portion on the other part of the pattern. If you are uncertain if the irregular curve was drawn correctly, mark and project a few additional points on the circumference to check your work. Remember that the more segments you draw, the more accurate your construction will be.

Figure 10.131 (*a*) Top and front view of truncated cylinder. (*b*) Stretch-out line added. (*c*) Construction (projection) of first two heights shown. (*d*) Finished construction. Computer generated image.

Figure 10.132 (*a*) Construction of ellipse by connecting three or more points with irregular curve. (*b*) Irregular curve flipped for mirrored points.

Creating flat patterns with the computer is generally no more difficult than it is to do them by hand. A feature that is required of the CADD system is the ability to fit a curve to the points. However, once the flat pattern is created with a computer system, the drafter can draw the stock size of the material from which the pattern(s) will be cut. The patterns can then be moved around on the stock size so that there is the least amount of waste. In addition, some CADD systems have been interfaced (connected) to laser or mechanical cutting systems so that no programming is required to set up the production system once the pattern has been drawn. Figure 10.133 shows patterns developed on two CADD systems. Refer back to Figure 10.126 for an example of multiple patterns closely packed to minimize waste.

10.7.4 Cones

When dealing with both cones and pyramids, it is important to keep in mind the necessity for true-length lines in order to be able

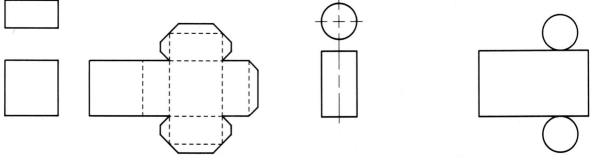

Figure 10.133 Pattern development done on CADD systems. (*a*) GraphiCAD, and (*b*) AutoCAD. Computer generated image.

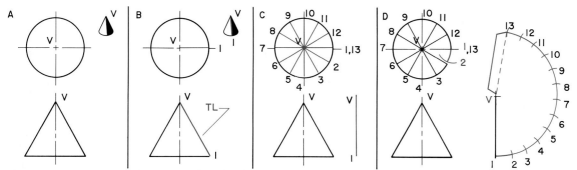

Figure 10.134 (*a*) Top, front, and pictorial views; of a right circular cone; (*b*) top, front, and pictorial views with *TL* line highlighted; (*c*) beginning of development with apex of cone and first radial line with arc for base; (*d*) chord system for laying off base arc and finished drawing.

to complete developments. It is easy to lose track of which side lengths are known and which are unknown. These shapes require a thorough understanding of the principles of revolution to obtain the true lengths of sides. As a reminder, if you rotate in one view, you translate in the other view and make the measurements in that view. Labeling is very important to keep track of the given views and the true-length segments.

Cones will be examined first because a view where the axis of a right-circular cone is true length provides a true-length edge or radial line. In Figure 10.134, the top view of the cone appears to be a circle and the front view, where the cone appears as a triangle, is the one that provides the true-length radial line from the top to the base of the cone. The radial line is highlighted in both views.

In Figure 10.134 the top and front views are given. The first step is to find a true-length line, which is shown highlighted in Figure 10.134*b*. The development drawing begins in Figure 10.134*c* by locating the apex of the cone and drawing the first radial line and the arc for the base. Two methods are available for finding the length of the base arc: the calculation method described in paragraph 10.7.3 and the chord method that is shown in 10.134*d*, which shows the completed pattern.

Figure 10.135 shows the same right-circular cone with the top having been truncated at an angle to the cone axis. In this case much additional work must be done. The base circle in the top view is divided into an even number of equal arcs. These arcs are labeled in both the top and front view. In the front view, each mark has two labels. One for the front of the cone and one for the back. These marks are connected to what would be the apex of the cone in both views. You can see that the radial lines have

Figure 10.135 Development of truncated cone.

varying lengths because of the truncation. Each radial line must be laid off individually in order to complete the pattern development. In this case only half of the pattern need be drawn because the object is symmetrical. After tracing the half pattern on the stock piece, the pattern can be flipped over and the second half of the material to be cut can be traced.

10.7.5 Pyramids and Tetrahedrons

The pyramid shown in Figure 10.136 has none of its inclined edges (legs) shown true length in either view. The first step is to number or letter the vertices of the pyramid. Again work from the inside and number clockwise. The next step is to revolve one of the legs about the axis of the pyramid in the top view and translate the lower end in the front view to get the leg's true length. Lay the leg's true length off near the two given views and begin the development. Use this length as the radius of an arc and strike an arc to the right. The pyramid base is shown true size in the top view. Measure the length of the first edge of the base with your compass and strike an arc with the lower end of the first leg as center. This second arc will intersect the large arc locating the second corner of the base. Continue this process until you have done each of the edges of the pyramid. Construct a tab along the

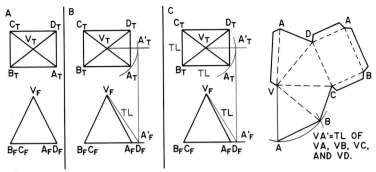

Figure 10.136 Development of full pattern for right pyramid.

finishing edge. Attach a base with tabs to the first side and darken the edges of the pattern that will be cut. Use a dashed line for the location of the folds. In this case half of the pattern is sufficient if the proper starting point is chosen.

Figure 10.137 shows the development of a truncated tetrahedron where the axis is not perpendicular to the base. In this case the true length of every line must be determined either through rotation or auxiliary views.

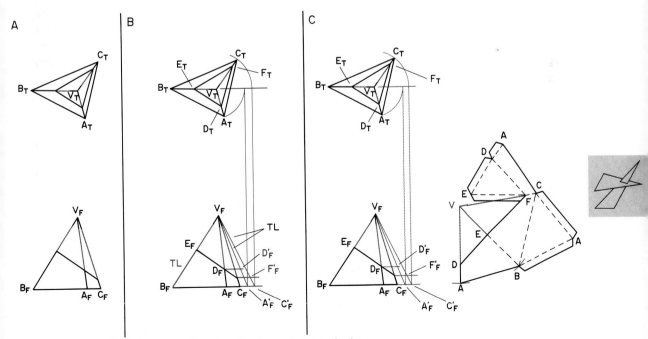

Figure 10.137 Development of pattern for truncated tetrahedron.

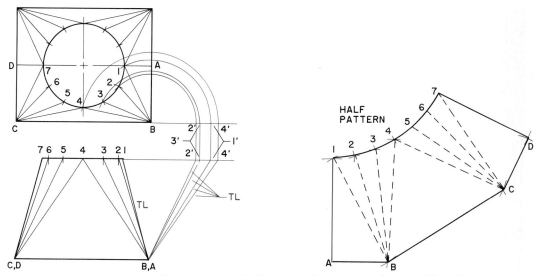

Figure 10.138 Half pattern development for transition piece.

10.7.6 Transition Pieces

There are occasions when one shape for tubing or ductwork must change to a different shape. One example of this is when a furnace's air outlet is round and the desired ductwork shape is rectangular. A piece of ductwork is needed to provide the transition from one shape to the other.

Figure 10.138 shows the steps for the development of a transition piece for a square to a circular cross section. The key here is to remember that corners are in effect part of a cone while the sides are triangles. A quarter of the pattern can be drawn and used four times to get the full pattern. In the example shown, half of the pattern was constructed. You can see the work that could be saved by clever planning.

10.7.7 Applications

Take time to observe the number of items that you use daily that are three-dimensional shapes that were created from patterns drawn on flat stock. Some examples are grocery sacks, french fry containers, waxed cardboard drink containers, ice cream cartons, and boxes for toys. The floor mats in cars, metal desks, bicycle fenders, wing section panels for airplanes, and some newspaper stands are also the product of development methods used in design (Figures 10.139 and 10.140).

Figure 10.139 Container pattern.

Figure 10.140 Container pattern.

10.8 *MAPS AND CONTOUR PLOTS*

One of the more important graphics topics for all engineers is the subject of maps and plans that depend on maps. Any construction project or transportation project involves looking at space or distances. If you have flown on an airplane or looked down at the ground from a high building or hill, you have seen what needs to be represented on paper. Figure 10.141 shows the view from an airplane. Your eyes have provided the information about what items were present and whether they had some height or altitude. The colors and textures also provided your brain with information about the presence of each of the items such as buildings, trees, streams, roads, bridges, and other items of interest.

Graphics provides the rules for representing the details of a particular section of terrain on paper and the symbology so that all items present can be correctly interpreted. It also provides the conventions for direction, distance or scale, and altitude, elevation, or height. Most of the critical information needed is provided on the map or in the margin around the map and is called the legend. Figure 10.142 shows a section of a geological survey map.

10.8.1 Map Introduction

When we look down from a high point, everything appears to be smaller than it really is. A two-dimensional representation or

Figure 10.141 View from an airplane. Courtesy of OSU Engineering Publications

Figure 10.142 Section from geological survey map. Courtesy of USGS

Figure 10.143 (*a*) Typical lot with house (1 in. = 20 ft); (*b*) typical subdivision (1 in. = 500 ft); (*c*) typical highway section (1 in. = 100 ft). Computer generated image.

Figure 10.144 An engineer's scale and a metric scale.

drawing must be much smaller than real objects, and we have to make very small distances represent very large distances. This is called scaling. Typical scales for maps might be 1 in. on the paper equals 1000 ft on the ground or 1 in. equals 1 mile or several miles. Figure 10.143 shows three portions of maps or plans at different scales.

Engineer's scales provide 1 in. divided into 10, 20, 30, 40, or 50 parts. The parts can represent 1 ft, 10 ft, 20 ft, 2000 ft, and so on. Metric scales are available and are used in Europe and Canada. Figure 10.142 shows a section of a contour map where the scale is 1:50,000. Figure 10.144 shows the ends of two scales. One is an engineer's scale based on the inch and the other is a metric scale.

Unless the sun is shining and we know what time it is and therefore where shadows are located, we do not know whether

UTM GRID AND 1971 MAGNETIC NORTH
DECLINATION AT CENTER OF SHEET

The 1976 MAGNETIC BEARING is 1°38' (29 mils) EAST of GRID NORTH.

ANNUAL CHANGE INCREASING 2.3'

GRID NORTH is 1°45 (31 mils) WEST of TRUE NORTH for centre of map.

Le REPÈRE MAGNÉTIQUE en 1976 est à 1°38' (29 mils) EST du NORD DU QUADRILLAGE.

VARIATION ANNUELLE CROISSANTE 2.3'

NORD DU QUADRILLAGE est 1°45' (31 mils) à l'ouest du NORD GÉOGRAPHIQUE au centre de la carte.

Figure 10.145 Declination diagram from USGS map and written information from Canadian map. Courtesy of USGS. This map is based on information taken from 52-I-14 © 1976. Her Majesty the Queen in Right of Canada with permission of Energy, Mines and Resources Canada

Figure 10.146 Compass points.

Figure 10.147 Bearing convention east or west of north or south.

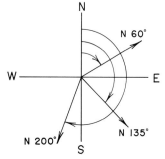

Figure 10.148 Azimuth convention angle to 360° clockwise of north.

a direction is north or south or some other compass point. With a compass, we can determine where north is. The convention or custom in drawing maps is to locate north at the top or near the top of the paper or map. North is indicated by an arrow with the letter N or the word *north*. Compasses can provide the location of magnetic north, but this is not the basis for map directions. We need to know where true north is, and this can be determined from magnetic north depending on where we are in the world. Figure 10.145 is a copy of a declination diagram from the U.S. Geological Survey (USGS). This figure also includes the written declination information from a Canadian map. The diagram provides the information on the difference between magnetic north and true north.

South is the opposite of north, obviously, and at the bottom of most maps. East is to the right and west to the left. Directions, when expressed as bearings, are always measured from north or south and are designated by the angle and whether it is east or west. Examples might be N 30 E or S 85 W. Another convention, called azimuth, is to assume that all directions are measured clockwise from north. Examples would be N 30 or N 330. Figures 10.146–10.148 show the compass points and the conventions for giving directions.

There are several different conventions for indicating direction depending on whether you are working for industry or the military or just going for a hike. However, the key is always to know where to find north.

A map drawn on a flat sheet of paper must provide the information about a section of the earth. The earth is not flat, and in order to represent the differences in height or elevation, map-makers have chosen to show lines of constant elevation or contour lines. Some or all of these lines can be labeled with the numbers that represent the elevation above sea level or just the elevation above or below some known point. See Figures 10.149 and 10.150.

Figure 10.149 Map without contours. Courtesy of AAA of Franklin County, Ohio

Figure 10.150 Map with contours. Courtesy of USGS

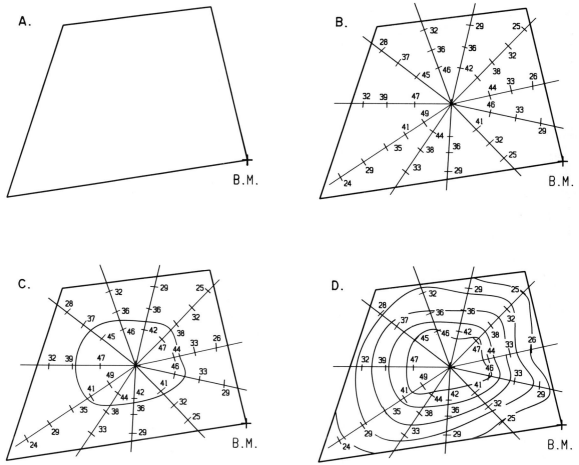

Figure 10.151 (*a*) Plot to be surveyed. (*b*) Drawing from surveyor's notes showing directions, distances, and elevations. (*c*) Single contour finished. (*d*) Completed contour map. Computer generated image.

These lines of constant elevation are determined by surveyors, people who know how to determine direction and measure distances accurately. They have used their instruments to determine the heights of specific points on the ground. This information is plotted on paper and then the contours are drawn in at constant elevations. Figure 10.151*a*, *b*, and *c* show the original area and preliminary drawing from a surveyor's notes used to create a contour map, and Figure 10.151*d* shows the map with the contour lines finished.

Symbols can be used to represent specific details. Rectangles can represent buildings. If the symbol for a building is modified with a flag, it represents a school. A cross on the rectangle symbolizes a church. Figure 10.152 shows the section of a map that represents a small town. Figure 10.153 shows a legend from a map giving the symbols for such features as a bridge, a dam, a railroad, and so on.

When maps are colored, the color can be used to represent a

Figure 10.152 Section showing small town from USGS map.
Courtesy of USGS

LANDMARK FEATURES	POINTS DE REPÈRE		
HOUSE, BARN	MAISON, GRANGE		
CHURCH, SCHOOL	ÉGLISE, ÉCOLE		
POST OFFICE	BUREAU DE POSTE		
HISTORICAL SITE	LIEU HISTORIQUE		
TOWERS: FIRE, RADIO	TOURS: FEU, RADIO		
WELL: OIL, GAS	PUITS: PÉTROLE, GAZ		
TANK: OIL, GASOLINE, WATER	RÉSERVOIR: PÉTROLE, ESSENCE, EAU		
TELEPHONE LINE	LIGNE TÉLÉPHONIQUE		
POWER TRANSMISSION LINE	LIGNE DE TRANSPORT D'ÉNERGIE		
MINE	MINE		
CUTTING, EMBANKMENT	DÉBLAI, REMBLAI		
GRAVEL PIT	GRAVIÈRE		

Figure 10.153 Legend from map. This map is based on information taken from 52-I-14 © 1976. Her Majesty the Queen in Right of Canada with permission of Energy, Mines and Resources Canada

variety of features. In some cases, color represents elevation on a map of a very large area. In other cases the colors represent land features. For example, blue may represent water; green, parks or vegetation; and black or gray, roads and bridges. Again, the legend on a map should describe the symbols or colors.

Any construction project starts with identification of the location for construction. The location is part of a larger area on a map. The information on a large map normally is not sufficient for construction, and surveyors go in and remeasure the boundaries to verify the deed. They also check elevations and draw a contour map or verify an existing one. Construction projects normally require changes to the ground either by removing soil for basements, adding soil to fill a low spot, or shifting soil on the location to level ground. This movement of soil costs money, and inaccurate maps or plans can cost the contractor if estimates are not made correctly.

When a map is used for a construction project, it is often referred to as the plan or plan view. That is, the plan is the same as the horizontal view or top view discussed earlier in the chapter on orthographic construction. Figures 10.154 and 10.155 show the contours of an area before and after road construction.

Although it might seem that maps and plans may not be important for engineering disciplines other than mining, civil, or petroleum engineering, they are the heart of many projects. For example, an industrial engineer involved in plant layout and design is interested in the plan view of the floors of a plant. The location of the sun relative to the building is important for calculating heat loads. See Figure 10.156. The relative elevations of the shipping and receiving departments as well as the manufacturing area can determine the power required to move materials. Gravity is very nice to have as the force to move raw materials or manufactured assemblies or subassemblies. All chemical processing plant designs require plan views of the plant area. These plan views are used to locate the piping and vessels that make up the plant. Here, as in the manufacturing plant, elevation can play an important role.

An engineer involved in the maintenance of a large facility must know how to use building plans to locate key features. Electrical engineers involved in the production and transmission of electrical power also get involved in drawing and creating plans for location of generation, conversion, and transmission hardware. Figure 10.157 shows a plan for an electrical substation. As engineers become managers, they may find themselves in the position of supervising personnel involved in the interpretation of plans and maps and in construction projects. They need to know how to read maps and make decisions based on information that is derived from maps and plans.

Figure 10.154 Section of map showing contours before road construction. Computer generated image.

Figure 10.155 Section of map showing new contours after road construction. Computer generated image.

Figure 10.156 Floor plan for office layout. Courtesy of American Electric Power Service Corporation

Figure 10.157 Electrical substation. Courtesy of American Electric Power Service Corporation

10.8.2 Concept of Representing Elevation

When discussing a map with contour lines in the preceding we talked about the fact that contour lines represent constant elevations. This would mean that you could follow the contour line around a hill and never go up or down. However, when you walk up the hill, if it is a steep hill, you get tired very rapidly. This means that you are crossing contour lines frequently or that they are close together. If, on the other hand, the walk up the hill does not tire you, the contour lines are further apart and the slope is rather shallow.

Figure 10.158 Contour map showing hill with steep and shallow slopes. Computer generated image.

PROFILE OF
STEEP SLOPE

PROFILE OF
SHALLOW SLOPE

Figure 10.159 Diagrams for rise and run showing portion of contour map for run and profile showing rise and run. Includes calculations. Computer generated image.

Study the contour map shown in Figure 10.158. Note the places that have contour lines close together and those that have contour lines far apart. These represent steep and shallow slopes. What does slope mean? It means that for every unit of distance that you go in a particular direction, you either go up or go down a certain distance. In other words, you are changing elevation. Slope is the change in elevation (rise) divided by the change in horizontal distance traveled (run). Figure 10.159 shows two different slope calculations. The slope is represented as a ratio and can be converted to an angle measured from the horizontal.

10.8.3 Sections or Profiles and Applications

As in orthographic drawings, there are times when we may need to look underground. Or we may need to plan for a road, dam, water supply system, or sewer system. In these cases, we have to draw a cross section of the earth along a particular direction. This is called a sectional view or, many times, the profile. The contours become parallel horizontal lines in the profile or sectional views. Figure 10.160 shows a portion of a contour map and a cross section through the ground where the cutting plane is shown.

Figure 10.160 Section of contour map (plan) and profile. Computer generated image.

Special paper is printed so that you can see the plan and the profile at the same time. However, whereas the scale in orthographic drawings of mechanical parts is the same in all three directions, the scales in plan–profile drawings are frequently different. That is, the plan, X and Y directions, is drawn to one scale, while the vertical distance in the profile is drawn at a greatly expanded scale. This allows more detail to be represented. Such detail may be the manholes in a storm sewer system or the relative sizes of pipes. Figure 10.161 shows a plan for a section of sewer for a residential development.

Another application of plan profiles is in road construction projects where earth must be removed, an operation called ''cut,'' or hauled in, an operation called ''fill''. In this case, there are two sections required. One is the profile that was discussed in the preceding, and the other is the typical cross section of the road down the center line. Normally, a typical cross section provides the cut and the fill plan. Because earth being removed generally is more stable than earth being filled in, the cut slope is normally steeper than the fill slope. When earth is being filled in, it must be compacted so that it will support the loads being applied. See Figure 10.162. Improperly compacted earth will not

Figure 10.161 Typical plan profile of sewer. Computer generated image.

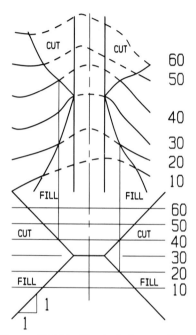

Figure 10.162 Typical plan profile for road construction. Shows typical cross section. Computer generated image.

support roads and the roads will sink. You may have noted this when you approach some bridges on the highway. The bridge requires pilings to be driven into the earth or special footings to be poured so that the structure will not move after construction. When the road sinks, there are large bumps just before and after the bridges. This is due to the settlement or sinking of the soil behind the bridge pier.

PROBLEMS

In the following problems, the grid divisions represented by the calibrated axes associated with each figure may be assigned any value per grid division. Suggestions: 1 grid unit = 0.20 in., 0.25 in., 5 mm, 10 mm. Label any true length lines, TL, edge views of planes, EV, and true size views of planes, TS.

Problems 10.1 through 10.6
Sketch the given orthographic views and the right profile view of each figure on grid paper, and color or crosshatch the principal planes in green, inclined planes in blue, and oblique planes in red. Use a double-width colored line for the edge views of planes.

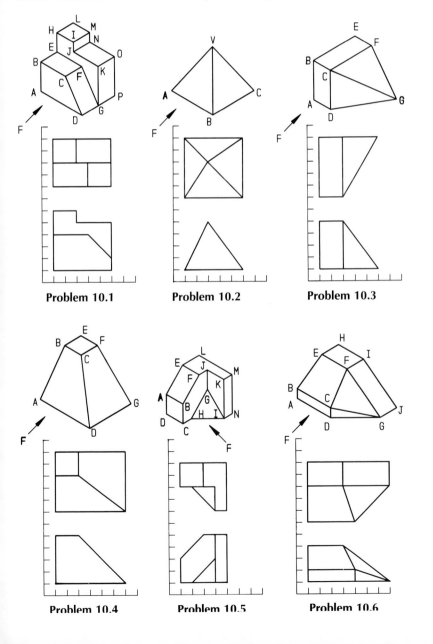

Problem 10.1   Problem 10.2   Problem 10.3

Problem 10.4   Problem 10.5   Problem 10.6

Problem 10.7
Referring to the figure of problem 10.2, draw the given orthographic views and find (or construct the necessary views to find) the true length, of:
> (a) line *VA*.
> (b) line *VB*.
> (c) line *VC*.

Problem 10.8
Referring to the figure of problem 10.3, draw the given orthographic views and find (or construct the necessary views to find) the true length of:
> (a) line *DG*.
> (b) line *CG*.

Problem 10.9
Referring to the figure of problem 10.4, draw the given orthographic views and find (or construct the necessary views to find) the true length of:
> (a) line *BA*.
> (b) line *CD*.

Problem 10.10
Referring to the figure of problem 10.6, draw the given orthographic views and find (or construct the necessary views to find) the true length of:
> (a) line EB.
> (b) line CG.
> (c) line FC.

Problem 10.11
Referring to the figure of problem 10.1, draw the given orthographic views and construct the necessary view(s) to find the true size of plane *CFGD* and label the edge view, EV, and true size view, TS.

Problem 10.12
Referring to the figure of problem 10.2, draw the given orthographic views and construct the necessary view(s) to find the true size of:
> (a) plane *VBC*.
> (b) plane *VAB*.

Problem 10.13
Referring to the figure of problem 10.3, draw the given orthographic views and construct the necessary view(s) to find the true size of:
> (a) plane *CFG*.
> (b) plane *CDG*.

Problem 10.14
Referring to the figure of problem 10.6, draw the given orthographic views and construct the necessary view(s) to find the true size of:
> (a) plane *CDG*.
> (b) plane *FGHI*.
> (c) plane *BEFC*.
> (d) plane *CFG*.

Problem 10.15
Referring to the figure of problem 10.5, draw the given orthographic views and construct the necessary view(s) to determine whether line *HG* is perpendicular to the line *GI*. If they are not perpendicular, determine the angle between them.

Problem 10.16
Referring to the figure of problem 10.3, draw the given orthographic views (you may draw just the given views of the points, lines, or planes specified rather than the entire figure in this and all succeeding problems) and construct the necessary view(s) to determine the shortest distance between:
    (a) *F* and line *CG*.
    (b) *E* and line *DG*.
    (c) *A* and line *CG*. Show the line representing the shortest distance in all views.

Problem 10.17
Referring to the figure of problem 10.6, draw the given orthographic views and construct the necessary view(s) to determine the shortest distance between:
    (a) *B* and line *DG*.
    (b) *B* and line *FG*.
    (c) *I* and line *CG*. Show the line representing the shortest distance in all views.

Problem 10.18
Referring to the figure, draw the given orthographic views and construct the necessary view(s) to determine the shortest distance between the following pairs of lines:
    (a) *AB* and *CD*.
    (b) *AB* and *HK*.
    (c) *EG* and *CD*.
    (d) *EG* and *HK*.
    (e) *CD* and *HK*.

(Optional: Show the line representing the shortest distance in all views.)

Problem 10.18

Problem 10.19

Referring to the figures, draw the orthographic views of the plane, the given view of the dashed line through the lettered point, and the lettered point in the adjacent view. Construct the adjacent view of the dashed line so that the line is parallel to the plane:

  (a) the first (*a*) figure and letter *A*.
  (b) the first (*a*) figure and letter *C*.
  (c) the second (*b*) figure and letter *E*.
  (d) the second (*b*) figure and letter *G*.

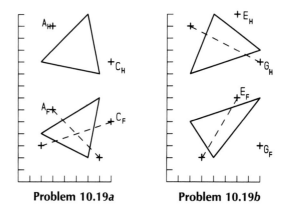

| Problem 10.19*a* | Problem 10.19*b* |

Problem 10.20

Draw the figure on linear grid paper and construct a plane *ABC* parallel to plane *XYZ*.

Problem 10.20

Problem 10.21

Draw the figure on linear grid paper and construct a plane *GHI* parallel to plane *STU*.

Problem 10.21

Problem 10.22
Referring to the figure, draw orthographic views of the plane shown and the point specified on linear grid paper. Construct the necessary view(s) to determine the shortest distance between the point and the plane. Draw the line representing that distance in the true length view and label the point where the line strikes the plane:
    (a) point A.
    (b) point B.
    (c) point C.

Problem 10.22

Problem 10.23
Referring to the figure of problem 10.5, draw plane HGI and point F in two orthographic views and construct the necessary view(s) to find the distance from point F to plane HGI.

Problem 10.24
Referring to the figure of problem 10.5, draw the two orthographic views of plane GHI and the line specified and construct the necessary view(s) to find the angle between the specified line and plane GHI:
    (a) line LM.
    (b) line MN.

Problem 10.25
Referring to the figure of problem 10.6, draw the two orthographic views of plane CFG and the line specified and construct the necessary view(s) to find the angle between the specified line and plane CFG:
    (a) line BF.
    (b) line HF.

Problem 10.26
Referring to the figure of problem 10.6, draw the two orthographic views of plane CFG and the plane specified and construct the necessary view(s) to find the dihedral angle between the specified plane and plane GHI:
    (a) plane $GFIJ$.
    (b) plane CDG.

Problem 10.27
Referring to the figure of problem 10.5, draw the two orthographic views of plane *GHI* and the plane specified and construct the necessary view(s) to find the dihedral angle between the specified plane and plane *GHI*:

    (a) plane *GIKJ*.
    (b) plane *ABCD*.

General Note: Label all points of intersection and show visibility in the following problems.

Problem 10.28
Referring to the figure, draw the two orthographic views of the prism *ABCDE* and the line *XY* and construct the necessary view(s) to find the intersection between line *XY* and the specified plane:

    (a) plane *ABC*.
    (b) plane *ADE*.

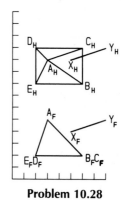

Problem 10.28

Problem 10.29
Referring to the figure, draw the two orthographic views of the prism *ABCDEF* and the line *ZW* and construct the necessary view(s) to find the intersection between line *ZW* and the specified plane:

    (a) plane *ABCD*.
    (b) plane *CDEF*.

Problem 10.29

Problem 10.30
Referring to the figure, draw the two orthographic views of the prism containing plane ABC and the line ST and construct the necessary view(s) to find the intersection between line ST and the plane ABC.

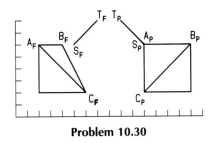

Problem 10.30

Problem 10.31
Referring to the figure, draw the two orthographic views of the prism containing KLM and the line VR and construct the necessary view(s) to find the intersection between line VR and the plane KLM.

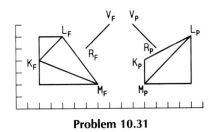

Problem 10.31

Problem 10.32
Referring to the figure, draw the two orthographic views of the prism $VABC$ and the plane KLM and construct the necessary view(s) to find the intersection between plane KLM and the plane(s) specified:
 (a) plane VAB.
 (b) plane VAC.
 (c) entire prism $VABC$.

Problem 10.32

Problem 10.33

Referring to the figure, draw the two orthographic views of the prism *ABCDEGH* and the plane *XYZ* and construct the necessary view(s) to find the intersection between plane *XYZ* and the plane(s) specified:

(a) plane *ECG*.
(b) plane *ABC*.
(c) plane *BEC*.
(d) entire prism *ABCDEGH*.

Problem 10.33

Problem 10.34

Referring to the figures, draw the two orthographic views of the prism and plane combination specified and construct the necessary view(s) to find the intersection between the prism and plane specified:

(a) prism *VABCD* and plane *STU*.
(b) prism *VKLM* and plane *PQR*.

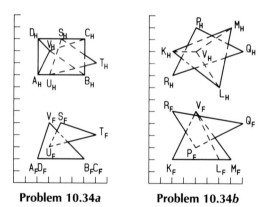

Problem 10.34a   **Problem 10.34b**

Problem 10.35

Referring to the figures, draw the given orthographic views of the figure specified and find the intersections between the line and the cylinder or cone and label each intersection point. Show visibility.

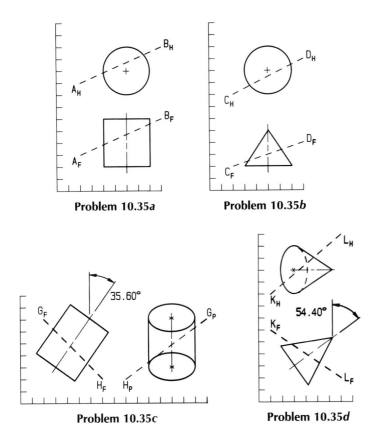

Problem 10.35a

Problem 10.35b

Problem 10.35c

Problem 10.35d

Problem 10.36

Referring to the figures, draw the given orthographic views of the figure specified and find the line of intersection between the plane and the cylinder or cone and label key intersection points.

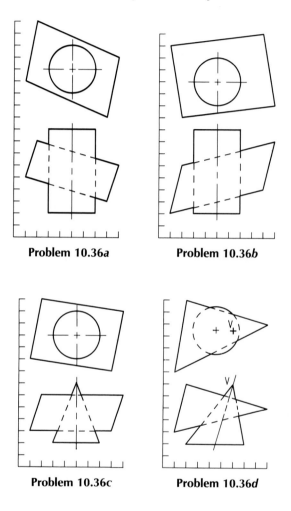

Problem 10.36a

Problem 10.36b

Problem 10.36c

Problem 10.36d

Problems 10.37 to 10.44
For the following figures, and the figure of problem 2, develop the inside full, half, or quarter patterns:

 (a) With the figure only with no tabs or bottom or top "lids."
 (b) With tabs for assembly without the bottom or top lids.
 (c) With the tabs and the bottom and top lids.
 (d) With tabs and lids and then lay out the pattern on heavy paper, cut it out, and assemble it.

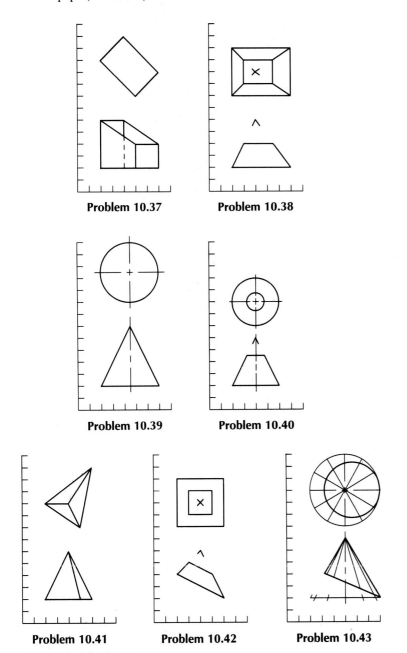

Problem 10.37   **Problem 10.38**

Problem 10.39   **Problem 10.40**

Problem 10.41   **Problem 10.42**   **Problem 10.43**

Problem 10.45

For the figure, write out the (*a*) bearing and the (*b*) azimuth for the vectors assigned relative to the north arrow:

(a) *A,C,E,H*.

(b) *B,D,F,G*.

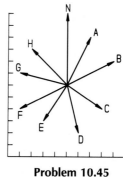

Problem 10.45

Problem 10.46

Copy the elevations in the figure to a piece of linear grid paper and draw the contours at 2 ft intervals.

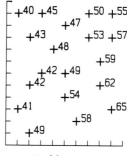

Problem 10.46

Problem 10.47

Copy the contour map shown in the figure to a piece of linear grid paper. Grid scale is 10 ft per division. Plot a 20 ft wide level roadway along the given centerline with a cut and fill of 1:1 at one of the following elevations as assigned:

    (a) 20 ft.
    (b) 30 ft.
    (c) 40 ft.
    (d) 50 ft.

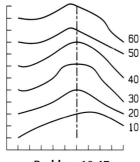

Problem 10.47

Problem 10.48

Copy the contour map shown in the figure to a piece of linear grid paper. Grid scale is 10 ft per division. Plot a 20 ft wide level roadway along the given centerline with a cut and fill of 1:1.5 at one of the following elevations as assigned:

    (a) 10 ft.
    (b) 20 ft.
    (c) 30 ft.
    (d) 40 ft.

Problem 10.48

TERMS YOU WILL SEE IN THIS CHAPTER

AXES   Scales along which data are plotted.

DATA POINT   Symbol used on a graph to show the location of a specific value relative to the axes.

GRAPH TITLE   Name of the chart or graph.

AXIS CALIBRATION   Units and markings on an axis that allows all data to be plotted.

AXIS TITLE   Name and units for the data to be plotted along the axis.

INTERCEPT   Where the curve crosses the Y axis at a specified X value, normally 0.

SLOPE   Angle that the line on a graph makes with the horizontal or vertical axis.

GRID   Coordinate system that allows consistent plotting of data points.

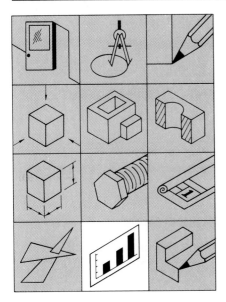

11

GRAPHICAL MATHEMATICS

11.1 INTRODUCTION

Graphical mathematics is the creation and use of charts and graphs to record and display data. It also includes the use of graphical methods to analyze the data. In the first part of this chapter you will study several types of charts and graphs and the methods used to create them. In the next part, you will learn about empirical equations and ways to determine whether the data on a graph can be represented by one of three common types of equations: linear, power, and exponential. Then in the third part, you will learn how to use graphical means to integrate and differentiate the graph of a function. Examples of graphical mathematics are shown in Figure 11.1.

$$\Sigma y = \Sigma f(x) dx$$

$$Y = f(x)$$

$$\frac{\Delta Y}{\Delta X} = \frac{\Delta f(x)}{\Delta X}$$

Figure 11.1 Sample illustrations for graphical mathematics.

11.2 CHARTS AND GRAPHS

Engineers use charts and graphs to present information in graphical or pictorial form. They know that information that is presented graphically is read and understood much more quickly by most people than the same information presented in words or tables of numbers. Either of the terms *chart* or *graph* can be used, although the former is generally used when referring to a map or presentation display and the latter when presenting technical information. This chapter will cover some typical types of charts and graphs and their uses.

11.2.1 Classes of Charts and Graphs

Charts and graphs can be classified according to their intended use. Some are used to provide technical information, such as the horsepower being developed by an automobile engine in a laboratory test, while others are used to provide general information, such as a report of weather conditions. Those used for technical purposes are generally prepared with more attention to detail and precision. Those used for general information purposes may provide less detail but more attention may be paid to readability and appearance. Additional detail can make the chart or graph more difficult to read.

11.2.2 Typical Uses of Charts and Graphs

One of the many ways in which charts and graphs are used is in the analysis of data. Figure 11.2 is a graph of U.S. foreign trade from 1965 to 1985. Both total exports and total imports are plotted at five-year intervals over the period. By looking at the graph, it is easy to determine (1) the years in which exports exceeded imports, (2) the years in which imports exceeded exports, (3) whether exports are increasing or decreasing relative to imports, and (4) other important information such as the amount by which exports and imports differed in any year. Because data are plotted at five-year intervals, it is not possible to tell the exact year when imports began to exceed exports. Plotting data points for every year would aid in the analysis of the data.

Figure 11.3 is a graph from a recorder attached to an environmental chamber. It shows the temperature and the humidity of the process on a continuous basis as a function of the time of day. This graph tells at a glance whether the chamber is maintaining the temperature and humidity within acceptable limits or whether adjustments need to be made.

We have all seen weather maps on television. These maps are a form of chart. They are used to present a wide variety of weather

Figure 11.2 Example of line graph.

Figure 11.3 Strip chart from recording instrument. Courtesy of Ohio State Engineering Publications

information that would be nearly impossible for the weather forecaster to present in words. Examples of other commonly used maps are road maps and survey maps showing land parcel locations and uses.

Other ways in which charts and graphs are used to display

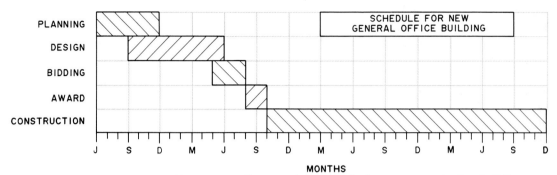

Figure 11.4 Construction schedule for new general office building.

Figure 11.5 Pie chart showing family budget.

information include using them to show schedules (Figure 11.4), to describe the makeup of something such as a budget (Figure 11.5), and to show trends such as in the stock market (Figure 11.6).

11.2.3 Types of Graphs and Their Uses

Line Charts

Line charts relate one quantity to another. For instance, a sales chart might show the sales of an item with respect to time (Figure 11.7). The type of line chart with which we may be most familiar has rectilinear or linear coordinate scales along its **axes,** that is,

Figure 11.6 Line graph showing trend of stock market from July 8 through August 4, 1987.

Figure 11.7 Line chart: Sales of item over period of time.

scales with even spacing between units, as represented by Figure 11.7. Some line charts used for data analysis or problem solving may use a logarithmic scale on one axis and a rectilinear scale on the other axis (Figure 11.8). These are often referred to as semilogarithmic, or semilog, scales. In other cases they may use a logarithmic scale on both axes (Figure 11.9). These are referred to as logarithmic, or log–log, scales. Applications where semilog or logarithmic scales would be used will be covered later in this chapter.

Multiple-Line Graphs

Many times it is desirable to plot two or more different relationships on the same line graph. In that case the plotted points are marked with different symbols for each relationship, and different line styles or colors are used for the lines (Figure 11.10).

Multiple-Scale Graphs

Sometimes, when we wish to plot more than one relationship on a single graph, we find that we cannot use the same scale for both relationships. For instance, we may wish to plot both the acceleration and velocity of a cam follower as a function of time but find that the values are so different that different scales are needed. The problem can be solved by creating two scales on the vertical axis (Figure 11.11). The two scales can be placed on the same side of the graph, but it is often more convenient to put one on the left side and the other on the right, as shown in Figure 11.11.

Figure 11.8 Example of graph that uses logarithmic scale along ordinate and arithmetic scale along abscissa: Semilogarithmic graph.

Figure 11.9 Example of graph that uses logarithmic scale along both axes: Logarithmic graph.

Figure 11.10 Multiple-line graph.

TIME AND ACCELERATION
OF A CAM FOLLOWER FOR
CAM ROTATING AT 60RPM

Figure 11.11 Multiple-scale graph.

Pie Charts

Pie charts, as shown in Figure 11.5, show the relative size of each part of a whole. A pie chart is drawn as a circle that represents the entire item, such as a budget or the cost of a manufactured product. The circle is then divided into sectors (Figure 11.12) to show each component that makes up the total and how much each component contributes to the total. Finally, the chart and each of its sectors are labeled. A typical use for a pie chart would be to show what portion of the total weight of an automobile is steel, plastic, rubber, glass, and other materials. Another typical use is to show how much of a budget is used for each of the spending categories in the total. Pie charts can display absolute amounts (Figure 11.13), percentages (Figure 11.14), or both.

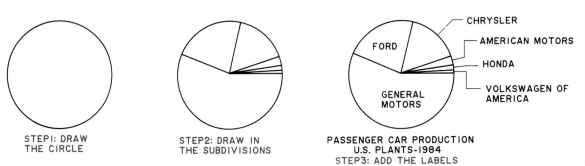

STEP1: DRAW
THE CIRCLE

STEP2: DRAW IN
THE SUBDIVISIONS

PASSENGER CAR PRODUCTION
U.S. PLANTS-1984
STEP3: ADD THE LABELS

Figure 11.12 Step-by-step construction of pie chart.

Figure 11.13 Pie chart that displays absolute amounts.

Figure 11.14 Pie chart that displays percentages.

Polar Charts

A type of line chart not often seen in presentations or reports but often encountered in equipment monitoring and control applications is circular and uses polar coordinates (Figure 11.15). This

Figure 11.15 Example of chart that uses polar coordinates. Courtesy of Ohio State Engineering Publications

Figure 11.16 Recording instrument that draws polar charts. Courtesy of Ohio State Engineering Publications

type of chart is often used when one of the variables being plotted is time. A recorder that operates similar to a clock turns the chart and one or more pens draw the line(s) (Figure 11.16). Another use for polar graphs is to represent intensity levels or patterns such as illumination patterns from a light source or radiation patterns from a radio or television broadcasting station.

Bar Charts

Bar charts can be classified as either vertical bar, often called column charts, or horizontal bar charts. Vertical bar charts are often used to present the magnitudes of quantities. (Figure 11.17). Horizontal bars are more often used when periods of time are being presented or when the lengths of the bars are too great to draw them vertically (Figure 11.18).

More than one relationship can be plotted on the same graph by shading or crosshatching the bars (Figure 11.19) or by drawing multiple bars (Figure 11.20). Note that Figure 11.19 shows the total sales from all product lines as the total height of the bar. Figure 11.20 shows more clearly the relative contribution of each of the three product lines to total sales.

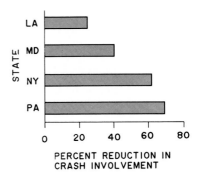

Figure 11.17 Vertical bar chart.

Figure 11.18 Horizontal bar chart. Courtesy of The Insurance Institute for Highway Safety

Figure 11.19 Vertical bar chart that shows more than one relationship using multiple shading of bars.

Figure 11.20 Vertical bar chart that shows more than one relationship using multiple bars and shading.

Figure 11.21 Surface chart.

Surface Charts

Surface charts are drawn in a manner similar to line graphs. However, the area under the line is shaded to represent a surface or area. If the chart contains more than one line, the areas between each of the lines are shaded differently (Figure 11.21). The shading helps the viewer to determine the relative size of each shaded area. Since the exact area of each shaded surface is difficult to determine, surface charts are not generally intended to be used for careful analysis of data but rather to give an overall impression or feel for the information.

Three-Coordinate Charts

Three-coordinate charts, often called volume charts, are used when it is necessary to relate three variables on one chart. The third coordinate is plotted by adding a third axis (the Z axis) to the chart (Figure 11.22). Volume charts are generally difficult to construct by manual methods but are easy to do on a computer. As a result, most three-coordinate charts are computer generated (Figure 11.23). Since each **data point** has X, Y, and Z coordinates, the chart can be displayed in any of several three-dimensional forms: oblique (Figure 11.23), dimetric (Figure 11.24), trimetric (Figure 11.25), or isometric (Figure 11.26).

Trilinear Charts

The trilinear chart is a valuable tool used in the fields of chemical engineering, metallurgy, and agriculture. It can be used to analyze

A THREE COORDINATE (X,Y,Z) PLOT

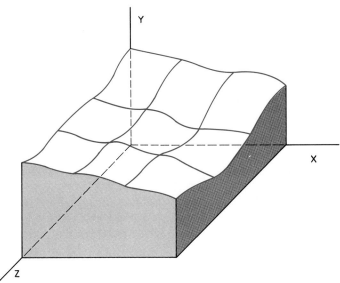

Figure 11.22 Three-coordinate chart or volume chart produced by manual drawing methods.

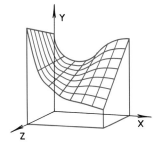

Figure 11.23 Three-coordinate chart or volume chart produced by computer. Computer generated image.

Figure 11.24 Computer-generated dimetric three-coordinate chart. Computer generated image.

Figure 11.25 Computer-generated trimetric three-coordinate chart. Computer generated image.

mixtures containing three variables. For example, chemical compounds consisting of three elements can be studied using a trilinear chart.

The chart is formed by drawing an equilateral triangle. The altitude of this triangle represents 100% of each of the three variables that make up the compound, mixture, or alloy. Using this type of chart depends upon the geometric principle that the sum of the perpendiculars to the sides of an equilateral triangle from any point within the equilateral triangle is a constant and is equal to the altitude of the triangle. Figure 11.27 is an example of a trilinear chart.

11.2.4 Plotting a Graph

Selecting a Size for a Graph

When preparing to plot a chart or graph, it is important to do some planning before putting any marks on the paper. First, the amount of space surrounding the plot itself should be considered. Most charts employ a horizontal axis (the abscissa) and a vertical axis (the ordinate). These axes will need to be labeled with the name of the quantity being recorded and the units being used. They will also need to be graduated and calibrated (Figure 11.28). It is important to reserve enough space for those purposes. A **graph t.** will also be needed to identify the purpose of the chart

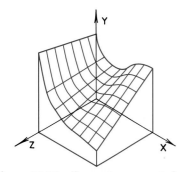

Figure 11.26 Computer-generated isometric three-coordinate chart. Computer generated image.

Figure 11.27 Trilinear chart. Courtesy of The USDA-Soil Conservation Service

Figure 11.28 Important elements to consider when constructing graphs.

and important information regarding it (Figure 11.28). If the chart title will be outside of the plotting area, space will need to be reserved for it also. Many times the chart title is placed in an unused part of the plotting area.

Selecting and Calibrating a Scale

When the approximate location and length of each axis line has been determined, it is necessary to work out a scale so that the axis line can be subdivided into increments. The first step is to determine the range of data values to be recorded for that axis. The scale is normally chosen so that the final plot uses nearly the entire plotting area. That is because it is easier to read a large plot than a small one and because the large plot presents a better appearance. To choose a scale, you must first determine the maximum and minimum values that you need to plot. It is important that the scale be such that the plotted points can all be contained within the plotting area.

After choosing the scale, it is necessary to decide what size increments to use, that is, how many subdivisions to have (Figure 11.29). The increments should be chosen for maximum readability. Too many increments will make the axis markings appear crowded and difficult to read. Too few will make it difficult to visually determine the values of the plotted points. The increments should be chosen so that the values at the increment points are even numbers. For instance, a scale of 0, . . . ,100 might be subdivided as 0, 20, 40, 60, 80, 100 or 0, 10, 20, 30, 40, 50, 60, 70, 80, 90, 100 but not 0, 12.5, 25, 37.5, 50, 67.5, 75, 87.5, 100 (Figure 11.29). It is not necessary or even desirable to show the value of each subdivision on the axis. For example, a scale with subdivision markings every 5 units might have values shown every

Figure 11.29 Examples of correct and incorrect scale calibrations.

$$m = \frac{L}{f(V_{max}) - f(V_{min})}$$

$$l = m(f[v] - f[V_{min}])$$

Figure 11.30 Formulas for computing scale modulus and scale length when calibrating scales.

20 units. As a rule, try to use increments that are some multiple of 2, 5, or 10.

Before we can place the tick marks, or **axis calibration** markings, on the scale, we must determine where each one is to go. To do this, we first calculate a scale modulus. The scale modulus is a factor that relates the length of the scale in inches or millimeters to the range of the data. It is calculated as the scale length divided by the scale range. The scale range is the difference between the maximum scale value and the minimum scale value:

$$m = (\text{scale length})/(\text{maximum} - \text{minimum})$$

If we think of the data as being represented by a function $y = f(v)$, then

$$m = (\text{scale length})/[f(v_2) - f(v_1)]$$

where $f(v_2)$ is the maximum scale value and $f(v_1)$ is the minimum scale value (Figure 11.30).

When the scale modulus has been calculated, the location of any data point v on the scale can be determined by rearranging the preceding equation to read

$$\text{Length} = m[f(v) - f(v_1)]$$

where length is the distance in inches or millimeters from the beginning of the scale to the location of v.

This method of calibrating a scale can be used for all scales, nonlinear as well as linear. The only requirements are that the function $f(v)$ is known and that it is continuous over the range of the scale. Figure 11.31 shows the use of the method to calibrate a logarithmic scale along the ordinate axis for a range of values from 1 to 10. In this case $f(v) = \log(10)v$. Note that $f(1) = 0$ and $f(10) = 1$. Therefore, $m = (\text{scale length})/(1 - 0)$. If the scale length is 6, then $m = 6/(1 - 0) = 6$. For any value of v, the distance from the start of the scale is length $= 6 \, [f(v)]$. For example, the distance to the calibration point for $v = 4$ is length $= 6[\log 10(4)] = 6(0.602) = 3.612$ in.

You can also calibrate a scale using graphical methods. Chapter 2 discusses a method in which you can divide a line into a required number of parts. The scale along the abscissa in Figure 11.31 was calibrated using this graphical procedure.

When more than one relationship is shown on a single chart, it is necessary to mark the points for each relationship differently so that there will be no question of which data points belong to which relationship. If the chart is to be done in color, different colors can be used. Otherwise different symbol types should be

Figure 11.31 Using formulas to calibrate logarithmic scale for range of values from 1 to 10 along ordinate. Using graphical procedure to develop arithmetic scale for range of values from 1 to 10 along abscissa.

used. The symbol types should be as different as possible. A chart might use circles, squares, and triangles to plot three sets of relationships. Use of circles and polygons with more than five sides on the same chart should be avoided because they might look too much alike.

The lines or curves connecting the points should also be different for each relationship. Again, color can be used effectively to set them apart. Line style can do the same thing. A chart showing three relationships might use a solid line for one, a dashed line for the second, and a dotted line for the third (Figure 11.32).

A question that is frequently asked is whether the line connecting the plotted points should be a continuous smooth curve or a series of straight-line segments. The answer depends on the nature of the data that is being graphed. If it is known that the relationship would provide a smooth curve if you plotted enough points, then draw a smooth curve. However, if the data are only available as specific points and there is no information about the relationship between the plotted points, then straight lines should be used.

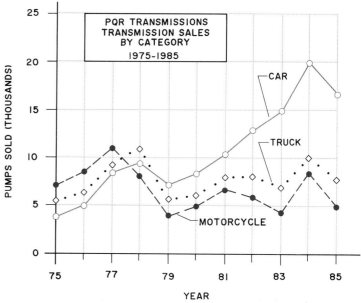

Figure 11.32 Multiple-line graph showing three relationships using different line styles to set them apart.

11.2.5 Modeling Empirical Data (Data Scatter)

Best-Fit Methods

Much of the engineering data that are plotted result from measurements taken in the laboratory or in the field. These data are called empirical data, meaning that they are measured rather than computed. Such data are often not precise due to inaccuracies in the measurement procedure or the equipment being used or due to uncontrolled changes taking place in the process being measured while the data are being taken. As a result, a plot of the data may show an erratic relationship known as data scatter (Figure 11.33). When it is known that the data should conform to some known relationship, then an attempt is usually made to fit a smooth curve describing that known relationship.

Many different methods have been developed to fit known curve forms to empirical data. One of the best known is the method of least squares. A curve of the known shape is drawn in a position that minimizes the sum of the distances or "errors" between the plotted points and the curve. Since all error values must be positive, the square of each error is used to avoid the problem of "negative errors," that is, points where the mathematical difference between the plotted point and the curve is a negative value (Figure 11.34).

Figure 11.33 Scatter diagram.

Figure 11.34 Least-squares fit of curve through experimental data.

Types of Data

Empirical data may be taken from either a discrete process or a continuous process. Discrete processes are those that produce data at certain times. The Yellowstone National Park Geyser Old Faithful only erupts about once an hour, so data regarding the eruption can only be taken at those times. Electrical current, on the other hand, flows continuously, so data regarding voltage, current, waveform, and so on, can be taken on a continuous basis. Discrete data would not be plotted using a smooth curve. The multiple-line graph shown in Figure 11.32 is an example showing how discrete data should be plotted.

11.2.6 Adding Titles and Labels

Chart/Graph Title

In order for the plotted data to be read and understood, it is necessary to complete the chart with descriptive information. A chart title is needed. The chart title should be placed in a box along with notes stating what the graph is for, dates, locations, and other identifying information. Other notes may be included to help explain the chart. The chart title should answer the questions who, what, where, when, how, and why.

Labeling the Axes

If the chart is a line or bar graph, then the axes need titles. Each **axis title,** or legend, should state the variable that is being plotted for that axis and the units of measurement being used (Figure 11.35). The vertical axis labels may be lettered sideways on the sheet to read from the right side or they may be computer plotted one below the other in an upright position (Figure 11.36). The

Figure 11.35 Axes labels show variable being plotted as well as units of measurement being used. Ordinate labeled sideways on sheet so it can be read from right side.

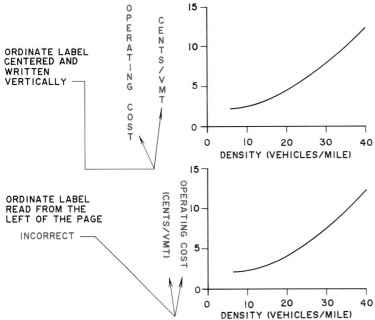

Figure 11.36 Example of ordinate axis labeled so it is read vertically. Letters printed one below the other in upright position. Axis labels are never read from left of page.

letters should never be printed to read from the left side of the sheet (Figure 11.36).

11.2.7 Creating Charts by Computer

You can draw pie, line, and bar charts with CADD systems. However, CADD systems are not designed for that purpose, and the process is not very efficient. There are many specialized software packages for creating charts and graphs. They fall into two general categories. The first is statistical packages. Statistical packages will accept your data, perform various statistical analyses, and plot both the data and the results of the analyses. The second category is presentation graphics packages. These do not generally do any data analysis. They will accept your data and create the plot. Both the statistical and presentation packages will do most or all of the work of determining scales, drawing, calibrating and labeling axes, and plotting the data. You may be given some choices regarding locations of titles and other drawing features, but the system will make most of the decisions.

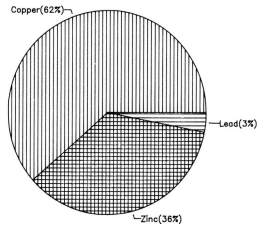

COMPOSITION of FREE CUTTING BRASS

Copper(62%)

Lead(3%)

Zinc(36%)

Figure 11.37 Computer-generated pie diagram. Computer generated image.

Creating a Pie Chart with CADD

As with any drawing to be done with a CADD system, the first step is to make a rough sketch on paper. Do the finished drawing on the computer. Start by setting up the work area so that the origin is in the center of the screen. Draw a circle centered at that origin. The circle should be as large as possible while still allowing enough space to place the needed text information around it. Next, draw a horizontal line from the center to the right edge of the circle. Working counterclockwise, draw in the rest of the dividing lines by starting each line at the circle's center (the origin) and using your system's distance and angle features to locate the other end of the line. The included angle for each segment of the pie will be the percentage for that segment divided by 360, and the angle for each line will be the sum of the angle for that segment and the angle of the previous line. When the pie is complete, label the segments and add the chart title using the system's text features. A pie chart generated by the computer is shown in Figure 11.37.

Creating a Line or Bar Chart with CADD

Start by studying the data and determining scales for the axes. Make a rough sketch of the chart and place the axis labels and chart title. Start a new drawing on your CADD system and locate the origin where you would like the X and Y axes to meet. This will generally be about one-quarter of the way up from the bottom and one-quarter of the way in from the left side of the screen. Select a drawing scale that will allow you to draw and calibrate the Y axis so that most of the vertical dimension of the screen is used. Move the origin slightly if necessary to accommodate the X- and Y-axis scales. Note that you will not be able to apply separate drawing scales to each axis, so the X-axis scale will probably not match your drawing scale. Typically, the Y-axis values are more critical, and that is why the scale should be set to accommodate the Y axis.

Draw in the two axis lines from the origin, and using the computer's drawing scale as a guide, calibrate the Y axis, drawing tick marks for the divisions. Now, calibrate the X axis. If the X axis will have numerical calibrations, you will need to select a location for the largest value on the axis and then calculate a conversion factor to locate the other values. The conversion factor is obtained by dividing the system's X-coordinate value for the location of this point by the chart value. Then, determine the system's value for each additional calibration point by multiplying the chart value by the conversion factor. Draw tick marks at the calibration points. Use the text feature to label the calibration points on both axes.

Plot the data on your chart by moving the cursor along the X axis to the approximate X location (determined by eye from the scale calibrations) and then moving the cursor vertically until the Y location is reached. Watch the system's coordinate readouts to find the correct Y location. Draw a symbol at the data point. When that data is plotted, select the line drawing mode and connect the points with lines of the desired color and style. If you are creating a bar chart, you may wish to eliminate the point plotting and simply draw rectangles from the X axis up to the Y value for the data point. Again, having set the drawing scale to the chart scale will make finding the correct Y location very easy. Figure 11.38 shows a computer-generated line graph.

11.3 *EMPIRICAL EQUATIONS*

11.3.1 Definition of Empirical Equations

Empirical equations are equations that are developed from observing natural phenomena or from data obtained through controlled laboratory experiments. These equations are useful in the

Figure 11.38   Computer-generated line graph. Computer generated image.

Figure 11.39   Pressure distribution due to wind on highway sign.

design and analysis of structures, machines, and electrical and electronics equipment. For example, the pressure on a traffic sign can be related to the velocity of the wind by an empirical equation. The pressure is carefully measured using an instrument called a pitot tube in a controlled experiment in a wind tunnel. The pressure data are observed as the velocity of the wind is varied in a controlled manner. The experimenter can develop an equation to calculate the pressure on the sign based on the velocity of the wind (Figure 11.39). This information helps engineers to design signs that resist wind damage and last longer. This section will relate the types of empirical equations that engineers use and the methods used to develop them.

11.3.2   Types of Empirical Equations

Empirical equations generally are classified into one of three categories: linear, power, and exponential. The type of equation that is developed depends on the shape of the curve obtained when the data from the controlled experiment are plotted on rectangular graph paper.

Linear Equations

An equation is linear if the curve through the data points is a straight line on rectangular graph paper. A linear equation takes the form

$$y = mx + b$$

where x is the independent variable, y is the dependent variable, m is the **slope** of the line, and b is the location of the **intercept** of the line and the y axis of the graph.

Before proceeding, a discussion of independent and dependent variables is in order. The independent variable x is the variable that is controlled in the experiments. The dependent variable y is not controlled but is measured as the independent variable is changed. For example, the length of a metal rod will increase as its temperature increases. By measuring the length of the rod as its temperature is increased, data that relates these variables can be obtained (Figure 11.40). The temperature can be controlled by the experimenter; therefore, it is the independent variable. The length of the rod is the variable that is being measured as the temperature increases; therefore, its magnitude depends on the temperature and is the dependent variable (Figure 11.41). The

Figure 11.40 Linear expansion of steel rod due to rise in temperature.

Figure 11.41 Relationship between change in temperature and change in length of steel rod. Change in length is dependent variable and change in temperature is independent variable.

temperature as the independent variable is plotted along the x axis, or abscissa, of the graph. For each temperature observed, a length of the rod is recorded. The change in length of the rod is the dependent variable and is plotted along the y axis, or ordinate, of the graph (Figure 11.41). The dependent variable "depends" on the independent variable for its magnitude. For example, as the independent variable changes, the dependent variable also changes, which forms the relationship that is described by the equation. The relationship between temperature and change in length of a metal rod is linear (Figure 11.41). An equation that describes this relationship is a linear equation.

Figure 11.41 shows a straight line drawn through the data. The data is said to have "scatter" as a result of instrument error or human error in measuring the independent and dependent variables. The straight line drawn on the graph represents a "best-fit" curve of the relationship between the variables. This "fit" can be done using judgment or it can be done using a mathematical approach that is easily analyzed by a computer, as described in Section 11.2.5. The mathematical procedure is called a least-squares fit. Several sophisticated software packages are commercially available that will do this analysis.

Power Equations

When the curve through the plotted data is not a straight line on rectangular graph paper, the equation form might be either the power form or the exponential form. It may be neither, in which case a more sophisticated means for equation development must be used. These methods are beyond the scope of this book. The power equation is the equation for a hyperbola or a parabola (Figure 11.42). When the exponent of the equation is positive, the curve is a parabola. When the exponent is negative, the curve is a hyperbola with the asymptotes being the coordinate axes.

The empirical equation is difficult to develop graphically when the curve is not linear. The parabolic curve plotted using rectangular coordinates in Figure 11.42 shows this. The curve can be rectified or made linear by replotting on logarithmic graph paper. Logarithmic paper is a type of graph paper that has a logarithmic scale for the abscissa and ordinate. The log scale, as it is called, can be one cycle (divisions from 1 to 10) or more than one cycle in length (two or more divisions from 1 to 10). A two-cycle scale is read by beginning at the first cycle (the origin). The number 1 can be 1, 10, 100, 1000, ... It also can be 0.1, 0.01, 0.001, 0.0001, ... Its value determines the scale for the entire axis. For example, suppose the value of the origin for the abscissa were 1.

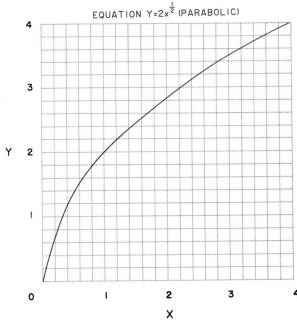

Figure 11.42 Rectangular plot of parabola. Power equation is form of empirical equation that defines parabola or hyperbola.

The first cycle (1, . . . ,10) would indicate values from 1 to 10. The second cycle would indicate values from 10 to 100. If a third cycle exists, it would indicate values from 100 to 1000. Each cycle is increased by a factor of 10 from the previous cycle. When a parabolic curve is plotted on logarithmic graph paper, it will appear as a straight line or is said to be rectified (Figure 11.43). From this logarithmic graph, the power equation can be developed. The power equation is of the form

$$y = bx^m$$

where m is the slope of the rectified curve and b is the y intercept.

The dynamic pressure on a traffic sign discussed previously is an example of the power equation. This equation relates the dependent variable P, the pressure on an area perpendicular to the direction of the wind in pressure units of pounds per square foot (pascals), with the independent variable V, the velocity of the wind in units of feet per second (meters per second). The resulting equation in English units is

$$P = 0.0036V^2$$

This equation is based on the units for wind velocity being feet

Figure 11.43 Logarithmic plot of parabola. Curve has been rectified or is straight line as logarithmic plot.

per second. You can see that this equation was derived through empirical means rather than mathematical means. The value of the intercept (0.0036) is the key in identifying this equation as an empirical equation.

Exponential Equations

Curves of exponential equations look similar to curves of power equations when they are plotted on rectangular coordinates. Sometimes it is very difficult to see any difference between these curves. Figures 11.44 and 11.45 show typical examples of parabolic curves and exponential curves, respectively, on rectangular coordinate graph paper. A rectangular plot of an unknown curve can be compared to the curves in the figures to give a clue as to whether the data fits the power form of the equation or the exponential form of the equation. The exponential curve will approach the x axis or abscissa as an asymptote. The power curve will not be asymptotic if it is a parabola; however, the hyperbola is asymptotic with the coordinate axes. If the curve appears to be exponential in nature, it can be rectified by plotting the data on semilog paper. The resulting equation will have the form

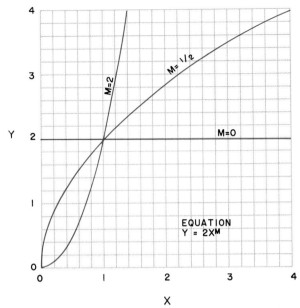

Figure 11.44 Typical examples of power equation curves plotted using rectangular scales.

Figure 11.45 Typical examples of exponential curves plotted using rectangular scales.

$$y = b(10)^{mx} \quad \text{or} \quad y = b(e)^{mx}$$

where again b is the y intercept and m is the slope. The value in parentheses depends on the logarithm base desired. The value 10 is used for base 10 logarithms while e (2.718 . . .) represents the base of the Naperian logarithm system. If the base is changed, the value of the slope will also change.

11.3.3   Typical Uses for Empirical Equations

Empirical equations are used to determine relationships between variables when the data have been measured during experiments. In these situations, there are two possibilities. The first is that a particular set of data is being measured where you know the type of relationship, the appropriate **grid** for plotting, and the equation that applies. The goal in these cases is to find the equation parameters so that they can be used in other work. The second case is where you have a set of data and you would like to know the relationship between the independent and dependent variables.

In courses such as strength of materials or mechanics of materials, data are taken during tensile tests to determine the modulus of elasticity, the yield strength, and the ultimate strength. A tensile test is defined as a test in which a specimen (a cylindrical rod of steel, aluminum, or other material) is pulled by a machine that records the applied load and the amount the specimen elongates. The specimen is pulled with increasing loads until it breaks. A graph can be plotted from the data gathered during the test (Figure 11.46). The modulus of elasticity of the material is measured as the slope of the linear portion of the graph. When the loads are such that a linear curve is produced, the specimen will react as a rubber band and recover or snap back to its original length after the load is removed. The yield strength for the material is the point on the graph where the curve is no longer linear but begins to curve. At this point, the specimen will be permanently deformed or elongated if the load is removed. It will not recover or snap back to its original length. The ultimate strength of the material is the highest point on the curve. All of these points can be seen in Figure 11.46. Figure 11.47 shows the graph of an actual tensile test of a steel rod. In this case the intercept is always known because it is zero, but the slope of the linear portion of the curve provides critical information about the modulus and the ultimate strength. A rectangular grid is used for plotting the data; therefore, the linear portion of the curve can

Figure 11.46 Typical curve showing relationship between applied load and elongation of metal bar during tensile test. Computer generated image.

Figure 11.47 Plot of laboratory data taken during tensile test of steel rod. Computer generated image.

Figure 11.48 Rectangular plot showing bearing wear measurements. Computer generated image.

Figure 11.49 Same data as in Figure 11.48 rectified using semilogarithmic coordinates. Computer generated image.

be used to develop an empirical equation relating the modulus of elasticity (the slope of the linear portion of the curve) with parameters based on the applied load to the specimen and elongation of the specimen.

Another example where empirical equations can be used involves bearing life. The plot of bearing life in hours versus the bearing operating temperature shows a nonlinear relationship (Figure 11.48). Here we want to know what the relationship is so that we can determine the bearing life at any operating temperature and be able to predict when bearings should be replaced. Either the equation that can be derived or the plot itself can be used to give us this information. The data for a bearing are shown on rectangular and on semilog grids (Figures 11.48 and 11.49).

A third example deals with water flowing in pipes. The pressure decreases in a pipe the further the water has to travel. When the velocity of the water flowing through the pipe is increased, the pressure loss becomes more severe. This relationship also is dependent on the diameter of the pipe. Engineers are interested in having a plot of this data so that they can extrapolate values, if necessary, to predict what would happen in cases outside those for which the experimental data has been gathered. Engineers are also interested in plotting the data so that they can make decisions about the size of pipe to use given the amount of water that must be delivered to the customer. This information would be plotted on a logarithmic graph. Figures 11.50 and 11.51 show the data for two pipes on rectangular grid paper and on logarithmic grid paper, respectively. From these plots an empirical equation can be developed as an aid in pipe design.

Figure 11.50 Rectangular plot showing head loss as water flows through pipe at various velocities. Computer generated image.

Figure 11.51 Same data as in Figure 11.50 rectified using logarithmic coordinate scales. Computer generated image.

11.3.4 Solutions of Empirical Equations

Empirical equations are generally solved using one of three methods. These methods are the slope-intercept method, the method of selected points, and the method of averaging. The slope-intercept method and method of selected points are graphical in nature, while the method of averaging is based on the values of the experimental data and not on a graph of the relationship between the variables. The slope-intercept method and method of selected points will be emphasized in the solutions of empirical equations. The method of averages will not be presented in that it is not a graphical method.

Empirical equations are generally solved using a step-by-step approach. The steps involved in the solution of any empirical equation are outlined in what follows.

1. Plot the experimental data on rectangular graph paper.

2. If a straight line can be drawn connecting the data points, proceed with the solution as described in step 4. When a significant amount of scatter in the experimental data is shown, a best-fit curve should be used or the least-squares method can be used.

3. If a curve is required to connect the data points on the rectangular graph, then compare the curve to the curves in Figure 11.44 or 11.45 before attempting to rectify the curve. If the curve appears to be parabolic or hyperbolic, as shown in Figure 11.44, plot the data on logarithmic paper and draw a straight line through the data points. If the curve appears to be exponential, as shown in Figure 11.45, plot the data on semilog paper and draw a straight line through the data points. When the curve is rectified, the development of the equation can proceed.

4. Using the rectified curve, develop the empirical equation using either the slope-intercept method or the method of selected points. These two methods of solutions are discussed in the following sections.

Linear Equation Solutions

The best way to present the methods of developing empirical equations is to show examples of each and follow the step-by-step procedure outlined in the preceding. Figure 11.52 shows a table of x and y values that are plotted as a rectangular graph. A linear relationship between the independent and dependent variables is shown because a straight line can be fit to the data points on the graph. The graph suggests that the relationship between the variables can be represented by the equation

$$y = mx + b$$

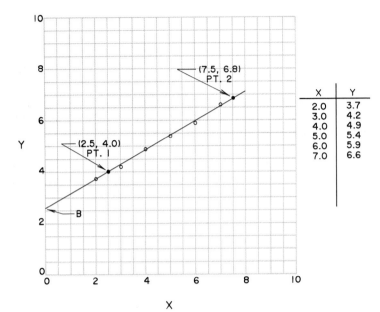

X	Y
2.0	3.7
3.0	4.2
4.0	4.9
5.0	5.4
6.0	5.9
7.0	6.6

Figure 11.52 Graph showing linear relationship between variables.

where x and y are the independent and dependent variables, respectively, m is the slope of the line, and b is the y intercept of
the line. The slope-intercept method of solution for this equation
is a graphical solution. The y intercept is read from the graph
(point B in Figure 11.52). This value is read as 2.6. The slope m
is determined by locating two points on the line (1 and 2). The
values X_1, Y_1, X_2, Y_2 are read from the graph. The slope is determined by the calculation

$$m = (Y_2 - Y_1)/(X_2 - X_1)$$

In this case the values from Figure 11.52 are $X_1 = 2.5$, $Y_1 = 4.0$,
$X_2 = 7.5$, and $Y_2 = 6.8$, yielding a slope value of 0.56. The
empirical equation representing this data based on the slope-intercept method is

$$y = 0.56x + 2.6$$

The empirical equation can be developed using an approach
that combines a graphical procedure with an analytical procedure.
This approach is the method of selected points. In this method,
two points are selected on the line (points 1 and 2 in Figure 11.52).
Points 1 and 2 are not specific data points used to plot the curve.
Again the values X_1, Y_1, X_2, Y_2 are read from the graph. This time
the values are used in the linear equations

$$Y_1 = mX_1 + b$$
$$Y_2 = mX_2 + b$$

The equations are solved simultaneously for m and b to yield the empirical equation relating the variables. For this example, $X_1 = 2.5$, $Y_1 = 4.0$, $X_2 = 7.5$, and $Y_2 = 6.8$. The resulting empirical equation is

$$y = 0.56x + 2.6$$

In this example, the resulting equations using the two methods of solution were identical; however, in most cases the equations will be slightly different. For example, the y intercept may be 2.5 or 2.7 using the method of selected points. This slight variation will not affect the results when this equation is used for design purposes. Many times the empirical equation is developed using both procedures and compared using a method called residual analysis to determine which equation best represents the experimental data from which it was derived. (Residual analysis is beyond the scope of this book.)

Power Equation Solutions

Figure 11.53 shows data plotted on rectangular graph paper. The resulting curve is not a straight line; therefore, the linear equation as discussed in the preceding cannot be used. The curve shown on the rectangular graph is hyperbolic and is rectified when plotted on logarithmic paper (Figure 11.54). Figure 11.55 shows the data

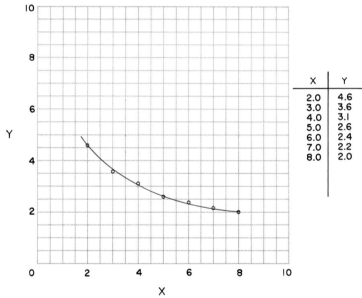

X	Y
2.0	4.6
3.0	3.6
4.0	3.1
5.0	2.6
6.0	2.4
7.0	2.2
8.0	2.0

Figure 11.53 Rectangular plot of some experimental data. Results do not indicate linear relationship between variables.

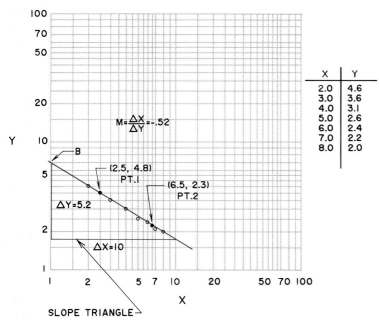

X	Y
2.0	4.6
3.0	3.6
4.0	3.1
5.0	2.6
6.0	2.4
7.0	2.2
8.0	2.0

Figure 11.54 Same data as in Figure 11.53 plotted as logarithmic graph indicates rectified curve.

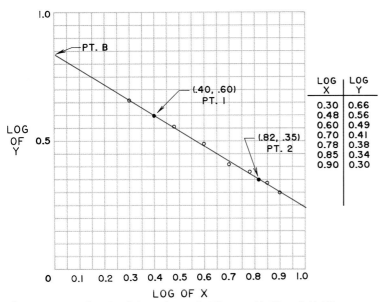

LOG X	LOG Y
0.30	0.66
0.48	0.56
0.60	0.49
0.70	0.41
0.78	0.38
0.85	0.34
0.90	0.30

Figure 11.55 Graph of data shown in Figures 11.53 and 11.54 converted to logarithms and plotted on rectangular coordinate scales.

converted to logarithms and plotted on rectangular paper with the resulting curve a straight line. The slope-intercept method or the method of selected points can be used to develop the power equation from either rectified curve.

The power equation is of the form $y = bx^m$ and is the equation for a parabola or a hyperbola. This equation is not linear, and developing the empirical relationship from it is difficult. The power equation is transformed into a linear power equation simply by taking the logarithm of both sides of the equation. The logarithmic form of the power equation is

$$\log y = \log b + m \log x$$

This form of the equation shows that the curve is rectified (linear) when plotted on logarithmic paper or when the logarithmic values of the data points are plotted on rectangular paper.

The slope-intercept method of analysis uses a graphical means to develop the power equation. From the rectified curve, the y intercept, b is read where the rectified curve crosses the y axis (point B in Figure 11.54 or 11.55). The y axis on a logarithmic graph is always located where the value of x is 1. This is extremely important since on a logarthmic scale the x axis may begin with a value of 0.01 or 0.001. The y axis is not necessarily located where the x axis begins but where $x = 1$. The value of the intercept in this example is read to be 7.0. The slope is determined by using a slope triangle (see Figure 11.54 or 11.55). Since both scales (X and Y) are logarithmic scales, the slope can be determined by looking at the difference in the Y values and the difference in the X values defined by the slope triangle or by measuring the sides of the slope triangle directly. The difference in the Y values is 5.2, and the difference in the X values is 10. The slope is determined by the equation:

$$m = (Y_2 - Y_1)/(X_2 - X_1)$$

For this example, the slope $m = -0.52$. This slope equation is identical to the equation for slope presented for linear empirical equations. With the value of the y intercept b and the value of the slope m known, the empirical equation is

$$y = 7.0x^{-0.52}$$

The negative exponent indicates that the curve is a hyperbola rather than a parabola.

As in the case of linear empirical equations, the method of selected points for determining the power empirical equation is more mathematical than the slope-intercept method. Two points are selected on the rectified curve (1 and 2 in Figure 11.54 or 11.55). The values are read from the curve (Figure 11.54 for this example). The values are $X_1 = 2.5$, $Y_1 = 4.0$, $X_2 = 6.5$, and $Y_2 = 2.3$. The logarithmic form of the equation is then used to develop the equation. Two equations are written substituting the values of X_1, Y_1, X_2, and Y_2:

$$\log 4.0 = \log b + m \log 2.5$$
$$\log 2.3 = \log b + m \log 6.5$$

These equations are solved simultaneously for the y intercept b and the slope m. In this example, the y intercept $b = 6.79$ and the slope $m = -0.58$. These values are inserted into the power equation, which is written as

$$y = 6.79x^{-0.58}$$

Compare the two equations obtained using the slope-intercept method and the method of selected points. They are not exactly the same and you should not expect them to be the same. For engineering purposes, the error associated with using either equation is not significant.

Exponential Equation Solution

Figure 11.56 shows data plotted on rectangular graph paper forming a curve that is exponential. This curve is rectified by plotting the data on semilogarithmic paper or by plotting the logarithmic values of the dependent variable (y axis). Figures 11.57 and 11.58

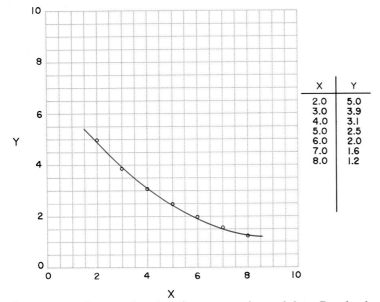

X	Y
2.0	5.0
3.0	3.9
4.0	3.1
5.0	2.5
6.0	2.0
7.0	1.6
8.0	1.2

Figure 11.56 Rectangular plot of some experimental data. Results do not indicate linear relationship between variables.

Figure 11.57 Same data as in Figure 11.56 as semilogarithmic graph indicates rectified curve.

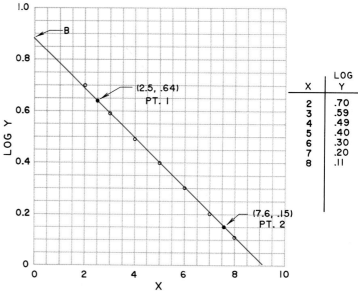

Figure 11.58 Graph of data shown in Figures 11.56 and 11.57 converted to logarithms (dependent variable only) and plotted on rectangular coordinate scales.

show the semilog plot and the logarithmic y-axis plot, respectively. A curve that is rectified in this manner can be used to develop an exponential empirical equation.

As before, the exponential equation will be developed using the slope-intercept method and the method of selected points. The exponential equation can be of the form $y = b(10)^{mx}$ or $y = b(e)^{mx}$ depending on the logarithm system desired. As in the case of the power equation, the logarithmic form of the equation is linear. The logarithmic form of the equation (base 10 logarithms) is

$$\log y = \log b + mx \log 10$$

Of course, the value of log 10 is 1, so the equation is simplified to

$$\log y = \log b + mx$$

The equation for base e, or Naperian, logarithms is

$$\ln y = \ln b + mx$$

As in all cases of empirical equations the slope-intercept method of solution is a graphical method. The y intercept is read from the graph of the rectified curve (point B in Figures 11.57 and 11.58). The slope is determined by drawing a translation line parallel to the rectified line through the data. This translation line crosses the Y axis at the end of a cycle (the grid labeled 10 in Figure 11.57). This translation line is extended through one cycle on the Y axis until it crosses the X axis. In a similar way, as in the case of the slope triangle presented for power equations, the difference in the Y values of the translation line is always 10. The value of log 10 is 1. The difference in the X values with regard to the translation line is read and the slope $m = 1/(X_2 - X_1)$.

For this example, the y intercept is read as $b = 7.7$ and the slope using the translation line is $- 1/10$ or -0.10. Using these values, the exponential empirical equation is

$$y = 7.7(10)^{-0.10x}$$

for base 10 logarithms.

To determine the empirical equation using Naperian (base e) logarithms, the translation line is drawn from the value of e (2.718. . .) on the Y axis parallel to the line through the data points until it crosses the X axis (Figure 11.57). The value of ln e is 1 and the difference in the X values with regard to the translation line is read. The slope is determined by the same relationship shown previously. For this example, the Naperian translation line crosses the X axis at 4.3; therefore, the slope m is $-1/4.3$, or -0.23. The equation based on Naperian logarithms is

$$y = 7.7e^{-0.23x}$$

The method of selected points in this example uses the logarithmic form of the equation. The X and Y values of the selected points (points 1 and 2 in Figure 11.57 or 11.58) are read and substituted into the logarithmic form of the exponential empirical equation. The X and Y values from the rectified graph (Figure 11.57) are $X_1 = 2.5$, $Y_1 = 4.3$, $X_2 = 7.5$, and $Y_2 = 1.4$. Two simultaneous equations are used to solve this problem. The two equations are

$$\log 4.3 = \log b + m(2.5)$$
$$\log 1.4 = \log b + m(7.5)$$

These equations are solved simultaneously to yield values for the y intercept b and the slope m. In this example, the values for the intercept b and the slope m are calculated to be 7.54 and -0.097, respectively. Using the Naperian-based logarithm system yields values of 7.65 and -0.23 for the intercept and the slope, respectively.

11.4 GRAPHICAL CALCULUS

11.4.1 Definition of Graphical Calculus

Graphical calculus is a tool that can be used by the engineer to do computations involving integration and differentiation. The advantage of graphical calculus over analytical methods is that it can be used for very difficult problems (discontinuous functions) that may be nearly impossible to solve by another means. Also, the method provides a graphical picture of the integration and differentiation processes.

Integration is the process by which a number of small elements are summed between limits to give a total result. Differentiation is the process that determines the rate of change of a variable of a function with respect to another variable of the function. Integration and differentiation processes are the reverse of one another, and the graphical procedures are worked inversely to obtain the desired result. For any given function that is represented by a graph, it can be integrated or differentiated. The result is a graph showing the function integrated or differentiated.

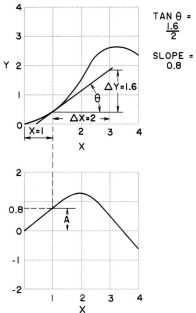

Figure 11.59 Slope at any point of given curve relating two variables is numerically equal to ordinate at corresponding point on derived curve of next lower degree (derivative curve).

11.4.2 Principles of Graphical Calculus

The procedures involved in doing graphical integration and differentiation are based on two principles. The first principle states that the slope at any point of a given curve relating two variables is numerically equal to the ordinate at the corresponding point on the derived curve of the next lower degree (derivative curve). Figure 11.59 demonstrates this principle. The top curve in Figure 11.59 is considered the integral curve of the bottom curve, and the bottom curve is considered the derivative curve of the top curve. The slope of the top curve at $x = 1$ is simply the tangent of the angle Θ or 0.8. This slope is numerically equal to the ordinate value on the bottom curve at $x = 1$.

The second principle states that the area bounded by a curve, any two coordinates, and the abscissa is numerically equal to the difference between the two corresponding ordinates of the curve of the next higher degree (integral curve). Figure 11.60 demon-

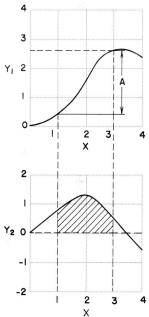

Figure 11.60 Area bounded by curve, any two coordinates, and abscissa is numerically equal to difference between two corresponding ordinates of curve of next higher degree (integral curve).

strates this principle. The area under the bottom curve in Figure 11.60 from $x = 1$ to $x = 3$ is numerically equal to the distance A_1 shown on the top curve.

The procedures for doing graphical integration and graphical differentiation are based on these principles. Study these principles and the corresponding figures so that you have these concepts in mind before attempting to solve any graphical calculus problems.

11.4.3 Graphical Differentiation (Chordal Method)

Graphical differentiation is the inverse process of graphical integration. The higher degree curve is given and the differentiated curve is desired. At any point on the given curve, the corresponding point on the differentiated curve represents the value of the slope on the given curve. The best way to understand graphical differentiation is through a step-by-step procedure with an example problem. Figure 11.61 shows a given curve to be differentiated and the solution. The following steps are used to solve this problem.

1. Divide the given curve into small elements using vertical strips. If the curve is relatively flat, the strips can be fairly wide. If the curve is steep, then the strips should be fairly narrow to ensure the differential curve is as accurate as possible (Figure 11.61).

2. Determine the point on the given curve where a tangent can be drawn that represents the maximum slope of the curve. Draw this tangent and determine its slope. On the ordinate scale for the differential curve, locate the value of the slope that you have just determined. Draw a line parallel to the tangent line on the given curve from the ordinate point on the differential representing the slope of the tangent line through a point on the abscissa (line 1 in Figure 11.61). The point where this line crosses the abscissa is the focus, or pole point (Figure 11.61).

3. In each elemental strip on the given curve, draw a chord connecting the intersection points of the curve with each side of the element. The tangent line through a point on the curve is the true slope line; however, the chord is easier to find, and its slope is very close to the slope of the tangent line over small distances.

4. For each chord drawn in step 3, draw a line parallel through the focus intersecting the ordinate of the differential curve (lines 2–11, Figure 11.61). At each intersection of the ordinate, draw a horizontal line between the appropriate elements.

5. When all the chords are shown as horizontal lines in their proper position on the differential curve, use the center point on

Figure 11.61 Graphical differentiation using chordal method.

the horizontal between each element to establish a point for the differential curve. Then complete the curve using an irregular curve to draw the differential curve.

These steps will ensure that an accurate curve will result. The procedure is easy to follow and does not take much time to show a differentiated function.

11.4.4   Graphical Integration (Chordal Method)

Graphical integration is done using areas under the given curve to get the tangents on the integral curve. The solution is exactly the inverse process of differentiation. A key to solving this type of problem is knowing that the ordinate values of the given curve are slopes of the integral curve (the first principle described in Section 11.4.2). As in the case of graphical differentiation, the best way to present graphical integration is through a step-by-step procedure involving an actual solution of a problem. Figure 11.62 shows a curve that is to be integrated and the solution. The steps in solving this problem follow.

1. Divide the given curve using vertical strips. The width of these vertical strips can vary. Where the curve is relatively flat, the strips can be fairly wide. Where the curve is steep, the strips should be narrower. The more elemental strips you use, the more accurate the integral curve will be (Figure 11.62).

2. The focus, or pole point, is the link between the given curve and the integrated curve. The focus to be used to get the chords or tangents for the integral curve must be determined. This is done by simply drawing a line that represents a tangent on the integral graph (line 1, Figure 11.62). Any line will do so long as its slope can easily be determined. The slope of line 1 in Figure 11.62 is 4.0. The slope of this tangent should be equal to or greater than the maximum slope of the integral curve. An estimate of the maximum slope of the integral curve is found by approximating the area under the given curve. After establishing the maximum slope on the integral curve, draw a line parallel to this slope (line 1, Figure 11.62) from the ordinate value on the given curve to the abscissa (line 2, Figure 11.62). Where this line crosses the abscissa is the focus, or pole point. This focus will be used to develop the integral curve.

3. On the given curve draw horizontal lines within each elemental strip so that the curve is cut in half over this small section (Figure 11.62). The shaded areas should be equal to ensure the accuracy of the integral curve. The average heights of the strips projected to the ordinate are the slopes for the chords drawn on the integral curve.

Figure 11.62 Graphical integration using chordal method.

4. Transfer each of the horizontal lines drawn in step 3 to the ordinate scale of the given curve.

5. From each point where the transferred line intersects the ordinate, draw a line through the focus. Then draw a line within the proper element on the integral parallel to the focus line just drawn. Connect these straight-line segments end to end. These straight-line segments are the chords of the small curve segments represented by each elemental strip.

6. Draw with an irregular curve the arc portions of the integral curve for each elemental strip.

11.5 *VECTOR GRAPHICS*

Many engineering problems can be solved by using **vectors.** These problems can be solved with or without the graphical representation of the vectors; however, a picture of the problem makes it easier to understand. Typical problems that can be solved using vectors include the analysis of simple structures in statics, dynamics, strength of materials, machine design, and fluid mechanics. These subject areas will become familiar to you as you advance in your engineering education.

Understanding descriptive geometry is required to solve three-dimensional vector problems. You may wish to review the auxilary view material presented in Chapter 10 before proceeding with this section.

Vector quantities have a direction and a magnitude. To represent vector quantities graphically, we draw a line and put an arrowhead at one end. The end of the vector that has the arrowhead is called the "head," or "tip," of the vector. The end without the arrowhead is called the "tail" of the vector. The length of the line is proportional to the magnitude of the vector and the arrowhead indicates the vector's direction. Accuracy in drawing the vector is important. A good protractor, straight edge, and scale are required to solve vector problems.

Vectors can represent forces, velocities, accelerations, displacements, and other quantities that have both magnitude and direction. The vector is drawn from a location in a direction specified by an angle. A scale is used to mark the length and an arrowhead is placed on the end of the vector to indicate direction (Figures 11.63*a–d*).

11.5.1 Vector Addition and Subtraction

Vectors can be added or subtracted. For these processes, we assume that the vectors are concurrent–coplanar. This means that

Figure 11.63 Examples of vector quantities.

Figure 11.64 Vector addition: Parallelogram method.

all the vectors (two or more) act at the same point and are contained in the same plane. Also, assume the plane on which the vectors are acting is the plane of the paper or the plane of the computer screen (Figure 11.64a). When adding two concurrent–coplanar vectors, first lay out the vectors to scale and in their respective directions. Then, using the two vectors as two sides of a parallelogram, lay out the full parallelogram (Figure 11.64b). Addition of two vectors is accomplished by drawing the diagonal of the parallelogram from the common point on which the two vectors act to the opposite end of the parallelogram (Figure 11.64c). This is called the resultant vector. Another approach with regard to vector addition is to visualize the tail of the second vector being located at the head of the first vector. The sum is then the vector drawn from the tail of the first vector to the head of the second vector. This is called the triangle method of vector addition (Figure 11.65). The triangle method can be expanded to the polygon method when more than two concurrent–coplanar vectors are to be added. Figure 11.66a shows four vectors that can be added to give a resultant vector. Simply connect the head of one vector to the tail of another until all vectors are connected. Then draw a resultant vector from the tail of the first vector to the head of the last vector to complete the polygon (Figure 11.66b). The magnitude of the resultant vector can be determined by measuring with the scale used to lay out the parent vectors. Its direction can be determined using a protractor.

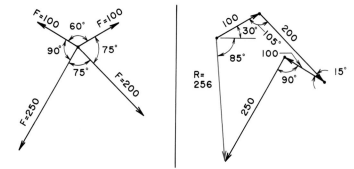

Figure 11.65 Vector addition: Triangle method.

Figure 11.66 Vector addition: Polygon method.

Vectors can also be subtracted using the same principles discussed in vector addition. Figure 11.67*a* shows vector subtraction. The opposite diagonal of the parallelogram from the diagonal used in vector addition is used in vector subtraction. Vector subtraction shows relative relationships between vectors. For example, Figure 11.67*b* shows a top view of a river flowing with a current. On one shore is a canoe the canoeist is going to use to cross the river and arrive at point *X*. The canoeist must start at an angle upstream in order to arrive at the designated point. This example illustrates the concepts of vector subtraction.

11.5.2 Applications

Statics: Two-Dimensional, Single Joint Problem
When objects are said to be in static equilibrium (there is no movement), it means that the sum of all of the forces on an object or any part of the object must be zero. The vector diagram of such an object is shown in Figure 11.68. Note that there is no resultant force shown in the vector diagram. When the vectors are connected in head-to-tail fashion, the vector diagram closes to form a polygon. This means that the resultant vector for equi-

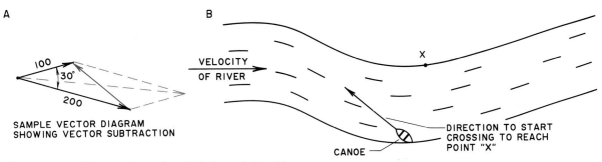

Figure 11.67   Vector subtraction: Relative relationships.

librium is zero. This makes it convenient to solve problems with no more than two unknown values. At this point, you need to remember that vectors have two properties: direction and magnitude. When solving problems with vectors, typical situations might be one vector (magnitude and direction) is unknown, the magnitudes of two vectors are unknown, or two of the directions are unknown.

Figure 11.69 illustrates a situation where the load on a connection point or joint is known and the direction of the forces provided by the rigid support and the cable are known. When solving such problems, you must remember that cables, ropes, and chains always pull on the joint and their forces are directed away from the joint. Rigid members can either pull or push on a joint, and their vectors can be represented as either toward the joint or away from the joint.

The solution step where the initial vector directions are assumed is called a free-body diagram. This diagram is drawn show-

Figure 11.68   Vector diagram showing forces in equilibrium. Vector diagram closes when vectors are connected in head-to-tail fashion.

A   LOAD DIAGRAM

B   FREE BODY DIAGRAM OF RIGID SUPPORT

C   FREE BODY DIAGRAM OF JOINT B

Figure 11.69   Two-dimensional force analysis of joint: (*a*) load diagram; (*b*) free-body diagram of rigid support; (*c*) free-body diagram of joint at *B*.

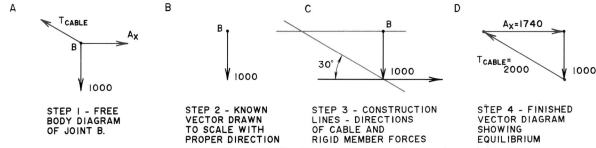

STEP 1 - FREE BODY DIAGRAM OF JOINT B.

STEP 2 - KNOWN VECTOR DRAWN TO SCALE WITH PROPER DIRECTION

STEP 3 - CONSTRUCTION LINES - DIRECTIONS OF CABLE AND RIGID MEMBER FORCES

STEP 4 - FINISHED VECTOR DIAGRAM SHOWING EQUILIBRIUM

Figure 11.70 Step-by-step procedure to determine unknown quantities related to vectors using graphical means. (*a*) Draw free-body diagram of joint. (*b*) Lay out to scale the known vector quantities in proper directions. (*c*) Draw directions of forces in cable and in rigid member using light construction lines. (*d*) Complete vector diagram by connecting vectors in head-to-tail fashion until polygon closes indicating equilibrium. Measure unknown values at selected scale to determine their magnitudes.

ing that the load, the chain, and the rigid member have all been replaced by their forces on a single point (Figure 11.70*a*).

The next step is to draw the known vector with the proper direction and magnitude (Figure 11.70*b*). Next from the head end of the known vector draw a construction line in the direction of the cable, and from the tail end of the known vector draw a second construction line in the direction of the rigid member (Figure 11.70*c*).

The final step is to draw the arrowheads for the two vectors that were unknown and to measure their length using the scale chosen at the beginning of the problem. The completed solution is shown in Figure 11.70*d*. Recall that when a joint is in equilibrium, the vector diagram of the joint closes when the vectors are connected head to tail.

Now try to solve this problem first using your drawing instruments and then sketching on grid paper. When using grid paper, assign a value to each grid square and then use your pencil or the edge of a second piece of paper to determine the magnitudes of any vectors that do not lie on either the vertical or horizontal axes.

Statics: Two-Dimensional Truss Analysis

The forces in the structural members of complicated trusses (those with many joints) can be solved finding the joints that have only two unknown values. These unknowns can be magnitudes or directions. From the previous discussion, we know that these types of problems can be solved graphically using vectors. Once the unknowns for a joint have been solved, subsequent joints can be solved because the number of unknowns for that joint will be reduced to 2. Figures 11.71a–d show the steps involved to create a vector diagram to aid in solving for the unknown forces acting at joint A. Figures 11.71e–h show the vector diagrams for the remainder of the truss structure. The black vectors are the known forces and the vectors shown in color are the unknown forces.

Statics: Three-Dimensional Analysis

It is also possible to solve three-dimensional problems with vectors. To do this, you need to remember some of the information presented in Chapter 10. Specifically, you should know that by drawing any two adjacent orthographic views of a line, you can find the true length of any vector (line). Sometimes you will need an auxiliary view to determine the true length.

Figure 11.72a shows a top and front drawing of the rigging of a tow truck. The vector labeled as 1000 lb represents the cable of the tow truck that is loaded in the direction indicated. Assuming

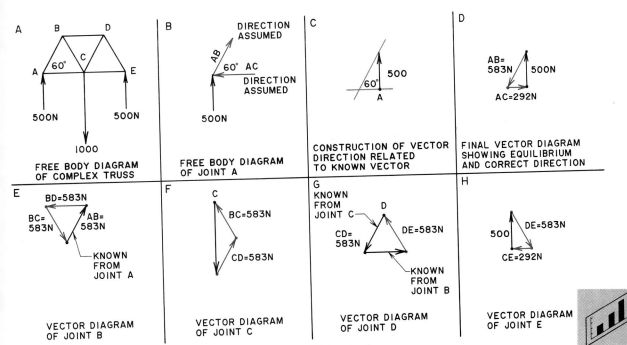

Figure 11.71   Two-dimensional force analysis of truss: (a) free-body diagram of truss with applied loads and reactions; (b) free-body diagram of joint A; (c) construction of vector diagram using directions of unknown vectors related to known vectors; (d) final vector diagram showing equilibrium and magnitudes of unknown vectors; (e–h) procedure repeated to get vector diagrams for each joint in truss. Magnitude of each unknown force vector is found.

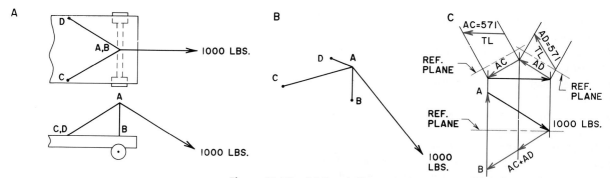

Figure 11.72 (*a*) Load diagram of three-dimensional force system. (*b*) Free-body diagram of tow truck rigging. (*c*) Top and front views of vector diagram of forces in rigging drawn to scale. Using methods of descriptive geometry, unknown force vectors in members are solved.

that all of the forces go through a single joint (joint *A*), a free-body diagram can be drawn (Figure 11.72*b*).

Figure 11.72*c* shows a vector diagram with a top and front view. The front view is solved first because the load (the car being towed) is known (the vector is true length) and there are only two unknowns. Again the black vectors represent the known forces and the colored vectors represent the unknown forces. Once the front view has been solved, the auxilary views can be used to determine the magnitudes of the unknown forces in the top view (*AC* and *AD*).

11.6 SUMMARY

This chapter serves as an introduction to basic methods of graphical mathematics. Graphical mathematics includes the production of charts and graphs, development of empirical equations, methods of graphical calculus, and methods of vector analysis. Each of these areas was presented in detail. This summary presents the highlights of each section.

Engineers use charts and graphs to present information in graphical and pictorial form. Information can be easily understood if presented as a chart or graph. Trends can be analyzed and functions of mechanism can be documented.

Charts and graphs can be classified according to their intended use. Some are used to provide technical information with regard to the design of a product while others present complex information in an understandable way.

Various types of charts and graphs are used to present data in an understandable way. Pie charts show the relationship between the various parts of a whole. Line graphs show trends or the relationship between two variables. Bar charts are used to compare various quantities. Multiple-line graphs, surface charts, three-coordinate charts, map charts, and trilinear charts are other types of charts that can be used to display data.

When developing a chart or a graph, it is important to take into consideration chart size, scale calibration, axis labels, tic mark labels, and chart title. Scale calibration is a very important concern when designing a chart or graph. The primary consideration is to be sure that data can be interpolated or read for the chart without having to do time-consuming calculations.

Empirical equations are equations that are developed from observing natural phenomena or from data obtained through controlled laboratory experiments. There are three types of empirical equations with which we can deal: linear, power, and exponential. The form that describes these types of empirical equations are as follows: linear, $y = mx + b$; power, $y = bx^m$; and exponential, $y = b(10)^{mx}$ or $y = b(e)_{mx}$.

Empirical equations are developed by using a graph of the experimental data. In order to use one of the preceding equation forms, the graph of the data must first be rectified or made linear. If the curve of the data is linear using rectangular coordinates, then the linear form of the equation is used. If the curve of the data is rectified using logarithmic coordinates, then the power equation is used. Lastly, if the curve of the data is rectified using semilogarithmic coordinates, then the exponential equation is used.

Graphical calculus is a tool that can be used by the engineer to do computations involving integration and differentiation. Its main advantage over analytical methods is that it can be used to solve very complex problems that may be impossible by any other means. The method also provides a graphical picture of the integration and differentiation processes.

Both integration and differentiation are accomplished using the chordal method. Several elemental strips are drawn through the given curve and various widths to do the integration or differentiation. If the curve is flat, the elemental strips can be wide; if the curve is steep, the elemental strips should be thinner so that the plotted curve will be accurate. From these elemental strips, chords are developed that relate the slope between the integral and differential curves.

Many engineering problems can be solved using vector graphics. Vectors offer a picture of the problem that makes the problem easier to understand. Vectors are important in the solution of problems related to statics, dynamics, strength of materials, machine design, and fluid mechanics because they can represent forces, velocities, accelerations, displacements, and other quantities that have magnitude and direction.

Vectors can be added or subtracted. When vectors are added, a resultant vector is obtained. A resultant vector is a single vector that can replace a series of vectors without changing the effect of the vectors on an object.

Vectors are used in solving two-dimensional and three-dimensional static equilibrium problems. When an object is in equilibrium, the resultant force vector on the object is zero. Vector solutions to these type of problems can be less time consuming and easier to understand than analytical methods.

PROBLEMS

Solve each problem either by manual or computer methods as defined by your instructor.

Problem 11.1
Draw and calibrate the axes for a line graph that will be used to plot data with abscissa (X axis) values ranging from 3 to 96 and ordinate (Y axis) values ranging from 270 to 9500.

Problem 11.2
Draw a complete pie chart showing the proportion of a student's budget that is spent in each of the following categories. Show the percentage for each item.

Tuition	$1800	Transportation	$400
Books	400	Clothing	600
Housing	1200	Entertainment	600
Food	2200	Personal and miscellaneous	400

Problem 11.3
Draw a pie chart similar to the one for problem 11.2 but use your own budget items and amounts.

Problem 11.4
Calculate the scale modulus m for the function $f(v) = v$. Use a 6-in. line and values of $v_{min} = 0$ and $v_{max} = 10$.

Problem 11.5
Calibrate a 6-in.-long scale using the data from problem 11.4 and your calculated value of m.

Problem 11.6
Calibrate a 4-in.-long scale for the following data: $f(v) = \sin(v)$, $v_{min} = 0°$, and $v_{max} = 90°$.

Problem 11.7
Calibrate a 3-in.-long scale for the following data: $f(v) = \log 10v$, $v_{min} = 10$, and $v_{max} = 100$.

Problem 11.8
Calibrate a 5-in.-long scale for the following data: $f(v) = 2.5v$, $v_{min} = 0$, and $m = 0.2$.

Problem 11.9
ABLE Consulting Co. posted the following operating revenues and expenses (in thousands of dollars) for the years 1981 through 1985:

	1981	1982	1983	1984	1985
Revenue	1520	1716	2135	1987	2225
Expense	1215	1410	1530	1560	1923

Draw a surface chart for this data with the year on the abscissa (X axis). Calibrate and label the axes. Include a short title.

Problem 11.10
Draw a bar chart for the data of problem 11.9. Shade the bars to distinguish between revenue and expense.

Problem 11.11
Answer the following questions by examining your chart from problem 11.9 or 11.10:
(a) In what year was the most profit made?
(b) In what year was the highest percentage of profit made?
(c) In what year was total revenue the highest?
Confirm your answers mathematically.

Problem Problem 11.12
The Clausius-Clapeyron equation relates the vapor pressure (p) of a pure substance with the absolute temperature (T).

$$\ln p = -\frac{\Delta H}{RT} + B$$

where

p = the vapor pressure in millimeters of mercury
T = the temperature in °K
ΔH = the latent heat of vaporization
R = the universal gas constant
B = a constant

The vapor pressure of benzene is measured at two temperatures with the following results:

$$T_1 = 285.7 \text{ °K} \qquad p_1 = 52 \text{ mmHg}$$
$$T_2 = 298.5 \text{ °K} \qquad p_2 = 97 \text{ mmHg}$$

Estimate, using the graphic method, the vapor pressure of benzene at 320 °K.

Problem 11.13
The rate at which heat is transferred from a hot wall to its surroundings is given by the formula:

$$\frac{Q}{A} = H(T - TO)$$

where

T = temperature of the wall (°C)
TO = temperature of the surroundings (usually a constant)
A = surface area of the wall (m^2)
Q = rate of transference of heat (kJ/h)
H = heat transfer coefficient (kJ/hm² °C)

The following information was obtained from measurements made in a heat transfer experiment:

T (°C)	$\frac{Q}{A}$ (kJ/hm²)
25	0
30	250
35	500
40	750
45	1000

What is the value of TO?
Using the appropriate coordinates, graphically determine the value of H for these conditions.

Problem 11.14
The equation proposed by Thiesen has proved to accurately follow the variation of the heat of vaporization (Q) with temperature.

$$Q = K(TC - T)^n$$

The critical temperature for water (TC) is 647 °K and the values of Q in calories per mole are 3250 and 6600 when temperatures (T) are 632 °K and 550 °K respectively. What is the heat of vaporization for water at 347 °K? K and n are constants. Solve this problem graphically using logarithmic graph paper.

Problem 11.15
Determine the empirical equation that describes the relationship between the variables from the experimental data shown.

X	Y
20	6.4
30	5.5
40	4.8
50	4.0
60	3.1
70	2.3
80	1.5

Problem 11.16
Determine the empirical equation that describes the relationship between the variables from the experimental data shown.

R	T
10	24
20	27
30	31
40	35
50	38
60	42
70	46

Problem 11.17

The internal energy of a traffic stream on a freeway can be measured by a parameter known as acceleration noise (ft/sec/sec). This parameter deals with the variation in speeds of the vehicles in the traffic stream. The relationship between acceleration noise and average speed was studied and the following data were collected:

Mean Acceleration Noise (ft/sec/sec)	Mean Speed (mph)
0.04	57
0.10	56
0.18	54
0.33	49
0.65	39

Determine the equation that defines the relationship between acceleration noise and speed. Speed is the independent variable and acceleration noise is the dependent variable.

Problem 11.18

Determine the empirical equation that relates the variables shown in the following experimental data.

P	V
10.0	3200
30.0	1010
50.0	600
80.0	370
100.0	300
150.0	198
200.0	149
250.0	120

Problem 11.19

Determine the empirical equation that relates the variables shown in the following experimental data.

X	Y
0.20	9.75
0.40	12.10
0.80	15.00
1.20	17.00
2.00	20.00
4.00	24.80

Problem 11.20

The pressure of a gas at a constant temperature varies with volume as described in the experimental data shown. Determine the empirical equation that describes the relationship.

Volume (cm³)	Pressure (mmHg)
2.5	1500
3.0	1220
3.5	1050
4.0	915
4.5	810
5.0	745
5.5	690
6.0	620
7.0	520
8.0	480
9.0	410
10.0	390

Problem 11.21

Determine the empirical equation that relates the variables in the following experimental data.

XFTR	YFTR
2.0	2.20
3.0	2.55
4.0	2.85
5.0	3.25
6.0	3.75
7.0	4.20
8.0	4.82

Problem 11.22

Determine the empirical equation that relates the variables in the following experimental data.

W	S
10.0	32.0
20.0	20.5
30.0	12.9
40.0	8.1
50.0	5.1
60.0	3.2

Problem 11.23

Strength tests of a metallic material used under high temperature conditions were conducted, and the data from these tests are tabulated below. Determine the empirical equation.

Temperature (°C)	Stress (Mn/m²)
550	41.20
600	18.62
650	8.62
700	3.92
750	1.86

Problems 11.24–11.26

Plot the given data on rectangular graph paper. Draw the derivative curve for each of the plotted curves.

Problem 11.24

X	Y
0	0
10	10
20	20
30	10
40	−10
50	−10
60	−10

Problem 11.25

X	Y
0	0
2	20
4	33
6	42
8	47
10	50
12	47
14	41
16	32
18	18
0	0

Problem 11.26

X	Y
0	100
5	95
10	82
15	70
20	50
25	25
30	0
35	−25
40	−50
45	−70
50	−85
60	−100

Problems 11.27–11.29

Plot the given data on rectangular graph paper. Draw the integral curve for each of the plotted curves.

Problem 11.27

X	Y
0	12
1	8
2	4
3	0
4	−4
5	−8
6	−12
8	−12
10	0

Problem 11.28

X	Y
0	−10
2	−10
4	−10
5	0
6	10
7	20
8	16
9	10
10	8
14	0

Problems 11.29

X	Y
0	0
2	3
3	9
4	15
5	25
7	25, −15
8	−12.5
9	−10
10	−7.5
13	0

Problem 11.30

Determine the resultant force of the concurrent coplanar force system shown. Scale: 1 inch = 50 lbs.

10 mm = 50 Newtons (N)

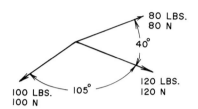

Problem 11.31

Determine the magnitude and direction of the force in each member of the frame shown. Direction can be indicated by a T for tension or a C for compression. Tension indicates the force is directed away from the joint; compression indicates the force is directed toward the joint.

Scale: 1 inch = 5000 lbs.

10 mm = 5000 Newtons (N)

Problem 11.32

For the three dimensional force system shown, determine the magnitude of the resultant force. Show the resultant force in the horizontal and frontal views. Scale: 1 inch = 50 lbs.

10 mm = 50 Newtons (N)

TERMS YOU WILL SEE IN THIS CHAPTER

ITERATIVE   Involving repetition.

STRESS   Load applied to an object. The load is divided by area to obtain values that can be compared.

$$S = \frac{F}{A}$$

STRAIN   Deformation resulting from an applied load. The deformation is divided by the length over which the deformation occurs to obtain values that can be compared.

$$\delta = \frac{d}{l}$$

TENSILE STRENGTH   Resistance to being pulled apart.

COMPRESSIVE STRENGTH   Resistance to being crushed.

SHEAR STRENGTH   Resistance to being cut.

COST ANALYSIS   Detailed investigation of all factors affecting the total cost of a product or system.

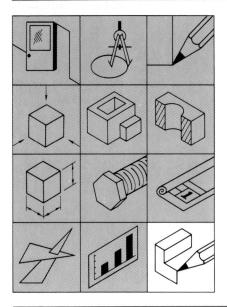

12

DESIGN PROCESS

12.1 INTRODUCTION

Design is the essence of engineering. It includes all activities involved in creation from the original idea to the finished product or system (Figure 12.1). An engineer may engage in many different activities during his or her career. Some of these are design, research, development, testing, manufacturing, technical service, construction, sales, and management (Figure 12.2). Among all these activities we consider design as the essence of engineering. Here all the technical knowledge, analysis, and creative thinking acquired in engineering education and experience are focused on solving problems to meet the needs of mankind. Design is the process by which we create and modify products and systems.

Design is a creative process that requires interaction between

Figure 12.1 Design includes concept from initial idea to finished product.

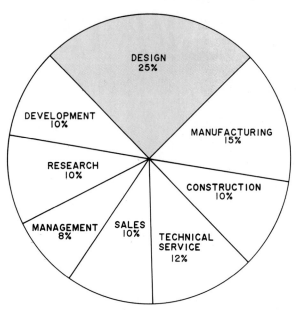

Figure 12.2 Engineering embraces many activities: *Design* is essence of engineering.

the engineer and other people or groups of people and also between the engineer and machines. (A former chief engineer of a large manufacturing company used to tell his engineers, "Listen to the machines; they will tell you what they need." Learning to do this kind of listening is a skill that takes practice; it also produces results.) The engineer uses interactive communications with the computer for technical data, analysis, the sharing of data, and computer graphics.

All design is an **iterative** process. We do not expect to achieve the best design on the first try. The steps of the design process are repeated until all concerned believe that an optimum solution has been reached within the time and money available.

Computer programs for design calculations and graphics have increased our ability to reach an optimum solution in less time. Computer graphics does not offer much of an advantage in speed over manual graphics when preparing a single drawing. However, once the data are entered, they can be modified and presented in many different ways very quickly. This is one of the advantages of computer graphics over manual graphics. In the "old days," when only manual graphics were available, we could seldom afford more than three trials to design a new part or machine. All the drawings were done by hand, and strength calculations were done by inputting the data separately to a calculator or computer. Today you can build a data base, display orthographic or pictorial drawings, calculate **stresses,** display stress patterns, estimate **strains,** calculate weights, and redesign using commercially available software packages. In many organizations, your data base can be sent directly to manufacturing to develop tools and to control the production machinery making the parts!

12.1.1 Product Design and Systems Design

Engineers are concerned with product design and systems design. The product may be an original concept or, more commonly, a refinement of an existing product. The system may be a manufacturing process, a construction time schedule, a computer system, or the communication structure for a large group of people working toward a common objective.

Product design is the process of developing new products or modifying existing ones. This process starts when someone recognizes a need for a product. For example, Velcro fasteners were developed to meet the needs of astronaut clothing in our space program. Product redesign centers activities around the redesign of an existing product to update or improve the product for its intended use. This type of design activity may be functional or aesthetic. A lawn care company produces a new product line that includes a redesigned garden hose. This development makes it easier for the user to attach an accessory to the garden hose and also prevents the user from damaging the threads on the hose or the accessory (Figure 12.3).

Both product design and product redesign require the engineer to have a firm knowledge of the use of the product, the market in which the product will be used, and technical data related to the product, its materials, and production. With product design or product redesign, the engineer will have to interact with people. Regardless of the source of the idea, the engineer must work with other creative people to further the scope of application of the product. These may include specialists in other engineering disciplines, manufacturing people, marketing representatives, and consultants. These encounters may be on a one-to-one basis or in group discussions.

System design is the process of developing engineered

Figure 12.3 Quick disconnect system for garden hose. Courtesy of OM Scott & Sons

systems that are used in manufacturing; in establishing relationships among people, machines and data; and developing complicated assemblies made of many parts. Chemical engineers create chemical processing manufacturing facilities to alter and/or manufacture chemicals. Mechanical engineers develop automated assembly lines that involve many mechanical processes. These are applications of manufacturing system designs. Industrial engineers design interfaces between people and machines and seek to optimize the productivity from this union. Complex products also require systems design. Auto manufacturers apply system design to build cars that contain simple and complex assemblies. Within an automobile there are many systems that require design such as the electrical system, the steering system, the drive train system, the heating and air-conditioning system, and the sound system.

Engineers working on system designs must also work with people. They may work with fewer people related to the user and are more likely to work with technical people from many disciplines during the design process. It is necessary that all the systems must be integrated with each other in order that the goals of the organization will be achieved.

12.2   *SCIENTIFIC METHOD*

The scientific method can be described as a vertical process. It is not peculiar to engineering design; it is used by all scientists and engineers. It includes a logical sequence of steps from start through completion. Following this logical order assures that we do not leave anything out. It results in a solution that can be reproduced and defended (Figure 12.4). This method has been described in many different words, but it always includes the following steps:

> Identify a problem.
>
> Collect data.
>
> Form a tentative hypothesis.
>
> Test the hypothesis.
>
> Accept, reject, or modify the hypothesis.
>
> Announce the results.

When you consider the method for engineering design you may remember the steps outlined here:

D	Determine a need.
E	Establish known information.
S	Stimulate design concepts.
I	Investigate design concepts with analytical tools.
G	Gather investigation data and review.
N	Now *test* the design that appears best.
P	Propose to accept, modify, or reject the design.
R	Render a final decision.
O	Optimize design by iteration.

If you follow these steps, you will arrive at a useful design solution. This is not the only way to arrive at a solution: Some

---

   I. Form a tentative hypothesis
  II. Determine methods for testing the hypothesis
 III. Test the hypothesis
  IV. Analyze results of tests
   V. Reformulate the hypothesis as a result of test reports
  VI. Retest and refine hypothesis
 VII. Announce results

Figure 12.4   One outline of the scientific method.

designers are intuitive, some use collective thinking, and some use the scientific method. Engineering solutions are never final: New information, new materials, and new tools mean that solutions can always be improved. So do not ignore the last step; you can continue to improve a good solution by repeating the process with new data. Keep it up and you will be a *design pro*.

12.2.1 Determine a Need

The process starts with the recognition of a need. The owner of a house overrun by mice might need a better mouse trap. A company that manufactures knives may have had complaints from people who have been cut. The need for safer packaging could result in fewer injuries to purchasers. The need for a car to appear different from last year's model is necessary in order that next year's model will keep up with the competition and sell well. A manufacturer of a high-quality item may identify the need to produce more products with lower labor cost in order to keep up with the competition. These are all examples of product and system designs that require change based on need.

At this stage of design, sketching will be one of your most valuable tools. As ideas and concepts for meeting the need are discussed, you and others can present sketches of possible ways to solve the problem. These may be pictorial or orthographic sketches of a product or a plan for a manufacturing process (Figures 12.5 and 12.6).

THREADS
THREE TURNS
TO SEAL

SNAP–RING
PUSH TO SEAL

Figure 12.5 Preliminary sketch of idea for product.

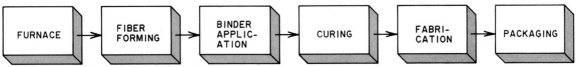

Figure 12.6 Preliminary flow chart for manufacturing facility.

Figure 12.7 Design information needs to be collected from many sources.

12.2.2 Establish Known Information

Once the need has been identified, the engineers must collect information. This information might include such items as sales history, availability of parts or materials (including new materials and new technology), market analysis, and financial and cost data (Figure 12.7). During this phase of the process, as much information as possible should be gathered so that later decisions will be based on the most complete collection of information possible.

At this step you will want to get as much expert opinion as possible. Do not attempt to design parts using new materials without advice from materials specialists. If your problem involves a new manufacturing method, get the manufacturing people involved; they know their existing process and may have many valuable ideas for changes. Interview computer systems experts before deciding that a new computer program will save time and money—a program to meet your needs may already exist.

12.2.3 Stimulate Design Concepts

At this stage you review your collected information and apply it to the need determined in step 1 (Section 12.2.1). Suppose it were a corrosion problem and the established information showed that you might:

Make the part of stainless steel.

Plate the part with cadmium.

Make the part of plastic material.

Change the corrosive environment.

A review of available data might show the following:

1. Cost information on the price of stainless steel and the cost of machining it might eliminate stainless steel as too expensive.
2. Cadmium vapors are hazardous to human health and environmental protection requirements would dictate a separate plating facility or having the plating done by a qualified outside supplier.
3. There are many plastics, some are reinforced with fibers and some are useful as structural materials without reinforcement; you may need more information on which materials to use. You would need to examine the cost and performance of both plain and reinforced materials, noting that there are several types of reinforcements that may be used with each plastic and what type of reinforcements will meet your needs.
4. Changing the environment is a possibility many people overlook. It is sometimes the best and most attainable solution. Do not overlook it! In this problem we shall assume that we do not change the environment, so we shall concentrate on the part.

Now you would truly be involved in the iterative nature of design. As you learn more, you would modify your design con-

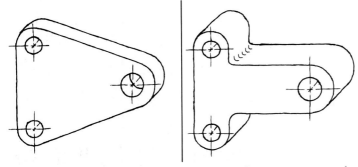

Figure 12.8 Two alternative design concepts. Computer generated image.

cepts. The information may even lead you to go back to the source of the problem and redefine the need more precisely. When you are confident that you have adequate information (not perfect information—no one ever has perfect information), make some preliminary design sketches or drawings. These may be done on the computer or manually. One thing you may be sure of: They will be changed. At this stage of design it is wise to consider several alternative designs. You have eliminated stainless steel, but how much would it cost to have a cast-iron part plated by an outside shop? Would polyester plastic reinforced with nylon be rigid enough? Would glass-fiber-reinforced polyester plastic be strong enough? During the designing can some material be taken out of the part or some manufacturing steps combined (instead of drilling holes, molding them in)? (Figure 12.8).

12.2.4 Investigate Design Concepts with Analytical Tools

The power of your computer allows you to enter the design for one material and with slight revisions have the design in a second material. Call up a finite-element analysis program and you can check for stresses throughout the part (Figure 12.9). Is the bending stress too much for cast iron? If the plastic part is not strong enough, can you make it large enough to carry the load and still fit in the assembly? Call up the assembly drawing into which this part must fit and check the space available. It is in this area of analysis that computer graphics and calculation programs give you power to do more iterations in less time. This means better design.

Figure 12.9 Finite-element analysis model for determining stresses in proposed design. Computer generated image.

If you are not using analytical tools on the computer, you still need to apply tools such as calculations of **tensile strength** and **compressive strength,** cost analysis, and producibility. Much valuable information on the design for efficient production has been developed in recent years by engineers, among them Boothroyd and Dewhurst.[1] You will need to consider their research in advanced design courses and in work projects.

12.2.5 Gather Investigation Data and Review

After you have investigated design concepts with various analytical tools, you need to review the results of the analysis. This process should include the input of representatives from every department or function concerned with the design. The results of the analysis should be given to these representatives before they are asked to comment on the designs.

Everyone should have had an opportunity to review the data before any meeting where the analysis is discussed (Figure 12.10). This is another example of where the engineer must interact with many people during the design process. Review meetings should reduce the number of choices and result in a decision to proceed with a particular design or sets of designs.

[1]Boothroyd, Geoffrey. *Automatic Assembly* (Mechanical Engineering Series: Vol 6). Dekken, 1982.

Boothroyd, Geoffrey. *Fundamentals of Metal Machining and Machine Tools.* Hemisphere Pub., 1975.

Boothroyd, G. and Dewhurst P. Series of articles on design for assembly. *Machine Design* V 55 n25 Nov. 10 '83 pp. 94–98 V 55 n28 Dec. 8 '83 pp. 140–145 V 56 n2 Jan. 28 '84 pp. 87–92 V 56 n4 Feb. 23 '84 pp. 72–76

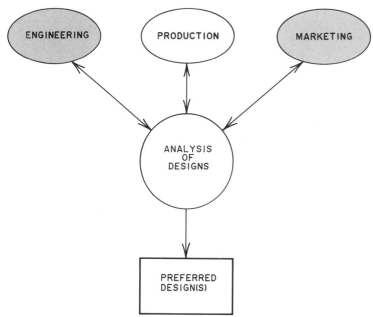

Figure 12.10 Everyone concerned reviews analysis of possible designs.

12.2.6 Now Test the Design

Test the design. This testing may be narrowed down to one design or conducted on several designs to see which is the most appropriate solution to the problem. Many techniques may be used to test a design (Figure 12.11). These could include, but would not be limited to, the following:

> *Numerical methods:* An extension of analysis procedures in which the performance of a design is analyzed under many conditions. Numerical answers are developed for the many conditions and statistical techniques applied to determine the most likely values and the probability of extreme values. A computer program is essential for a thorough analysis.

> Physical testing may be either nondestructive or destructive. A product requiring a certain amount of bending strength may be bent and the strain measured (nondestructive) or simply bent until it fails (destructive). Models may be bent and stress lines located on these models to determine areas of maximum stress.

Figure 12.11 Many kinds of testing are needed to review a design.

Market research will need to be conducted to estimate if there is a market for the product or system and how much of that market could be captured by the proposed product or system. If an existing product is being redesigned, you need to know if the redesign is as acceptable to the customers as the current design.

Economic analysis will need to be done to learn whether the product or system can be manufactured and distributed at a profit. An economic analysis is a mathematical model of the investment, the materials costs, the production costs, distribution costs, and overhead costs needed to bring the product to the customers. If done well, it considers various levels of production. In a capitalistic, free-enterprise economy, we do not produce goods at a loss (not for very long). An economic analysis may pinpoint certain areas of cost that need to be reviewed.

12.2.7 Propose to Accept, Modify, or Reject Design

The design team now proposes to the decision makers to accept, modify, or reject the design (Figure 12.12). The most common decision is to modify a design because we always learn more about a concept when we perform the testing. If the decision were to accept, then you could proceed to the next step. If the decision were to modify the design, then you go back to the appropriate design process level and work your way back to this level. Young engineers who contribute to a design are often mystified when this occurs. It may be that their personal contribution was flawless. If some part of the project is unacceptable, then it may be necessary to review every step of the design process and consider alternatives. Your computer data base makes this task a lot easier than doing everything by hand. If the decision were to reject the design, then the project team must decide whether the design process should be started again from scratch or abandoned.

12.2.8 Render the Final Decision and Prepare Working Drawings

When the design is accepted, it must be communicated to those who will do the manufacturing or construction. Working drawings for producing the product or system must be prepared, checked, and published. The information needed and the steps in preparing it are outlined in Chapter 9.

Figure 12.12 Three possible decisions for a design.

12.2.9 Optimize Design by Iteration

Again, during this process, constant feedback must be maintained. The iterative process continues within the production drawing phase for many decisions will need to be made. Here there may be the opportunity to eliminate or combine parts to reduce the complexity and cost of an assembly. This will often simplify the production process. Some decisions will require going back to earlier decisions to get the proper information or to modify information based on new knowledge. Feedback should be continuous even after the manufacture of the product or system begins so that the information will be available for later modification or redesign of the product.

12.3 *LATERAL THINKING*

The scientific method was referred to as a vertical process with each step following in sequence after another. Another technique is useful for design; it is called lateral thinking. The characteristic feature of this method is that it reaches out broadly for input to a design problem from many possible sources. Some of these sources may seem odd or unconventional. If you are designing a bridge, you might consider asking an entomologist to join your design team. Why? An entomologist studies insects and some insects produce highly efficient structures. Can we as engineers learn from other disciplines? Yes, of course, we can! The development of Velcro, noted earlier, borrowed the idea of little hooks from the burdock, a pesky weed whose burrs catch on clothing as you walk through fields.

First, we recognize a need that may arise from the market place or it may be the need to find a use for new technology. Then we assemble a team of people who can contribute to a design solution. People deeply involved in the technology and those from areas of expertise far from the problem are brought together. Ideas are generated to solve the problem. In the initial phase all ideas are acceptable, no matter how unrealistic they may seem. An odd idea from one person may inspire another to see a new way to solve the problem. After many ideas are presented, they are organized in patterns and resubmitted to the design team for elaboration and refinement. The process of iteration is used here also.

One method for lateral thinking is called brainstorming. People from diverse backgrounds get together, and the rules of the meeting are as follows:

1. Quantity of ideas is important, not quality.
2. No critical comments.

Figure 12.13 Think tank: Tool for lateral thinking.

3. Build on other ideas.

4. The wilder the idea, the better.

Others may be called just interactive design meetings. In some cases, individuals may be supplied with unusual inputs to stimulate creativity. One such device is the "think tank," a ball with a window (Figure 12.13). Inside the ball are 20,000 little tags with words on them. The designer stirs the tags inside the ball and then tries to relate each word showing in the window to the design problem. If you were an industrial engineer designing a new organizational plan, you might stir the tank and see the word *honey*. How does that relate to an organization plan?

> Honey is sweet—you might plan a reward system to make life under this plan sweeter.

> Honey is produced by bees. Many bees live in one colony under the dominion of one queen, and each bee seems to do its job with little or no direction: Could you structure a plan with only one manager for many people but with each person being self-directed to the goals of the organization?

> Honey is stored in hexagonal cells by bees: Would a hexagonal office layout promote efficiency?

These are some of the ideas an engineer might consider when relating the word *honey* to a problem.

The lateral thinking process will not solve all design problems, but it is a spur to creativity. The design process always requires creativity from the individual and from the design team. Good design includes learning and practicing creative techniques, analyzing carefully, and being willing to learn wherever you can. Engineers need to interact not only with technical people, but also with other creative people when searching for solutions to design problems.

12.4 *PRESENTATION*

Now that you have your design, how do you present it to your engineering management, to general management, or to customers? No matter how good the engineering design may be, it is worthless unless you are able to sell the design to others. If you consider what follows, you will be able to present your designs so that others can understand and accept them. The presentations may be oral, written, and/or graphic. A good overall presentation will probably include written reports, drawings, and oral reports to technical groups and management.

12.4.1 What Is Being Presented?

The engineer must make clear whether the design involves a product or a system. Also, you must make clear what you are about to present; you may be able to do this in a sentence or two. A rambling presentation with no clear subject will lose your audience quickly. This would be the opening paragraph of a summary report.

12.4.2 Why Is the Presentation Being Made?

You should identify the objective for making the presentation. Are you trying to convince the audience of the need for the product? Are you selling the need for a better way to produce the product? Are you trying to obtain approval for a better redesign of the product? Again, this should be stated at the beginning of the presentation so the observer or reader will know exactly what you are trying to accomplish.

12.4.3 Who Is the Customer?

As the presenter you must first identify whether your audience will be composed of people within your own group, within the same company, or consultants from outside your organization. Does your report go only to your supervisor or will it be sent to readers of diverse backgrounds? This could have a bearing on the understanding the audience may already have of the product or system. Also, the prior knowledge of the audience should be considered. Is the presentation to be made to production personnel, management level personnel, or the board of directors? Drawings in orthographic projection will be good for other engineers; management and marketing personnel may understand pictorial drawings and exploded assemblies more readily. Each of these groups has different interests and brings a different background and point of view to the problem.

12.4.4 When Will the Presentation Be Made?

Time constraints are important to consider for meetings. These include the time available to prepare the presentation and the time line needed for the design to be initiated. Graphic methods of presentation are useful here. The time span of a project might best be presented graphically using a PERT (Program Evaluation and Review Technique) chart that shows how the times for various parts of the project are related (Figure 12.14).

12.4.5 Where Will the Presentation Be Made?

Often you have the luxury of making the presentation in your conference room where you are very comfortable, where you know the availability of resources such as videotape or closed circuit television, and where you have unconditional access for set-up time. Other times, you must take your presentation "on the road," where some of these factors may not be available. If

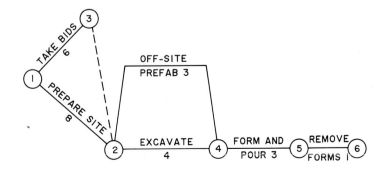

PLAN FOR CONCRETE STRUCTURE

Figure 12.14 PERT chart: Graphic time plan for project.

this happens, you have much less control over the presentation environment and must be able to make quick decisions concerning the presentation. One other consideration when presenting at a location away from your facility is the transportation of the materials needed to make the presentation. It is important to be able to move the objects to the new location (size of the object vs. size and availability of the vehicle to move it) and to have the needed personnel to move it. Also, weather can play a role since rain and snow can ruin some well-prepared illustrations.

12.4.6 How Will the Information Be Presented?

To make the greatest impact, the presenter must decide the method of presentation and what visual aids are necessary. A written report should have a short (one- or two-page) summary in the beginning. An easily read drawing of the basic concept should be a part of this summary. Experimental data, analysis, estimates and detailed drawings will follow as a major appendix.

Most readers will not examine the detailed information unless they are convinced by the summary that the concept is worthwhile. Many engineering design presentations consist of verbal presentations. As you learned in school, a plain lecture is a poor way to transmit information. However, if aids are used well, the impact of the presentation can be greatly enhanced. Simple drawings, charts, graphs, or other large graphics are always effective. The charts and graphs should be simple and easy to comprehend. Color can be very effective. Three-dimensional scale models are sometimes used when the design has been refined enough and the cost of the system or product warrants the cost of having the model built. As is often said, a picture is worth a thousand words. If this is true, then a moving picture (videotape) is worth a thousand pictures and a three-dimensional model is worth even more.

12.4.7 How Much Will the Presentation Cost?

As with all phases of design, the cost of the presentation is important. The cost should be relative to the level to which the design has progressed, the level of the audience, and the importance of the decision resulting from your presentation. Remember, no matter how good the design is, it is worthless if you cannot convince others of its value.

12.4.8 Notes on Content of Presentations

Each of the preceding paragraphs describing presentations included a key word. These key words are questions that need to be answered in any report. Good technical reports always answer them, good news reports always answer them, and you will want to be sure that you do too. The questions are

What? Why? Who? When? Where? How? How much?

12.5 PROTECTION OF DESIGNS

A design is a valuable property. Much time and talent is required to develop a good product or a useful system. Those who create these things should have the right to enjoy the fruits of their labor. This can be accomplished in several ways.

Some people protect their ideas by not disclosing them and by safeguarding them from others who might benefit from detailed information about the invention. This method works best for systems that are used within a private organization. If you develop a new assembly method for a product or a new chemical process, competitors see the result but do not know how you achieved it. Nondisclosure can protect a design under these circumstances.

More commonly, new designs are protected by patents. A patent is a license granted by a government that gives the owners the exclusive right to use the design for their benefit. Others wishing to use the design must pay a license or royalty fee to use it. The United States has a patent office as do many foreign countries.

The application to patent an invention must be filed on special forms with the patent office and accompanied by drawings that describe the device. The requirements for these drawings are very strict. A sample drawing is shown in Figure 12.15. Both products and manufacturing processes can be patented. Once the application is filed, the inventor may mark the device as "patent pending" and offer it to the public. The patent office searches all its records to learn if the proposed design is similar to others or is truly different. Sometimes it may be found to be similar in some respects and different in others; then a patent is granted only for the areas in which it is different. The search and final granting of a patent may take several years. Once it is granted, the owner of a U.S. patent has the exclusive right to use or license the invention for 17 years and the possibility of another 17-year renewal.

Patents are costly. Much time and effort of engineers, attorneys, and drafters is required to prepare an application. If the information is incomplete, the application will be rejected. One very important factor in patent records is time. If two people seek to patent similar devices at the same time, the one with the earliest record of conceiving the design will be granted the patent. For this reason written records, in ink and in bound notebooks, should

Figure 12.15 Example of patent drawing. Courtesy of Roylance, Abrams, Berdo & Goodman

be kept of creative discussions, meetings, research laboratory work, development trials, and tests. The notebooks should be checked weekly or monthly by a person not directly involved in the invention and witnessed. The reviewer should sign and date each page of notes to indicate that they existed at the time of the review.

Written material such as books, music, and computer software can be copyrighted. This is a form of license from the government that gives the creator the exclusive right to benefit from the written work. A copy of the work and an application must be submitted to the copyright office. The work can be labeled ''copyright'' or with the symbol ''©'' as soon as published and even before the application is submitted. A copyright prevents others from using the work for profit; there are regulations that allow copyrighted materials to be reproduced in small quantities for educational purposes without permission or payment of royalties.

12.6 A DESIGN PROBLEM

This design problem is not very difficult, yet it includes all the elements of the design process. What is it? To design an organizer for holding the supplies and materials needed by a student of graphics. As you follow the development of this project, you will see each step of the design process.

12.6.1 Determine a Need

Some of the problems that graphics students have include:

Forgotten or lost equipment.

Bent or wrinkled drawings.

Broken or crushed models.

Storage space not available in graphics laboratory.

Adequate drawing surface not available in residence.

Students are often seen scrounging in their backpacks for needed triangles, scales, or pencils; there is a need to keep graphics equipment so it can be easily found. Sometimes, laboratory drawings appear to have been through a war: wrinkled, wet, torn, or dirty. Finished work is valuable and needs protection. Students occasionally build models of a particular problem, but a crushed or broken model does not demonstrate a concept very well.

Some graphics laboratories are equipped with locked drawers

in each work station so that students can leave drawings and equipment in the laboratory. Not all graphics laboratories have these facilities. Some students live in crowded quarters where they have no smooth surface available on which to solve graphics problems, and it would be nice to have a portable drawing surface. With these needs in mind we can pursue the next step in the design process.

12.6.2 Establish Known Information

The organizer should hold all the equipment that a student usually needs to solve graphics problems. Typical equipment was shown in Chapter 1 and is shown again in Figure 12.16. The sheets to be carried in the organizer would be those from the workbook or those from the textbook, usually $8\frac{1}{2} \times 11$ in. (Figure 12.17). A graphics model might be of foam, clay, or cardboard; the model of a cut block used in Chapter 10 is shown in Figure 12.18.

Figure 12.16 Tools to be carried in graphics organizer.

Figure 12.17 Workbook sheets that should be in organizer.

Figure 12.18 Three-dimensional model to carry in organizer.

Figure 12.19 Quick analysis of space requirements to determine minimum width, height, and depth.

These materials establish the minimum dimensions of the organizer. The scale is the longest item; the model and cleaning pad have the greatest height. The flat pieces are the largest overall in combined width and depth. The minimum space requirements are sketched in Figure 12.19. You may have equipment that differs a little from these pieces, but these are typical. We should also allow a little extra space for tools a bit larger than those we have shown.

12.6.3 Stimulate Design Concepts

Here you can go into a "lateral thinking" mode. No ideas are unacceptable; unused ideas may stimulate a useful idea from another designer. Our design team may remember that some students in design and architecture use fishing tackle boxes for their equipment; most tackle boxes will not hold an $8\frac{1}{2} \times 11$- in. drawing without rolling or folding, but the idea is useful: The individual compartments in the tackle box keep the lures from getting mixed up and tangling the hooks; it would be good to have different compartments for different pieces of equipment.

The drawings and equipment need to be kept dry, so we need a case that closes tight. Should it fit in the backpack? Some quick sketches of "floor plans" show that if it holds all tools on one level, it would be too large for most backpacks (Figure 12.20). Figure 12.21 gives the overall dimensions of a one-level organizer with the second floor plan; note that the dusting brush handle would be toward the user and also that the scale would be handier than in the first plan. The lid has pockets for drawings in process or completed and a pocket for disks for computer graphics. Since this is too large for most packs, an alternative concept is developed in Figure 12.22. This model has a smaller floor plan accomplished by putting smaller parts in drawers under the large-area storage. The package is now 4 in. thick, perhaps too big for a backpack.

Figure 12.20 Two possible "floor plans" for organizer.

Figure 12.21 Concept sketch for organizer.

Figure 12.22 Another concept sketch smaller in width and depth but larger in height than first concept.

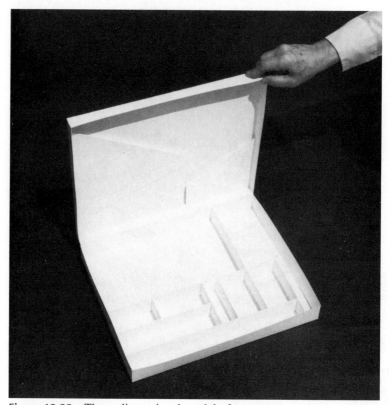

Figure 12.23   Three-dimensional model of a concept.

Since the organizer appears to have outgrown the backpack, it should be water resistant (rain resistant, not swimming pool resistant). At this point we may consider putting the workbook and the textbook into the organizer: It becomes much bigger and heavier. The team may decide to leave the books in the backpack.

Now we consider materials. Our model is of illustration board (Figure 12.23), but illustration board would not survive many rain storms. The participants in our concept session have steel and plastic tackle boxes. Both have survived several fishing expeditions, so plastics and steel are listed as candidate materials. One of the participants brings in a hand-made organizer made of walnut wood. It is beautiful, but walnut is too expensive (Figure 12.24). After considering some good ideas on the use of space in this example, wood is listed as a candidate material.

As the concept development progresses we discuss the idea of a drawing surface as a clipboard with a rectilinear grid on one side and an isometric grid on the other to aid in sketching with translucent media.

This is not an exhaustive list of concepts. You and your fellow students could produce many more. Nor are any of the concepts an ultimate or optimum design. These are the concepts developed so far and the next step is to investigate them.

12.6.4 Investigate Design Concepts with Analytical Tools

One of our first investigations is to determine if the designs will hold the needed tools. Figure 12.25 shows a model with the tools in place. Did they all fit? Is there enough space for each item? Could they be better arranged?

Another area would be strength calculations. How strong does the organizer need to be? What kind of strength: tensile, compressive, shear, bending, or crushing? There are no large loads on the box, but it may have books stacked on top of it or some one might sit on it. So we should check crushing of the box, which would be related to the compressive strength of the outside walls and the interior dividers and the beam strength of the top and bottom. We could assume that the worst load would be a 200-lb person sitting on the organizer; if they sat down hard, the impact might be 300 lb for an instant. These calculations would be done

Figure 12.24 Hand-made graphics kit box, fabric lined and beautifully finished.

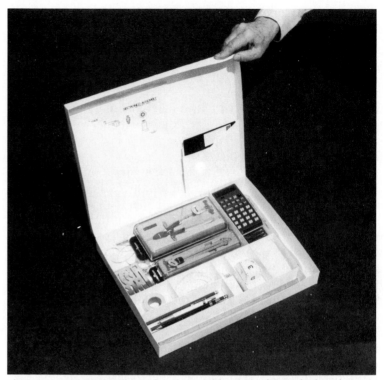

Figure 12.25 Three-dimensional model loaded with typical tools.

for different materials to determine minimum thicknesses for walls and partitions. This leads us to materials analysis.

Different materials have different costs, different strengths, and different resistance to moisture. As an aid in eliminating poor choices for materials, we can make a simple manufacturing cost estimate. This estimate would take into account only the direct costs of production (Figure 12.26). Manufacturing overhead, distribution costs, and profit would be considered later in a total economic model. Strength of materials and resistance to moisture can be found in engineering handbooks and current literature on materials. Tensile strength of three possible materials are shown here:

Material	Tensile Strength (psi)
Steel sheet	60,000
HD polyethylene	3100–5500
Polypropylene	5000

Steel has a much higher tensile strength than plastics, but it has a density of about eight times that of plastics. Considered as strength per unit weight, it is only $1\frac{1}{2}$ times stronger than plastics.

The investigative steps described here are only a few of those necessary for a thorough analysis. The quality of the investigation is dependent on the value of the final result and the cost of incomplete information: If our organizer project is delayed, some people may not realize the benefits of this device; if a flood control project is delayed or includes errors, lives and much property may be lost. The flood control project would require a thorough and exacting analysis.

12.6.5 Gather Investigation Data and Review

After the investigations are completed, the team would review the information. Suppose the cost analysis shows that the manufacturing cost of a steel box and a plastic box are about the same. Then we look at finishing costs. The steel box will need to be painted; a plastic box can be molded in color and will not rust. Latches for a steel box will be fabricated separately and probably spot welded to the basic box; latch parts can be molded directly into the plastic boxes. It appears that a plastic material is a better choice. Which one? This may be a time to recycle or iterate our analysis. Polypropylene has self-hinging properties (the top and bottom could be molded in one piece with a hinge strip in between). Exact material costs, mold costs, and production rates may need to be examined to make this decision.

The shallow design is larger in area but has no extra parts such as drawers. The drawers would require additional molds and some means of keeping them in place during transport. So our team may decide to pursue the shallow design.

After a review of the investigations, we decide on a shallow, injection-molded polyethylene container. Now we need to test the design. We could build one from polyethylene shapes or carve one from a block of polyethylene. In neither case would the physical properties be exactly like a molded shape. If this appears to be a good project, we might invest in a low-cost, temporary mold and mold a few boxes for testing.

12.6.6 Now Test the Design that Appears Best

Some of the tests that would need to be conducted include crushing strength, water resistance, and wear resistance (abrasion). (Figure 12.27).

GraphiComp, Inc.

PRELIMINARY COST ESTIMATE

PROJECT _GRAPHICS ORGANIZER_

DATE _9/1/88_ BY _F.D.M._

DIRECT COSTS $/Unit

    Materials .63

    Labor 1.20

    Sub-Total 1.83

SEMI-DIRECT COSTS

    Energy .20

    Tooling .15

    Supervision .40

    Sub-Total .75

DISTRIBUTED FIXED COSTS

    Equipment .13

    Plant services .40

    Sub-total .53

TOTAL UNIT COSTS 3.11

Figure 12.26 Brief manufacturing cost analysis.

CRUSH STRENGTH   WATER RESISTANCE   ABRASION

Figure 12.27 Three types of tests that need to be conducted for organizer.

A market test would also be a good idea. Some of the molded organizers could be finished and shown to graphics students and instructors to get their opinions about features, improvements, and willingness to buy such a device. Based on the willingness to buy the organizer at different selling prices, an economic model could be built on a computer to show profit or loss at various sales volumes, different prices, and different manufacturing concepts.

During our tests the molded box allows water to enter when exposed to a shower approximating a hard rain. A problem! A 200-lb weight dropped on the case 100 times from a height of 6 in. results in no noticeable damage. The prototype latches also stay closed during this test.

Some comments from potential users are favorable. However, some want a provision to carry a T-square. Simulated use show that the dry cleaning pad sheds particles that find their way throughout the organizer, even to the computer disk stored in the lid. Another problem!

Physical and market testing disclose at least two problems. Now it is time to review the project to date.

12.6.7 Propose to Accept, Modify, or Reject the Design

The design as tested is unacceptable. There is considerable interest among potential users if the price could be held below $15. So our team decides to modify the design. The water problem is resolved by changing the match line joint from a simple mating of the two sides to a U-channel that provides a sort of labyrinth seal (Figure 12.28). Investigation of some quality luggage finds a

Figure 12.28 Improved design for match line and interior corners of organizer.

similar joint used there. The dust from the dry cleaning pad could be contained if the pad were stored in a separate small container, such as a soap box sold for travel or a molded box such as the ones used to package screws. The internal corners of the organizer are also given generous fillets to improve cleanability (Figure 12.28).

If the modifications made are significant changes in terms of design or cost, we need to go back to testing the design and then reconsider our decision to accept, modify, or reject. Discussion with a mold designer may indicate that the U-seal slightly increases costs; the separate container for the dry cleaning pad also increases costs. These are judged not to be large enough to reject the project or go back further in the iteration process.

12.6.8   Render a Final Decision

This is it! Is the organizer as now conceived ready for production? A final review of the design, the market, and the economics will tell us if we should proceed.

Does the organizer meet the needs as determined in the beginning? Our sample of users says it would. Will enough people buy it to enable us to buy the tools to make it, buy the materials, pay the people, distribute the product, and make a profit? If our economic model says we have better than a 50% probability of doing this, our answer may be ''go.'' (If we are very conservative, we may want a higher probability of success.)

This is the time to make the detail and assembly drawings. From the detail drawings, the molds will be designed to mold the two sides of the case (Figure 12.29). The portable drawing board will be purchased as sheets of pressed board, cut, sanded, and faced with grid surfaces on both sides. Pockets for drawings and disks will be cut and installed in the top according to the assembly drawings (Figure 12.30). The bands to hold the drawings and disks will be installed according to the assembly drawings and the two sides of the organizer joined as shown in the assembly. The completed organizer is shown in Figure 12.30. This final decision will also trigger a packaging study that becomes a miniproject within the larger project. Go!

12.6.9   Optimize Design By Iteration

Are we done? Never! As the detail design and assembly drawings are made, questions will arise about the parts. Decisions about exact contours and locations will need to be answered. We may need to review some steps of the design process to get answers.

Figure 12.29 Detail drawing for lower half of organizer.

BILL OF MATERIAL				
NO.	REQ	DESCRIPTION	PART NO.	NOTES
1	1	UPPER ASSY	1021	DWG. 18-2
2	1	LOWER SHELL	1022	DWG. 18-3
3	1	HINGE PIN	1003	DWG. 18-7
4	1	BOX-CLEAN PAD	1025	DWG. 18-5
5	1	DRAWING BOARD	1026	DWG. 18-11

REVISIONS		
NO.	DESCRIPTION	DWN DATE

GRAPHICOMP, INC.
COLUMBUS, OHIO

DRAWN: DPO **DATE:** 1-11-88
CHECKED: CDD **DATE:** 1-13-88
APPROVED: FDM **DATE:** 1-5-88

PART: LOWER SHELL
PART NO: 1022
NEXT ASSY: 18-1

GEN. TOL: ±.020, ± $\frac{1}{2}$°
MATL: HDPE
SCALE: NOTED

DWG. NO. 18-3

Figure 12.30 Pictorial assembly of organizer.

When production begins, new information and new questions will be found. And the final judges of our work, the *users,* will tell us if the design meets their needs. They will note the failures and the improvements needed.

If we intend to continue to produce graphics organizers, the input from the users must be continually fed back to the design process. The product must be continually reevaluated through the steps of the design process to meet the users' need and to lower costs. If we fail to do this, we shall join the list of organizations that are extinct because they failed to meet their customers' needs while maintaining their own economic health.

12.7 *SUMMARY*

Design is the essence of engineering. It is a creative process and also an iterative process. The steps of the design process may be applied to any creative endeavor: a part, an assembled product, a process, or a system. Both vertical thinking and lateral thinking contribute to the design process. The steps may be remembered with the acronym DESIGN PRO:

D	Determine a need.
E	Establish known information.
S	Stimulate design concepts.
I	Investigate design concepts with analytical tools.
G	Gather investigation data and review.
N	Now test the design that appears best.
P	Propose to accept, modify, or reject the design.
R	Render a final decision.
O	Optimize design by iteration.

During the design process we need to communicate with others. Good communication requires that we answer basic questions in our written and oral presentations:

What? Why? Who? When? Where? How? How much?

A design is valuable property. Designs can be protected by patents and copyrights. If the design will be used in other countries, it must be protected under the laws of each country in which it will be used.

A design is never perfect. We go with the best we can conceive with the resources we have and then we iterate to improve it.

PROBLEMS

These problems are all presented verbally. You may receive problems in other situations as a combination of sketches and verbal requirements, or all sketches, or all verbal. Your instructor will specify how far you are to pursue the design process: ideas, concept sketches, plan for analysis, models or working drawings.

Problem 12.1
Need: a portable device to hold an $8\frac{1}{2} \times 11$ sheet of paper for reference while using a computer. Many computer stations do not have table space for reference material. The device must be portable, low-cost, lightweight, and inexpensive.

Problem 12.2
Need: a simple locking mechanism for automobile wheel covers. Thieves often steal wheel covers. Currently available locking devices require special key wrenches. The device must be low-cost, and it must be easy for the owner to unlock to change a tire.

Problem 12.3
Need: a holder for eyeglasses that will protect them from damage while the owner works at a computer station. Lenses are easily damaged; many users remove their glasses or use a second pair for computer work. The user should not have to fold the side supports—this would require a two-handed operation.

Problem 12.4
Need: a stadium pack for sports events that would serve as a seat with backrest and provide storage for a vacuum bottle, umbrella or rain cape, and programs. It should be easy to carry and attractive.

Problem 12.5
Need: a portable bookshelf for student housing. The shelf should be collapsible for easy moving and should provide built-in bookends.

Problem 12.6
Need: a small buffer/grinder stand for a basement workshop. A small electric motor ($\frac{1}{2}$ horsepower) would drive the tool through a V-belt drive. The tool should accept different size pulleys, up to 8 in. diameter, on one end, and different size wheel and brushes, up to 6 in. diameter, on the other end. The frame should be rigid (it might be a casting or a weldment).

Problem 12.7
Need: a candle holder that will immediately extinguish the flame if it is tipped over. A standard candle has a tapered base with a maximum diameter of $\frac{7}{8}$ in. The design should be aesthetically pleasing as well as safe.

Problem 12.8
Need: a single auto rack that can hold skis or a bicycle. Many people

like to ski in cold weather and bike in warm. A single rack would reduce changeover time and costs. The rack must be attractive and it would be a plus if it could be removed without damage to the auto.

Problem 12.9
Need: a dispenser for toilet paper that will limit the user to removing not more than 6 sheets at a time. A common roll of toilet paper is about 5 in. diameter and $4\frac{1}{2}$ in. long. The core diameter is approximately $1\frac{3}{8}$ in. Each sheet $4\frac{1}{2}$ in. square. Durability and ease of loading are prime considerations in this design.

Problem 12.10
Need: an entertainment center for a college room. It should be not more than 2 ft. square by 4 ft. high and should contain a television set, speakers, amplifier, AM/FM tuner, tape deck, record player (33 rpm or compact discs), and storage for tapes and records or discs. All wiring should be concealed and neatly run and power surge protection should be provided. Security and portability are major considerations (somewhat in opposition).

A
ADVANCED DIMENSIONING TOPICS

A.1 *INTRODUCTION*

You learned about dimensioning and tolerancing in Chapter 7. The material you studied there is an introduction to the fascinating world of production. It is only an introduction, for in actual production environments engineers must consider many other variables besides those described in Chapter 7. The standard for dimensioning practice, ANSI Y14.5M, includes information on topics beyond the scope of that chapter. It is one of the many references design and production engineers would use to develop a concept into useful products or processes.

The standard provides symbols for describing types of tolerances; this goes beyond just limits on numerical values. It also presents the concept of datum references: planes to which we can tie all the critical dimensions of a part. Geometric tolerancing includes tolerances for location, form, profile, orientation, and runout. These will be described in the section on geometric tolerancing.

Tolerancing for metric dimensions will also be reviewed. This information is based on ANSI B4.2 and B14.5. The methods for metric tolerancing are much different from those you have learned so far. Fortunately, the standards provide for transition steps in dimensioning so that designers, tool makers, and manufacturing personnel can learn the system gradually.

You learned in earlier dimensioning studies that the accuracy of measurements is related to the accuracy of a surface. (You cannot measure to 0.0001 in. between two surfaces rough as sandpaper.) Critical dimensions are always referenced to finished surfaces. The quality and character of finished surfaces varies. This variability is described in ANSI Y14.36 and B46.1. Material from the standards is included in this chapter.

If you have not worked in a manufacturing environment or had laboratory courses in manufacturing practices, the terms and concepts to be discussed will be unfamiliar to you. For this reason, this chapter includes a very short section on basic production processes before you get into the advanced topics.

A.2 MANUFACTURING PROCESSES

Manufacturing processes include methods for forming materials and methods for assembling parts made of various materials. Processes for forming can be classified as reformation, removal, and addition. Assembly processes include manual, mechanized, and automated.

A.2.1 Material Reformation

Figure A.1 Casting.

Figure A.2 Forging.

Historic methods for forming metals are casting and forging. In the casting process metal is heated until it is a liquid and the molten metal is poured into a mold. After it cools, the solid metal is taken from the mold. It will have the shape of the cavity in which it is cooled (Figure A.1). Casting involves making patterns of the object to be produced; enclosing the pattern in a mold of treated sand, ceramics, or metal of a higher melting point than the one to be cast; melting the metal; pouring it into the mold; cooling the cast part; and removing it from the mold. The part, as cast, may be very rough and have attached to it material needed for the casting process that is not a part of the finished object; this needs to be removed. All of these steps have been studied in detail, and there are texts, magazines, and technical papers providing information in depth. The casting process may also be used for nonmetallic materials such as ceramics and polymers. These plastic materials require less heat or no heat at all for casting. Some will melt at low temperatures while others will solidify through chemical action or without the addition of heat. They may be cast in precise metal molds and require little finishing after removal from the molds.

Forging is the method of the blacksmith (Figure A.2). Metal is heated until its structure changes and it becomes malleable (steel is usually red hot). Then it is hammered or pressed into the desired shape. The hammer may be wielded by a human arm or it may be powered by air or hydraulic fluid. The shape may be obtained by moving the forged part under the hammer to obtain the desired shape or it may be pressed into a mold known as a forging die. If the material is a thin sheet, it may be formed cold using only pressure. This related process is called pressing. Auto body parts

Figure A.3 Extrusion.

such as roofs, hoods, and doors are made by pressing. Forging and pressing are also complete engineering disciplines under constant study.

Extrusion is related to forging in that the material is softened but not usually liquid. The softened material is squeezed through a shaped hole (a die) and comes out as a continuous bar having the shape of the orifice through which it passed (Figure A.3). Extrusion may be used for metals, usually with heat to soften them. It may be used with plastic materials with heat or often without heat. Good extrusion processes can produce a very accurate shape with a smooth surface. The extruded bars are cut into pieces with each piece becoming an individual part.

Processes developed in more recent years include powder metallurgy and hot isostatic pressing. In the former the metal is ground to a powder and the powder pressed into a mold. Different metals or a mixture of metallic and nonmetallic materials may be used. The materials may be bonded by pressure only or they may be heated until they soften and fuse together. Powder metallurgy can produce a porous structure ideal for bearings: The cavities between the metal particles can be impregnated with oil or other lubricants that will be squeezed to the surface when the part is under load.

Hot isostactic pressing is a sophisticated process that immerses the heated raw material in a bath of high-pressure fluid that squeezes it uniformly from all sides into the desired shape. There are other reformation processes available. You will need to learn about new developments and the processes applicable to the products and process with which you work.

A.2.2 Material Removal

Material removal always produces scrap or waste material in addition to the desired part. These processes are still among the

Figure A.4 Drilling.

Figure A.5 Reaming.

most used and useful of manufacturing methods. They include a tool for removing material and a means of moving the tool against the part or the part against the tool or both. Material removal can produce a new geometric shape, for example, a round hole—a negative cylinder. It can also refine an existing shape—make the surface smoother or change a dimension. Classic processes include drilling, reaming, boring, turning, milling, broaching, shaping, and grinding. More recent processes include superfinishing, chemical milling, and electron discharge machining. Many other processes exist that are specialized for a single industry or are part of a production line making many units of a product; here they may be developed to perform a single operation very quickly with great precision.

Drilling uses a cylindrical tool with a pointed end and spiral cutting flutes (grooves) cut into the cylindrical surface. The tool is made of hardened steel. It is pressed against the surface to be formed and rotated (Figure A.4). The tool is usually rotated, but in some cases the part may be rotated. The drill enters the part and generates a round hole as it turns and advances.

Reaming is a finishing process: You start with a drilled hole and make it smoother and more accurately sized. The tool is a cylinder with a flat end slightly tapered or chamfered (you do not need a conical point to start the hole) and straight cutting flutes parallel to the axis of the reamer. It is rotated as it advances into the work. It removes only a small quantity of material as it smooths and sizes the hole (Figure A.5).

The process of boring is related to drilling and reaming. It uses a single-point tool to generate a negative cylinder. The tool is usually fixed and the work revolves (Figure A.6). The tool holder is fed axially into the work to the depth needed.

Figure A.6 Boring.

Figure A.7 Turning.

Figure A.8 Milling.

Figure A.9 Broaching.

Turning is the classic method for producing positive cylinders. The work is held between centers, or by one end, and rotated. As it turns, a single-point tool is pressed into the work and advanced axially to generate the cylindrical shape (Figure A.7). Tapered or conical shapes can be generated by offsetting the axis of the work from the path of travel of the tool. Multipoint tools may be used to generate special shapes.

Milling is one of the most versatile of machining processes. Shaped cavities, external shapes, or surfaces can be produced by milling. The work is fastened to a table that can be moved sideways, forward, backward, vertically, or tipped at an angle. The rotating tool generally moves vertically, but some move on a horizontal axis. The concept of milling is shown in Figure A.8.

The work is held still in the broaching process. A special tool, often resembling a large file, is pulled past the work. Each cutting edge on the broach removes a little material; with many teeth the broach can remove a lot of material in one pass (Figure A.9).

Shaping can produce flat parts or parts of a constant contour. The work is clamped to a fixed table that can be adjusted vertically while the tool is pushed across the work. The tool cuts on the forward stroke and is hinged to slide back over the work on the back stroke (Figure A.10).

Grinding uses an abrasive wheel or disk to remove material (Figure A.11). The abrasive tool may also have a special shape. The work is fed into the tool, and the abrasive removes small amounts of material. A course abrasive coupled with a rapidly moving work piece can remove a lot of material in a short time. A fine abrasive coupled with a slow movement of the work can result in a very fine and accurate finish.

Figure A.10 Shaping.

Figure A.11 Grinding.

A.2.3 Material Addition

Welding was discussed in Chapter 8 as a method for joining parts. Welding can also be used to add material to a single part. It is often used to repair or salvage a part that is broken or undersize in one area. A small amount of material may be deposited from the weld rod in one pass; multiple passes can be used to build up greater thicknesses.

Plating is another method for adding material to a part. This method adds a very thin layer of material to a part, often to provide new properties to the surface. Common plating materials are nickel, chromium, cadmium, silver, rhodium, and gold. Some plating is accomplished by dipping; many materials are applied by electroplating. The objective of plating may be to bring an undersize part back to specification, but more commonly it is to provide protection against corrosion. Plating with silver or gold is used to improve the electrical conductivity of electrical parts. These precious metals are also used to improve appearance. If a part is to be plated, the thickness of the plating must be taken into account in the design. The basic part must be finished undersize, so that when the plating thickness is added, the part is within needed tolerances.

Coating is another way to add material to a part. Some coating materials are nonmetallic and cannot be applied by electroplating. Coatings may be used to protect against corrosion or to improve appearance or both. A common method of coating is the blueing applied to the steel parts of firearms. It is wiped on the steel surface and bonds to the metal to provide protection against corrosion. Aluminum is anodized. This is a method of electrical deposition that provides a thin, protective layer over the aluminum. It also can provide color. Coatings are relatively thin and most often applied to exterior surfaces rather than mating surfaces, so tolerances are not so critical. The important consideration of thickness is getting a coating thick enough to protect or color the part while minimizing cost.

Material addition by welding will usually produce a surface that needs to be machined: The surface will be rough and irregular. Chromium plating is often a two- or three-step process. A coating of nickel is usually deposited on the steel to seal it against corrosion and the chromium deposited over the nickel to provide the shiny, hard surface. Chromium plating, by itself, is porous and would allow pinpoint corrosion. Coatings are thin and usually duplicate the surface quality of the underlying part.

A.2.4 Assembly Processes

Over half of the work required to produce useful products may be in the assembly of the component parts. This is true for electronic equipment; farm machinery typically has less than a quarter of the work in assembly. It is an area that production engineers constantly study.

Originally all assembly work was done by hand. One craftsman conceived the device, made the parts, and put them together. If the parts did not fit, he would adjust or remake parts until the device functioned properly. Beginning in the nineteenth century, the concept of interchangeable parts was introduced (thus the need for limit dimensions). A worker or team of workers received a collection of parts and an *assembly drawing*. Their job was to put the parts together according to the drawing. This method is still used for one-of-a-kind machines or in small shops building special equipment.

At the beginning of the twentieth century the concept of assembly lines was introduced. Frederick Winslow Taylor, the *father of scientific management*, developed time and motion study. Henry Ford conceived the assembly line and applied it to the building of the Model T Ford car. The work was subdivided into individual tasks, and each worker did his task as the cars moved by on a conveyor. Parts were delivered to the worker and tools were provided for the needed tasks.

The next step was to mechanize the operations so that the operation was independent of the strength and skill of the worker. Powered wrenches tightened fasteners to exactly the same torque every time. Parts feeders supplied the fasteners and other parts to the point of use without need for the worker to reach for them. The worker was still there, but the need for high skills was reduced for many operations. Imagine the consternation if the parts supplied were out of tolerance or the power tools were not adjusted properly! The design and production engineers had to transfer their knowledge directly to the production process; the craftsman was no longer using his skills to correct their errors.

We are now in the era of automated assembly. Parts are delivered to an assembly line by automatic feeders and put in place by robotic-powered tools. Positioning, fastening, and welding are all performed by robotic elements that sense position and the presence of the part and perform the same task in the same way every time without tiring. Now the engineers have no one to correct their errors; they must design the parts for feeding and orienting and design the end product so that it can be assembled easily without human intervention. The workers on the line now become supervisors of the machines. Their responsibility in the best manufacturing operations is to monitor quality and note ways to improve methods and product quality.

Design for automatic assembly places greater demands on the engineers. Not only must the end product perform, but it must be capable of being assembled by machines. This requires a different perspective on component design and assembly methods. The good design engineer will need to study research and current literature in both areas.

A.3 STANDARD TABLES FOR FITS

The application of tolerances to metric dimensions is prescribed by the International Standards Organization and by ANSI standards Y14.5 and B4.2. Tolerances for cylindrical fits in inches are prescribed in USA Standard B4.1. These standards are based upon years of experience and should be used for selecting tolerances unless special conditions require other choices.

A.3.1 Preferred Limits and Fits in Inches

Standard B4.1 includes all of the terms and concepts you learned in Chapter 7. It includes some new terms and concepts you will need to understand for the development of the tables in Appendix B. However, you can apply the tables starting with the concept of *basic size*. Although the tables may be applied to any basic size you choose, the standards also list preferred basic sizes for fractional and decimal dimensions (Figure A.12).

As the designer develops an assembly, he or she knows what kind of fit must exist between adjacent parts. There are three general types of fits: running fits, locational fits, and force fits. Each of these is further subdivided and identified with a letter code:

RC: running or sliding clearance fit.

LC: locational clearance fit.

LT: transition clearance or interference fit.

LN: locational interference fit.

FN: force or shrink fit.

There are nine classes of RC fits ranging from the closest, RC1, to the loosest, RC9. Each letter group is similarly subdivided. You will need experience to do a good job of selecting which class is needed for a particular pair of mating parts. The classes are described in the standard; an example follows: RC 7 *Free running fits* are intended for use where accuracy is not essential or where large temperature variations are likely to be encountered or under both of these conditions.

A portion of one of the tables is shown in Figure A.13; more complete tables are included in Appendix B. Note that all limits are given in thousandths of an inch. You start with the basic size to apply the tables. Note the ABC agreements are mentioned; they were also referred to in Chapter 8 as an agreement among the United States, Great Britain, and Canada. The symbols below the words *hole* and *shaft* are taken from the ABC System and are related to metric tolerancing, which is reviewed in the next section.

Assume you have two mating parts that need to move freely and to fit precisely with little temperature difference. This is the description for an RC 3 fit. If the nominal size is 1.40, then the basic size is 1.4000. This is true because the table provides values to 0.0001 in. in the range we are using. Reading down the table we are in the RC 3 column. Reading across the table we are in the nominal-size row "over 1.19 to 1.97." This zeros us in on one cell of the table, which looks like this:

Preferred Basic Sizes
Decimal

0.010	2.00	8.50
0.012	2.20	9.00
0.016	2.40	9.50
0.020	2.60	10.00
0.025	2.80	10.50
0.032	3.00	11.00
0.040	3.20	11.50
0.05	3.40	12.00
0.06	3.60	12.50
0.08	3.80	13.00
0.10	4.00	13.50
0.12	4.20	14.00
0.16	4.40	14.50
0.20	4.60	15.00
0.24	4.80	15.50
0.30	5.00	16.00
0.40	5.20	16.50
0.50	5.40	17.00
0.60	5.60	17.50
0.80	5.80	18.00
1.00	6.00	18.50
1.20	6.50	19.00
1.40	7.00	19.50
1.60	7.50	20.00
1.80	8.00	

Figure A.12 Preferred basic sizes (in.). Courtesy of USAS (B4.1)

Nominal Size Range Inches		Class RC 1			Class RC 2			Class RC 3			Class RC 4		
		Limits of Clearance	Standard Limits		Limits of Clearance	Standard Limits		Limits of Clearance	Standard Limits		Limits of Clearance	Standard Limits	
Over	To		Hole H5	Shaft g4		Hole H6	Shaft g5		Hole H7	Shaft f6		Hole H8	Shaft f7
0	− 0.12	0.1 0.45	+ 0.2 0	− 0.1 − 0.25	0.1 0.55	+ 0.25 0	− 0.1 − 0.3	0.3 0.95	+ 0.4 0	− 0.3 − 0.55	0.3 1.3	+ 0.6 0	− 0.3 − 0.7
0.12	− 0.24	0.15 0.5	+ 0.2 0	− 0.15 − 0.3	0.15 0.65	+ 0.3 0	− 0.15 − 0.35	0.4 1.2	+ 0.5 0	− 0.4 − 0.7	0.4 1.6	+ 0.7 0	− 0.4 − 0.9
0.24	− 0.40	0.2 0.6	+ 0.25 0	− 0.2 − 0.35	0.2 0.85	+ 0.4 0	− 0.2 − 0.45	0.5 1.5	+ 0.6 0	− 0.5 − 0.9	0.5 2.0	+ 0.9 0	− 0.5 − 1.1
0.40	− 0.71	0.25 0.75	+ 0.3 0	− 0.25 − 0.45	0.25 0.95	+ 0.4 0	− 0.25 − 0.55	0.6 1.7	+ 0.7 0	− 0.6 − 1.0	0.6 2.3	+ 1.0 0	− 0.6 − 1.3
0.71	− 1.19	0.3 0.95	+ 0.4 0	− 0.3 − 0.55	0.3 1.2	+ 0.5 0	− 0.3 − 0.7	0.8 2.1	+ 0.8 0	− 0.8 − 1.3	0.8 2.8	+ 1.2 0	− 0.8 − 1.6
1.19	− 1.97	0.4 1.1	+ 0.4 0	− 0.4 − 0.7	0.4 1.4	+ 0.6 0	− 0.4 − 0.8	1.0 2.6	+ 1.0 0	− 1.0 − 1.6	1.0 3.6	+ 1.6 0	− 1.0 − 2.0

Figure A.13 Portion of standard fit table (USA Standard). Courtesy of USAS (B4.1)

$$1.0 \quad +1.0 \quad -1.0$$
$$2.6 \quad \ \ 0 \quad -1.6$$

The first column gives us the limits for the fit, the second gives us the limits for the hole, and the third gives us the limits for the shaft. Applying these limits to the *basic size* we get:

Limits for hole: $\dfrac{1.4000}{1.4010}$ Difference = 0.0010

Limits for shaft: $\dfrac{1.3990}{1.3984}$ Difference = 0.0006

Tolerance of system = clearance$_{max}$ − clearance$_{min}$
= 0.0026 − 0.0010 = 0.0016

Tolerance of system = sum of individual tolerances
= 0.0010 + 0.0006 = 0.0016 *Check!*

You can apply the tables to solving any tolerance problem for a pair of parts in the same manner.

A.3.2 Preferred Metric Limits and Fits

Standard B4.2 describes the system for tolerancing metric dimensions. It includes an abbreviated way for designating tolerances that assumes that those using the information have access to the tables in the standard. A 40-mm hole can be completely toleranced by the description 40H8; a mating shaft could be described as 40f7.

Some terms you will need to know to understand the system follow:

Deviation: The algebraic difference between a size and the corresponding basic size.

Upper deviation: The algebraic difference between the maximum limit of size and the corresponding basic size.

Lower deviation: The algebraic difference between the minimum limit of size and the corresponding basic size.

Fundamental deviation: That one of the two deviations closest to the basic size. The letter H in the hole specification 40H8 designates the fundamental deviation.

International Tolerance Grade (IT): A group of tolerances that vary depending on the basic size but provide the same relative level of accuracy within a given grade. It is designated by the number 8 in 40H8; it would be described as IT8.

Basic Size, mm		Basic Size, mm		Basic Size, mm	
First Choice	Second Choice	First Choice	Second Choice	First Choice	Second Choice
1		10		100	
	1.1		11		110
1.2		12		120	
	1.4		14		140
1.6		16		160	
	1.8		18		180
2		20		200	
	2.2		22		220
2.5		25		250	
	2.8		28		280
3		30		300	
	3.5		35		350
4		40		400	
	4.5		45		450
5		50		500	
	5.5		55		550
6		60		600	
	7		70		700
8		80		800	
	9		90		900
				1000	

Figure A.14   Preferred basic sizes (mm). Courtesy ANSI (B4.2)

Note again that the basic size is the starting point for all these concepts. Preferred sizes also exist in the metric system (Figure A.14). The combination of a fundamental deviation and an international tolerance grade describes a *tolerance zone:* H8 or f7. Note that the fundamental deviation is a capital letter for cavities and a lowercase letter for external shapes. There are also preferred tolerance zones for both internal dimensions and external dimensions. Although there are over 100 tolerance zones for both holes and shafts, only 13 zones are preferred zones for each. Combinations of preferred holes with preferred shafts reduce the usual choices to 10 preferred fits (Figure A.15). Now the system begins to look very manageable and easier to specify than the inch system.

Assume that your design calls for a nominal size of 40 mm for a shaft and hole. If it is to be a free-running fit, you can see from Figure A.15 that the fit would be specified as 40H9/d9. The hole would be dimensioned 40H9 and the shaft 40d9. When we read the numerical tolerances from the tables, it looks like this:

	Hole H9	Free Running Shaft d9	Fit
Maximum	40.062	39.920	0.204
Minimum	40.000	39.858	0.080

| ISO SYMBOL | | DESCRIPTION |
Hole Basis	Shaft[1] Basis	
H11/c11	C11/h11	*Loose running* fit for wide commercial tolerances or allowances on external members.
H9/d9	D9/h9	*Free running* fit not for use where accuracy is essential, but good for large temperature variations, high running speeds, or heavy journal pressures.
H8/f7	F8/h7	*Close running* fit for running on accurate machines and for accurate location at moderate speeds and journal pressures.
H7/g6	G7/h6	*Sliding* fit not intended to run freely, but to move and turn freely and locate accurately.
H7/h6	H7/h6	*Locational clearance* fit provides snug fit for locating stationary parts; but can be freely assembled and disassembled.
H7/k6	K7/h6	*Locational transition* fit for accurate location, a compromise between clearance and interference.
H7/n6	N7/h6	*Locational transition* fit for more accurate location where greater interference is permissible.
H7/p6	P7/h6	*Locational interference* fit for parts requiring rigidity and alignment with prime accuracy of location but without special bore pressure requirements.
H7/s6	S7/h6	*Medium drive* fit for ordinary steel parts or shrink fits on light sections, the tightest fit usable with cast iron.
H7/u6	U7/h6	*Force* fit suitable for parts which can be highly stressed or for shrink fits where the heavy pressing forces required are impractical.

(Left margin, top to bottom): Clearance Fits — Transition Fits — Interference Fits

(Right margin): More Clearance — More Interference

[1] The transition and interference shaft basis fits shown do not convert to exactly the same hole basis fit conditions for basic sizes in range from 0 through 3 mm. Interference fit P7/h6 converts to a transition fit H7/p6 in the above size range.

Figure A.15 Description of preferred metric fits. Courtesy of ANSI (B4.2)

No addition or subtraction is requ d. The tables are longer, but no arithmetic is needed. Now, wha ɔ you put on your drawing? The following notes are recommended in the standards.

(a) $\dfrac{40.000}{40.062}$ (40H9)   (b) 40H9 $\dfrac{40.000}{40.062}$ (c) 40H9

Note (a) is recommended when the system is new to an organization; as experience is gained, note (b) is suggested; and when everyone is used to metric notation, note (c) may be used.

Metric dimensioning may be applied in a basic shaft system as well as in the basic hole system. The basic hole system is preferred unless you are dealing with standard size shafts and must dimension the holes to fit the shafts.

A.4 SURFACE CONTROL

We need to control surface characteristics for two reasons: Rough surfaces cause friction and wear with moving parts and it is hard to get precise dimensions when the surfaces are irregular. Surface control is addressed in ANSI Standard B46.1; the symbols for surface texture are defined in ANSI Y14.36.

There are three surface characteristics defined in the standards. *Roughness* refers to the small hills and valleys that can be found on a surface. They may be obvious on a surface such as sandpaper, but they still exist in smaller form on smooth-looking chromium-plated hardware. *Lay* refers to the direction of the tool marks on a machined surface. This factor can be important on mating surfaces that move relative to each other. *Waviness* refers to surface irregularities greater than roughness; they can usually be seen as a pattern on a machined surface.

Roughness is defined on a drawing as the arithmetic average of the deviations above and below a mean height of a surface. It is expressed in microinches or micrometers. Preferred values for surface roughness are shown in Figure A.16. Surface roughness is measured with electrical instruments or it may be estimated by comparison with standard samples (Figure A.17). The manufacturing processes referred to earlier in this appendix produce

Micrometers—μm		Microinches—μin	
μm	μin	μm	μin
0.012	0.5	1.25	50
0.025*	1*	1.60*	63*
0.050*	2*	2.0	80
0.075	3	2.5	100
0.10*	4*	3.2*	125*
0.125	5	4.0	160
0.15	6	5.0	200
0.20*	8*	6.3*	250*
0.25	10	8.0	320
0.32	13	10.0	400
0.40*	16*	12.5*	500*
0.50	20	15	600
0.63	25	20	800
0.80*	32*	25*	1000*
1.00	40		

*Recommended

Figure A.16 Preferred numbers for expressing roughness. Courtesy of ANSI (Y14.36)

Figure A.17 Surface roughness scales for comparison.

surfaces of different roughness; Figure A.18 provides some comparative ranges. If we want a shaft to run freely in a bearing hole, we would not put a cast shaft in a plain drilled hole. The surfaces would be rough—friction would be high, wear would be fast, and the combination would generate a lot of heat. If you were to specify appropriate tolerances for a precision running fit, you could not even measure them with these kinds of surfaces.

Lay is defined by the direction of the tool marks on a surface. Standard symbols and their meanings are shown in Figure A.19. Parallel and perpendicular lay would be the result of milling or shaping operations. Facing the end of an object held in a lathe would produce a circular pattern. Some methods of grinding would produce a multidirectional lay. Sand casting would reproduce the texture of the sand in the surface of the part and the lay would be P. Lay is independent of roughness but can also be a criterion for surface quality.

Waviness is measured in inches or millimeters. It is an order of magnitude greater than roughness. Roughness is imposed on top of waviness. Figure A.20 shows the preferred values for expressing waviness.

Surface control is noted on drawings with standard symbols. The basic symbol is a 60° V with the right leg extended up. For a simple designation of surface roughness the maximum allowable

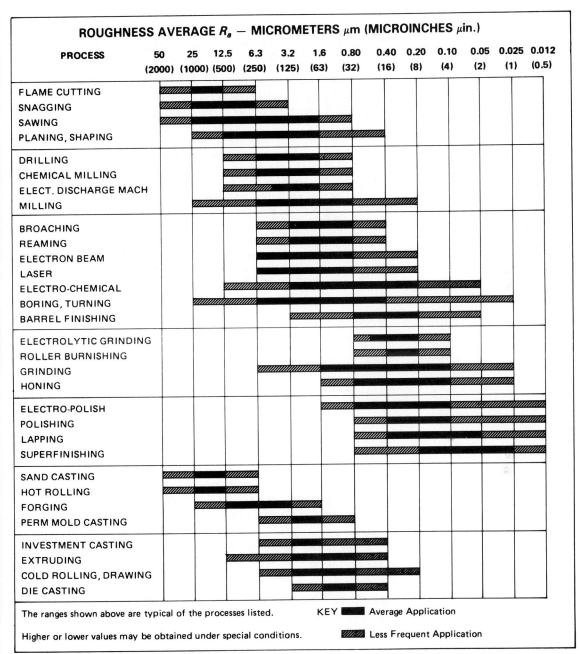

Figure A.18 Surface roughness produced by common production methods. Courtesy of ANSI (B46.1)

Lay Symbol	Meaning	Example Showing Direction of Tool Marks
—	Lay approximately parallel to the line representing the surface to which the symbol is applied.	
⊥	Lay approximately perpendicular to the line representing the surface to which the symbol is applied.	
X	Lay angular in both directions to line representing the surface to which the symbol is applied.	
M	Lay multidirectional.	
C	Lay approximately circular relative to the center of the surface to which the symbol is applied.	
R	Lay approximately radial relative to the center of the surface to which the symbol is applied.	
P	Lay particulate, non-directional, or protuberant.	

Figure A.19 Standard symbols and examples for lay. Courtesy of ANSI (Y14.36)

average value is entered in the open area of the extended V. For more complete specifications a horizontal bar is added to the extended leg and additional information given above and below the bar. A small circle in the V means that material removal is prohibited; closing the V to form a triangle means that material removal by machining is required. Several applications of the symbol are shown in Figure A.21.

mm	in.	mm	in.	mm	in.
0.0005	0.00002	0.008	0.0003	0.12	0.005
0.0008	0.00003	0.012	0.0005	0.20	0.008
0.0012	0.00005	0.020	0.0008	0.25	0.010
0.0020	0.00008	0.025	0.001	0.38	0.015
0.0025	0.0001	0.05	0.002	0.50	0.020
0.005	0.0002	0.08	0.003	0.80	0.030

Figure A.20 Preferred numbers for expressing maximum waviness. Courtesy of ANSI (Y14.36)

Figure A.21 Applications of surface texture symbol. Courtesy of ANSI (Y14.36)

The beginning designer should not apply surface control symbols frequently. Surface control is expensive, and it should only be required where needed. It is best left to the judgment of experienced engineers and designers. In industries producing equipment such as rough machinery and conveyor frames, you can inspect thousands of drawings and never find a symbol for surface texture. Use only where needed!

A.5 GEOMETRIC TOLERANCING

Geometric tolerancing includes specification of location and form. These are factors we have not considered yet; Chapter 7 considered lengths, diameters, and radii. We also applied tolerances to obtain maximum and minimum values for critical dimensions.

Now we can look at the geometric relationships among the elements of a part.

A.5.1 Definitions and Symbols

Some of the terms you will need are those you already know; some of the new terms follow:

> *Basic dimension:* A numerical value used to describe the theoretically exact size or location. Permissible variations from this theoretical value are a part of geometric tolerancing.
>
> *True position:* The theoretically exact location of a feature established by basic dimension.
>
> *Datum:* A theoretically exact point, axis, or plane used as the origin from which location or geometric characteristics of features are located.
>
> *Datum target:* A specified point, line, or area on a part used to establish a datum.
>
> *Datum feature:* An actual feature of a part used to establish a datum.
>
> *Maximum material condition (MMC):* The condition in which a feature of size contains the maximum amount of material within the stated limits of size, for example, minimum hole diameter, or maximum shaft diameter.
>
> *Least material condition (LMC):* Opposite of MMC, the feature contains the least material, for example, maximum hole diameter or minimum shaft diameter.
>
> *Regardless of feature size (RFS):* The term indicates that a geometric tolerance or datum reference applies at any increment of size of the feature within its size tolerance.

Geometric tolerances are specified with symbols (Figures A.22 and A.23). Several symbols and numbers may be combined in a sort of shorthand to describe what is required. The collection of symbols is presented in a rectangular shape called a feature control frame (Figure A.24). The frame shown is to control perpendicularity of a circular feature at maximum material condition. The tolerance of the diameter is 0.05 mm; the perpendicularity is to be referenced to a datum marked C. A lot of information in a little rectangle!

	TYPE OF TOLERANCE	CHARACTERISTIC	SYMBOL
FOR INDIVIDUAL FEATURES	FORM	STRAIGHTNESS	—
		FLATNESS	▱
		CIRCULARITY (ROUNDNESS)	○
		CYLINDRICITY	⌭
FOR INDIVIDUAL OR RELATED FEATURES	PROFILE	PROFILE OF A LINE	⌒
		PROFILE OF A SURFACE	⌓
FOR RELATED FEATURES	ORIENTATION	ANGULARITY	∠
		PERPENDICULARITY	⊥
		PARALLELISM	//
	LOCATION	POSITION	⊕
		CONCENTRICITY	◎
	RUNOUT	CIRCULAR RUNOUT	↗
		TOTAL RUNOUT	⌰

Figure A.22 Basic symbols for geometric characteristics. Courtesy of ANSI (Y14.5M)

TERM	SYMBOL
AT MAXIMUM MATERIAL CONDITION	Ⓜ
REGARDLESS OF FEATURE SIZE	Ⓢ
AT LEAST MATERIAL CONDITION	Ⓛ
PROJECTED TOLERANCE ZONE	Ⓟ
DIAMETER	⌀
SPHERICAL DIAMETER	S⌀
RADIUS	R
SPHERICAL RADIUS	SR
REFERENCE	( )
ARC LENGTH	⌒

Figure A.23 Modifying symbols for geometric characteristics. Courtesy of ANSI (Y14.5M)

A.5.2 Datum Referencing

A datum indicates the origin of a dimensional relationship between a toleranced feature and a designated feature or features on a part. A theoretically exact datum would be difficult to establish, so very flat surfaces such as machine tables or stone surface plates are used. Enough datum references should be selected to control all needed features. An example is shown in Figure A.25. Cylindrical parts should have datums passing through the center (Figure A.26). Datums may be referenced to a surface on a part, a line,

Figure A.24 Feature control frame used to combine geometric tolerance information. Courtesy of ANSI (Y14.5M)

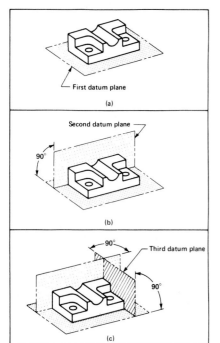

Figure A.25 Three datum reference planes at right angles. Courtesy of ANSI (Y14.5M)

a point, a single feature, or a collection of features (a group of holes).

A.5.3 Tolerances of Location

Location includes position, concentricity, and symmetry. Basic dimensions are used to establish true positions from datums. The basic dimensions are put in a box on the drawing. Datum references are usually designated by a letter preceded and followed by a dash. Examine Figure A.27, and you will find several basic dimensions, three datum references, and two feature control frames. The true position of the large hole is fixed relative to datums A and B; the true positions of the small holes are fixed relative to the center lines of the large hole.

Concentricity is specified by the small concentric circle symbol. An example is shown in Figure A.28. Symmetry is shown in Figure A.29.

Figure A.26 Cylindrical part with datum plane and datum axis. Courtesy of ANSI (Y14.5M)

Figure A.27 Example of positional tolerancing. Courtesy of ANSI (Y14.5M)

Figure A.28 Concentricity tolerance note and its meaning. Courtesy of ANSI (Y14.5M)

Figure A.29 Symmetry tolerance note and its meaning. Courtesy of ANSI (Y14.5M)

Figure A.30 Straightness note and its meaning. Courtesy of ANSI (Y14.5M)

Figure A.31 Flatness note and its meaning. Courtesy of ANSI (Y14.5M)

A.5.4 Tolerances of Form

Tolerances of form include form, orientation, profile, and runout. Form features that can be controlled include straightness, flatness, circularity, and cylindricity. Orientation tolerances control angularity, parallelism, and perpendicularity. Form and orientation are used together where tolerances for size and location do not provide enough control.

Straightness means being in a straight line. Figure A.30 shows a straightness requirement for the elements of a cylinder. The tolerance on the diameter is 0.11 mm, but the elements of the cylinder must all be straight within 0.02 mm. Flatness means all points on a surface are in the same plane. Flatness control is illustrated in Figure A.31. Note that the plane may not be truly horizontal, but all points on the top surface must be within 0.25 mm of the plane. Other tolerances for size are not shown; they must be met also. Figure A.32 shows control of circularity. A right section cut anywhere must contain all elements with the tolerance zone defined by the two concentric circles. Figure A.33 shows control of cylindricity.

Orientation control includes angularity, parallelism, and perpendicularity. Specification of these features is usually tied to a datum. Note that datum plane *A* is the reference for angular control in Figure A.34. Parallelism is demonstrated in Figure A.35 with datum *A* as the control reference. Perpendicularity is shown in Figure A.36; the bottom surface of the angle is specified as datum *A*.

A profile is the outline of an object in two dimensions: A pro-

THIS ON THE DRAWING

⃞ 0.25   ⃝ 0.25

MEANS THIS

A ←   90   A ←   90°

A ←   0.25 wide tolerance zone   A ←

SECTION A-A

Each circular element of the surface in a plane perpendicular to a common axis must lie between two concentric circles, one having a radius 0.25 larger than the other. Additionally, each circular element of the surface must be within the specified limits of size.

Figure A.32 Circularity note for cylinder or cone and its meaning.

THIS ON THE DRAWING

⌭ 0.25

MEANS THIS   0.25 wide tolerance zone

The cylindrical surface must lie between two concentric cylinders, one having a radius 0.25 larger than the other. Additionally, the surface must be within the specified limits of size.

Figure A.33 Cylindricity note and its meaning. Courtesy of ANSI (Y14.5M)

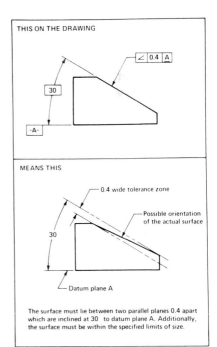

THIS ON THE DRAWING

∠ 0.4 A

30

-A-

MEANS THIS

0.4 wide tolerance zone

Possible orientation of the actual surface

30

Datum plane A

The surface must lie between two parallel planes 0.4 apart which are inclined at 30 to datum plane A. Additionally, the surface must be within the specified limits of size.

Figure A.34 Angularity note and its meaning. Courtesy of ANSI (Y14.5M)

jection or a section. The boundaries of the profile need to be shown on the drawing along with a datum and the control tolerance. Figure A.37 shows the application of profile tolerance to a curved shape. If the allowable deviation can occur on both sides of the true profile line, it is a bilateral tolerance and no special

Figure A.35 Parallelism note and its meaning. Courtesy of ANSI (Y14.5M)

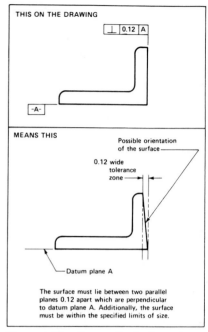

Figure A.36 Perpendicularity note and its meaning. Courtesy of ANSI (Y14.5M)

Figure A.37 Profile note given three ways and their meanings. Courtesy of ANSI (Y14.5M)

symbol is added. If the deviation can only be allowed on one side of the true profile line, a phantom line is added to show in which direction the deviation is allowed.

Runout is the deviation from an axis or datum surface along some given length. Runout for both a conical surface and a cylindrical surface is shown in Figure A.38. Note that runout control applies to only a 17-mm-long portion of the large cylinder (it probably has to mate with another part in this zone). A measuring tool with a dial reading parts of a millimeter or inch is used to take several readings in the control area as the part is rotated about axis A. In this example all the readings at each section must be within 0.02 mm.

A.6 *SUMMARY*

You have been introduced to a few of the basic manufacturing processes. Drilling and turning have been practiced for centuries. Swords and other weapons were made by forging since metals were discovered. New means for forming materials are being developed every year. If you become involved in manufacturing, you will need to be aware of them.

Specifying tolerances is not difficult when you use the standard tables. Tolerances should be selected from the tables based on the function of the parts. Checking tolerances can still be done by calculating the tolerance of the system. Metric tolerances may be specified in several ways depending on the familiarity of the people using them.

In addition to specifying tolerances, the designer may need to specify the quality of the surfaces of a part. Standard measures and symbols exist for controlling surface roughness, lay, and waviness.

Geometric tolerancing is a sophisticated method for specifying form and position. You have a detailed method and symbols for controlling all the geometric elements of a part. Datum references are an important part of the methods.

Tolerances of size, geometric tolerances, and surface control can improve the quality and performance of machinery. All of these controls add cost and time to the production of parts. They should be used always when necessary but only when necessary for the function of the part or assembly.

The information in this appendix is a very brief summary of the information needed to do a complete job of dimensioning. Most readers will not need to acquire total information in all of the areas mentioned; most of you who become practicing

Figure A.38 Circular runout note and its meaning. Courtesy of ANSI (Y14.5M)

engineers will need to learn more about some of the areas. For more information start with the standards noted with each topic; books and professional magazines and journals are available that cover many areas of manufacturing and design. When you take advanced courses in design, you may study some of these subjects in depth; as a practicing engineer you may need to take short courses or graduate work to gain the knowledge needed to fulfill your professional obligations.

B

TABLES

The following is a list of the tables in Appendix B. The numbers in the list are page numbers.

Table B.1 Preferred Limits and Fits (Inches)

Nominal Size Range Inches (Over)	(To)	Class RC 1 Limits of Clearance	RC 1 Hole H5	RC 1 Shaft g4	Class RC 2 Limits of Clearance	RC 2 Hole H6	RC 2 Shaft g5	Class RC 3 Limits of Clearance	RC 3 Hole H7	RC 3 Shaft f6	Class RC 4 Limits of Clearance	RC 4 Hole H8	RC 4 Shaft f7
0	0.12	0.1 / 0.45	+0.2 / 0	−0.1 / −0.25	0.1 / 0.55	+0.25 / 0	−0.1 / −0.3	0.3 / 0.95	+0.4 / 0	−0.3 / −0.55	0.3 / 1.3	+0.6 / 0	−0.3 / −0.7
0.12	0.24	0.15 / 0.5	+0.2 / 0	−0.15 / −0.3	0.15 / 0.65	+0.3 / 0	−0.15 / −0.35	0.4 / 1.2	+0.5 / 0	−0.4 / −0.7	0.4 / 1.6	+0.7 / 0	−0.4 / −0.9
0.24	0.40	0.2 / 0.6	+0.25 / 0	−0.2 / −0.35	0.2 / 0.85	+0.4 / 0	−0.2 / −0.45	0.5 / 1.5	+0.6 / 0	−0.5 / −0.9	0.5 / 2.0	+0.9 / 0	−0.5 / −1.1
0.40	0.71	0.25 / 0.75	+0.3 / 0	−0.25 / −0.45	0.25 / 0.95	+0.4 / 0	−0.25 / −0.55	0.6 / 1.7	+0.7 / 0	−0.6 / −1.0	0.6 / 2.3	+1.0 / 0	−0.6 / −1.3
0.71	1.19	0.3 / 0.95	+0.4 / 0	−0.3 / −0.55	0.3 / 1.2	+0.5 / 0	−0.3 / −0.7	0.8 / 2.1	+0.8 / 0	−0.8 / −1.3	0.8 / 2.8	+1.2 / 0	−0.8 / −1.6
1.19	1.97	0.4 / 1.1	+0.4 / 0	−0.4 / −0.7	0.4 / 1.4	+0.6 / 0	−0.4 / −0.8	1.0 / 2.6	+1.0 / 0	−1.0 / −1.6	1.0 / 3.6	+1.6 / 0	−1.0 / −2.0
1.97	3.15	0.4 / 1.2	+0.5 / 0	−0.4 / −0.7	0.4 / 1.6	+0.7 / 0	−0.4 / −0.9	1.2 / 3.1	+1.2 / 0	−1.2 / −1.9	1.2 / 4.2	+1.8 / 0	−1.2 / −2.4
3.15	4.73	0.5 / 1.5	+0.6 / 0	−0.5 / −0.9	0.5 / 2.0	+0.9 / 0	−0.5 / −1.1	1.4 / 3.7	+1.4 / 0	−1.4 / −2.3	1.4 / 5.0	+2.2 / 0	−1.4 / −2.8
4.73	7.09	0.6 / 1.8	+0.7 / 0	−0.6 / −1.1	0.6 / 2.3	+1.0 / 0	−0.6 / −1.3	1.6 / 4.2	+1.6 / 0	−1.6 / −2.6	1.6 / 5.7	+2.5 / 0	−1.6 / −3.2
7.09	9.85	0.6 / 2.0	+0.8 / 0	−0.6 / −1.2	0.6 / 2.6	+1.2 / 0	−0.6 / −1.4	2.0 / 5.0	+1.8 / 0	−2.0 / −3.2	2.0 / 6.6	+2.8 / 0	−2.0 / −3.8
9.85	12.41	0.8 / 2.3	+0.9 / 0	−0.8 / −1.4	0.7 / 2.8	+1.2 / 0	−0.7 / −1.6	2.5 / 5.7	+2.0 / 0	−2.5 / −3.7	2.2 / 7.2	+3.0 / 0	−2.2 / −4.2
12.41	15.75	1.0 / 2.7	+1.0 / 0	−1.0 / −1.7	0.7 / 3.1	+1.4 / 0	−0.7 / −1.7	3.0 / 6.6	+2.2 / 0	−3.0 / −4.4	2.5 / 8.2	+3.5 / 0	−2.5 / −4.7
15.75	19.69	1.2 / 3.0	+1.0 / 0	−1.2 / −2.0	0.8 / 3.4	+1.6 / 0	−0.8 / −1.8	4.0 / 8.1	+2.5 / 0	−4.0 / −5.6	2.8 / 9.3	+4.0 / 0	−2.8 / −5.3
19.69	30.09	1.6 / 3.7	+1.2 / 0	−1.6 / −2.5	1.6 / 4.8	+2.0 / 0	−1.6 / −2.8	5.0 / 10.0	+3.0 / 0	−5.0 / −7.0	5.0 / 13.0	+5.0 / 0	−5.0 / −8.0
30.09	41.49	2.0 / 4.6	+1.6 / 0	−2.0 / −3.0	2.0 / 6.1	+2.5 / 0	−2.0 / −3.6	6.0 / 12.5	+4.0 / 0	−6.0 / −8.5	6.0 / 16.0	+6.0 / 0	−6.0 / −10.0
41.49	56.19	2.5 / 5.7	+2.0 / 0	−2.5 / −3.7	2.5 / 7.5	+3.0 / 0	−2.5 / −4.5	8.0 / 16.0	+5.0 / 0	−8.0 / −11.0	8.0 / 21.0	+8.0 / 0	−8.0 / −13.0
56.19	76.39	3.0 / 7.1	+2.5 / 0	−3.0 / −4.6	3.0 / 9.5	+4.0 / 0	−3.0 / −5.5	10.0 / 20.0	+6.0 / 0	−10.0 / −14.0	10.0 / 26.0	+10.0 / 0	−10.0 / −16.0
76.39	100.9	4.0 / 9.0	+3.0 / 0	−4.0 / −6.0	4.0 / 12.0	+5.0 / 0	−4.0 / −7.0	12.0 / 25.0	+8.0 / 0	−12.0 / −17.0	12.0 / 32.0	+12.0 / 0	−12.0 / −20.0
100.9	131.9	5.0 / 11.5	+4.0 / 0	−5.0 / −7.5	5.0 / 15.0	+6.0 / 0	−5.0 / −9.0	16.0 / 32.0	+10.0 / 0	−16.0 / −22.0	16.0 / 42.0	+16.0 / 0	−16.0 / −26.0
131.9	171.9	6.0 / 14.0	+5.0 / 0	−6.0 / −9.0	6.0 / 19.0	+8.0 / 0	−6.0 / −11.0	18.0 / 38.0	+12.0 / 0	−18.0 / −26.0	18.0 / 50.0	+20.0 / 0	−18.0 / −30.0
171.9	200	8.0 / 18.0	+6.0 / 0	−8.0 / −12.0	8.0 / 22.0	+10.0 / 0	−8.0 / −12.0	22.0 / 48.0	+16.0 / 0	−22.0 / −32.0	22.0 / 63.0	+25.0 / 0	−22.0 / −38.0

Courtesy of USAS (B4.1)

Class RC 5			Class RC 6			Class RC 7			Class RC 8			Class RC 9			Nominal Size Range Inches	
Limits of Clearance	Hole H8	Shaft e7	Limits of Clearance	Hole H9	Shaft e8	Limits of Clearance	Hole H9	Shaft d8	Limits of Clearance	Hole H10	Shaft c9	Limits of Clearance	Hole H11	Shaft	Over	To
0.6	+ 0.6	− 0.6	0.6	+ 1.0	− 0.6	1.0	+ 1.0	− 1.0	2.5	+ 1.6	− 2.5	4.0	+ 2.5	− 4.0	0 − 0.12	
1.6	− 0	− 1.0	2.2	− 0	− 1.2	2.6	0	− 1.6	5.1	0	− 3.5	8.1	0	− 5.6		
0.8	+ 0.7	− 0.8	0.8	+ 1.2	− 0.8	1.2	+ 1.2	− 1.2	2.8	+ 1.8	− 2.8	4.5	+ 3.0	− 4.5	0.12− 0.24	
2.0	− 0	− 1.3	2.7	− 0	− 1.5	3.1	0	− 1.9	5.8	0	− 4.0	9.0	0	− 6.0		
1.0	+ 0.9	− 1.0	1.0	+ 1.4	− 1.0	1.6	+ 1.4	− 1.6	3.0	+ 2.2	− 3.0	5.0	+ 3.5	− 5.0	0.24− 0.40	
2.5	− 0	− 1.6	3.3	− 0	− 1.9	3.9	0	− 2.5	6.6	0	− 4.4	10.7	0	− 7.2		
1.2	+ 1.0	− 1.2	1.2	+ 1.6	− 1.2	2.0	+ 1.6	− 2.0	3.5	+ 2.8	− 3.5	6.0	+ 4.0	− 6.0	0.40− 0.71	
2.9	− 0	− 1.9	3.8	− 0	− 2.2	4.6	0	− 3.0	7.9	0	− 5.1	12.8	− 0	− 8.8		
1.6	+ 1.2	− 1.6	1.6	+ 2.0	− 1.6	2.5	+ 2.0	− 2.5	4.5	+ 3.5	− 4.5	7.0	+ 5.0	− 7.0	0.71− 1.19	
3.6	− 0	− 2.4	4.8	− 0	− 2.8	5.7	0	− 3.7	10.0	0	− 6.5	15.5	0	− 10.5		
2.0	+ 1.6	− 2.0	2.0	+ 2.5	− 2.0	3.0	+ 2.5	− 3.0	5.0	+ 4.0	− 5.0	8.0	+ 6.0	− 8.0	1.19− 1.97	
4.6	− 0	− 3.0	6.1	− 0	− 3.6	7.1	0	− 4.6	11.5	0	− 7.5	18.0	0	− 12.0		
2.5	+ 1.8	− 2.5	2.5	+ 3.0	− 2.5	4.0	+ 3.0	− 4.0	6.0	+ 4.5	− 6.0	9.0	+ 7.0	− 9.0	1.97− 3.15	
5.5	− 0	− 3.7	7.3	− 0	− 4.3	8.8	0	− 5.8	13.5	0	− 9.0	20.5	0	− 13.5		
3.0	+ 2.2	− 3.0	3.0	+ 3.5	− 3.0	5.0	+ 3.5	− 5.0	7.0	+ 5.0	− 7.0	10.0	+ 9.0	− 10.0	3.15− 4.73	
6.6	− 0	− 4.4	8.7	− 0	− 5.2	10.7	0	− 7.2	15.5	0	− 10.5	24.0	0	− 15.0		
3.5	+ 2.5	− 3.5	3.5	+ 4.0	− 3.5	6.0	+ 4.0	− 6.0	8.0	+ 6.0	− 8.0	12.0	+ 10.0	− 12.0	4.73− 7.09	
7.6	− 0	− 5.1	10.0	− 0	− 6.0	12.5	0	− 8.5	18.0	0	− 12.0	28.0	0	− 18.0		
4.0	+ 2.8	− 4.0	4.0	+ 4.5	− 4.0	7.0	+ 4.5	− 7.0	10.0	+ 7.0	− 10.0	15.0	+ 12.0	− 15.0	7.09− 9.85	
8.6	− 0	− 5.8	11.3	0	− 6.8	14.3	0	− 9.8	21.5	0	− 14.5	34.0	0	− 22.0		
5.0	+ 3.0	− 5.0	5.0	+ 5.0	− 5.0	8.0	+ 5.0	− 8.0	12.0	+ 8.0	− 12.0	18.0	+ 12.0	− 18.0	9.85− 12.41	
10.0	0	− 7.0	13.0	0	− 8.0	16.0	0	− 11.0	25.0	0	− 17.0	38.0	0	− 26.0		
6.0	+ 3.5	− 6.0	6.0	+ 6.0	− 6.0	10.0	+ 6.0	− 10.0	14.0	+ 9.0	− 14.0	22.0	+ 14.0	− 22.0	12.41− 15.75	
11.7	0	− 8.2	15.5	0	− 9.5	19.5	0	− 13.5	29.0	0	− 20.0	45.0	0	− 31.0		
8.0	+ 4.0	− 8.0	8.0	+ 6.0	− 8.0	12.0	+ 6.0	− 12.0	16.0	+10.0	− 16.0	25.0	+ 16.0	− 25.0	15.75− 19.69	
14.5	0	−10.5	18.0	0	−12.0	22.0	0	− 16.0	32.0	0	− 22.0	51.0	0	− 35.0		
10.0	+ 5.0	−10.0	10.0	+ 8.0	−10.0	16.0	+ 8.0	− 16.0	20.0	+12.0	− 20.0	30.0	+ 20.0	− 30.0	19.69− 30.09	
18.0	0	−13.0	23.0	0	−15.0	29.0	0	− 21.0	40.0	0	− 28.0	62.0	0	− 42.0		
12.0	+ 6.0	−12.0	12.0	+10.0	−12.0	20.0	+10.0	− 20.0	25.0	+16.0	− 25.0	40.0	+ 25.0	− 40.0	30.09− 41.49	
22.0	0	−16.0	28.0	0	−18.0	36.0	0	− 26.0	51.0	0	− 35.0	81.0	0	− 56.0		
16.0	+ 8.0	−16.0	16.0	+12.0	−16.0	25.0	+12.0	− 25.0	30.0	+20.0	− 30.0	50.0	+ 30.0	− 50.0	41.49− 56.19	
29.0	0	−21.0	36.0	0	−24.0	45.0	0	− 33.0	62.0	0	− 42.0	100	0	− 70.0		
20.0	+10.0	−20.0	20.0	+16.0	−20.0	30.0	+16.0	− 30.0	40.0	+25.0	− 40.0	60.0	+ 40.0	− 60.0	56.19− 76.39	
36.0	0	−26.0	46.0	0	−30.0	56.0	0	− 40.0	81.0	0	− 56.0	125	0	− 85.0		
25.0	+12.0	−25.0	25.0	+20.0	−25.0	40.0	+20.0	− 40.0	50.0	+30.0	− 50.0	80.0	+ 50.0	− 80.0	76.39−100.9	
45.0	0	−33.0	57.0	0	−37.0	72.0	0	− 52.0	100	0	− 70.0	160	0	−110		
30.0	+16.0	−30.0	30.0	+25.0	−30.0	50.0	+25.0	− 50.0	60.0	+40.0	− 60.0	100	+ 60.0	−100	100.9 −131.9	
56.0	0	−40.0	71.0	0	−46.0	91.0	0	− 66.0	125	0	− 85.0	200	0	−140		
35.0	+20.0	−35.0	35.0	+30.0	−35.0	60.0	+30.0	− 60.0	80.0	+50.0	− 80.0	130	+ 80.0	−130	131.9 −171.9	
67.0	0	−47.0	85.0	0	−55.0	110.0	0	− 80.0	160	0	−110	260	0	−180		
45.0	+25.0	−45.0	45.0	+40.0	−45.0	80.0	+40.0	− 80.0	100	+60.0	−100	150	+100	−150	171.9 −200	
86.0	0	−61.0	110.0	0	−70.0	145.0	0	−105.0	200	0	−140	310	0	−210		

Table B.1 *(Continued)*

Nominal Size Range Inches Over	To	Class LC 1 Limits of Clearance	Class LC 1 Standard Limits Hole H6	Class LC 1 Standard Limits Shaft h5	Class LC 2 Limits of Clearance	Class LC 2 Standard Limits Hole H7	Class LC 2 Standard Limits Shaft h6	Class LC 3 Limits of Clearance	Class LC 3 Standard Limits Hole H8	Class LC 3 Standard Limits Shaft h7	Class LC 4 Limits of Clearance	Class LC 4 Standard Limits Hole H10	Class LC 4 Standard Limits Shaft h9	Class LC 5 Limits of Clearance	Class LC 5 Standard Limits Hole H7	Class LC 5 Standard Limits Shaft g6
0 —	0.12	0 0.45	+0.25 −0	+0 −0.2	0 0.65	+0.4 −0	+0 −0.25	0 1	+0.6 −0	+0 −0.4	0 2.6	+1.6 −0	+0 −1.0	0.1 0.75	+0.4 −0	−0.1 −0.35
0.12—	0.24	0 0.5	+0.3 −0	+0 −0.2	0 0.8	+0.5 −0	+0 −0.3	0 1.2	+0.7 −0	+0 −0.5	0 3.0	+1.8 −0	+0 −1.2	0.15 0.95	+0.5 −0	−0.15 −0.45
0.24—	0.40	0 0.65	+0.4 −0	+0 −0.25	0 1.0	+0.6 −0	+0 −0.4	0 1.5	+0.9 −0	+0 −0.6	0 3.6	+2.2 −0	+0 −1.4	0.2 1.2	+0.6 −0	−0.2 −0.6
0.40—	0.71	0 0.7	+0.4 −0	+0 −0.3	0 1.1	+0.7 −0	+0 −0.4	0 1.7	+1.0 −0	+0 −0.7	0 4.4	+2.8 −0	+0 −1.6	0.25 1.35	+0.7 −0	−0.25 −0.65
0.71—	1.19	0 0.9	+0.5 −0	+0 −0.4	0 1.3	+0.8 −0	+0 −0.5	0 2	+1.2 −0	+0 −0.8	0 5.5	+3.5 −0	+0 −2.0	0.3 1.6	+0.8 −0	−0.3 −0.8
1.19—	1.97	0 1.0	+0.6 −0	+0 −0.4	0 1.6	+1.0 −0	+0 −0.6	0 2.6	+1.6 −0	+0 −1	0 6.5	+4.0 −0	+0 −2.5	0.4 2.0	+1.0 −0	−0.4 −1.0
1.97—	3.15	0 1.2	+0.7 −0	+0 −0.5	0 1.9	+1.2 −0	+0 −0.7	0 3	+1.8 −0	+0 −1.2	0 7.5	+4.5 −0	+0 −3	0.4 2.3	+1.2 −0	−0.4 −1.1
3.15—	4.73	0 1.5	+0.9 −0	+0 −0.6	0 2.3	+1.4 −0	+0 −0.9	0 3.6	+2.2 −0	+0 −1.4	0 8.5	+5.0 −0	+0 −3.5	0.5 2.8	+1.4 −0	−0.5 −1.4
4.73—	7.09	0 1.7	+1.0 −0	+0 −0.7	0 2.6	+1.6 −0	+0 −1.0	0 4.1	+2.5 −0	+0 −1.6	0 10	+6.0 −0	+0 −4	0.6 3.2	+1.6 −0	−0.6 −1.6
7.09—	9.85	0 2.0	+1.2 −0	+0 −0.8	0 3.0	+1.8 −0	+0 −1.2	0 4.6	+2.8 −0	+0 −1.8	0 11.5	+7.0 −0	+0 −4.5	0.6 3.6	+1.8 −0	−0.6 −1.8
9.85—	12.41	0 2.1	+1.2 −0	+0 −0.9	0 3.2	+2.0 −0	+0 −1.2	0 5	+3.0 −0	+0 −2.0	0 13	+8.0 −0	+0 −5	0.7 3.9	+2.0 −0	−0.7 −1.9
12.41—	15.75	0 2.4	+1.4 −0	+0 −1.0	0 3.6	+2.2 −0	+0 −1.4	0 5.7	+3.5 −0	+0 −2.2	0 15	+9.0 −0	+0 −6	0.7 4.3	+2.2 −0	−0.7 −2.1
15.75—	19.69	0 2.6	+1.6 −0	+0 −1.0	0 4.1	+2.5 −0	+0 −1.6	0 6.5	+4 −0	+0 −2.5	0 16	+10.0 −0	+0 −6	0.8 4.9	+2.5 −0	−0.8 −2.4
19.69—	30.09	0 3.2	+2.0 −0	+0 −1.2	0 5.0	+3 −0	+0 −2	0 8	+5 −0	+0 −3	0 20	+12.0 −0	+0 −8	0.9 5.9	+3.0 −0	−0.9 −2.9
30.09—	41.49	0 4.1	+2.5 −0	+0 −1.6	0 6.5	+4 −0	+0 −2.5	0 10	+6 −0	+0 −4	0 26	+16.0 −0	+0 −10	1.0 7.5	+4.0 −0	−1.0 −3.5
41.49—	56.19	0 5.0	+3.0 −0	+0 −2.0	0 8.0	+5 −0	+0 −3	0 13	+8 −0	+0 −5	0 32	+20.0 −0	+0 −12	1.2 9.2	+5.0 −0	−1.2 −4.2
56.19—	76.39	0 6.5	+4.0 −0	+0 −2.5	0 10	+6 −0	+0 −4	0 16	+10 −0	+0 −6	0 41	+25.0 −0	+0 −16	1.2 11.2	+6.0 −0	−1.2 −5.2
76.39—	100.9	0 8.0	+5.0 −0	+0 −3.0	0 13	+8 −0	+0 −5	0 20	+12 −0	+0 −8	0 50	+30.0 −0	+0 −20	1.4 14.4	+8.0 −0	−1.4 −6.4
100.9 —	131.9	0 10.0	+6.0 −0	+0 −4.0	0 16	+10 −0	+0 −6	0 26	+16 −0	+0 −10	0 65	+40.0 −0	+0 −25	1.6 17.6	+10.0 −0	−1.6· −7.6
131.9 —	171.9	0 13.0	+8.0 −0	+0 −5.0	0 20	+12 −0	+0 −8	0 32	+20 −0	+0 −12	0 80	+50.0 −0	+0 −30	1.8 21.8	+12.0 −0	−1.8 −9.8
171.9 —	200	0 16.0	+10.0 −0	+0 −6.0	0 26	+16 −0	+0 −10	0 41	+25 −0	+0 −16	0 100	+60.0 −0	+0 −40	1.8 27.8	+16.0 −0	−1.8 −11.8

Courtesy of USAS (B4.1)

Class LC 6			Class LC 7			Class LC 8			Class LC 9			Class LC 10			Class LC 11			Nominal Size Range Inches	
Limits of Clearance	Hole H9	Shaft f8	Limits of Clearance	Hole H10	Shaft e9	Limits of Clearance	Hole H10	Shaft d9	Limits of Clearance	Hole H11	Shaft c10	Limits of Clearance	Hole H12	Shaft	Limits of Clearance	Hole H13	Shaft	Over	To
0.3 / 1.9	+1.0 / 0	-0.3 / -0.9	0.6 / 3.2	+1.6 / 0	-0.6 / -1.6	1.0 / 3.6	+1.6 / 0	-1.0 / -2.0	2.5 / 6.6	+2.5 / 0	-2.5 / -4.1	4 / 12	+4 / -0	-4 / -8	5 / 17	+6 / -0	-5 / -11	0	0.12
0.4 / 2.3	+1.2 / 0	-0.4 / -1.1	0.8 / 3.8	+1.8 / 0	-0.8 / -2.0	1.2 / 4.2	+1.8 / 0	-1.2 / -2.4	2.8 / 7.6	+3.0 / 0	-2.8 / -4.6	4.5 / 14.5	+5 / 0	-4.5 / -9.5	6 / 20	+7 / -0	-6 / -13	0.12	0.24
0.5 / 2.8	+1.4 / 0	-0.5 / -1.4	1.0 / 4.6	+2.2 / 0	-1.0 / -2.4	1.6 / 5.2	+2.2 / 0	-1.6 / -3.0	3.0 / 8.7	+3.5 / 0	-3.0 / -5.2	5 / 17	+6 / 0	-5 / -11	7 / 25	+9 / -0	-7 / -16	0.24	0.40
0.6 / 3.2	+1.6 / 0	-0.6 / -1.6	1.2 / 5.6	+2.8 / 0	-1.2 / -2.8	2.0 / 6.4	+2.8 / 0	-2.0 / -3.6	3.5 / 10.3	+4.0 / 0	-3.5 / -6.3	6 / 20	+7 / 0	-6 / -13	8 / 28	+10 / -0	-8 / -18	0.40	0.71
0.8 / 4.0	+2.0 / 0	-0.8 / -2.0	1.6 / 7.1	+3.5 / 0	-1.6 / -3.6	2.5 / 8.0	+3.5 / 0	-2.5 / -4.5	4.5 / 13.0	+5.0 / 0	-4.5 / -8.0	7 / 23	+8 / 0	-7 / -15	10 / 34	+12 / -0	-10 / -22	0.71	1.19
1.0 / 5.1	+2.5 / 0	-1.0 / -2.6	2.0 / 8.5	+4.0 / 0	-2.0 / -4.5	3.0 / 9.5	+4.0 / 0	-3.0 / -5.5	5 / 15	+6 / 0	-5 / -9	8 / 28	+10 / 0	-8 / -18	12 / 44	+16 / -0	-12 / -28	1.19	1.97
1.2 / 6.0	+3.0 / 0	-1.2 / -3.0	2.5 / 10.0	+4.5 / 0	-2.5 / -5.5	4.0 / 11.5	+4.5 / 0	-4.0 / -7.0	6 / 17.5	+7 / 0	-6 / -10.5	10 / 34	+12 / 0	-10 / -22	14 / 50	+18 / -0	-14 / -32	1.97	3.15
1.4 / 7.1	+3.5 / 0	-1.4 / -3.6	3.0 / 11.5	+5.0 / 0	-3.0 / -6.5	5.0 / 13.5	+5.0 / 0	-5.0 / -8.5	7 / 21	+9 / 0	-7 / -12	11 / 39	+14 / 0	-11 / -25	16 / 60	+22 / -0	-16 / -38	3.15	4.73
1.6 / 8.1	+4.0 / 0	-1.6 / -4.1	3.5 / 13.5	+6.0 / 0	-3.5 / -7.5	6 / 16	+6 / 0	-6 / -10	8 / 24	+10 / 0	-8 / -14	12 / 44	+16 / 0	-12 / -28	18 / 68	+25 / -0	-18 / -43	4.73	7.09
2.0 / 9.3	+4.5 / 0	-2.0 / -4.8	4.0 / 15.5	+7.0 / 0	-4.0 / -8.5	7 / 18.5	+7 / 0	-7 / -11.5	10 / 29	+12 / 0	-10 / -17	16 / 52	+18 / 0	-16 / -34	22 / 78	+28 / -0	-22 / -50	7.09	9.85
2.2 / 10.2	+5.0 / 0	-2.2 / -5.2	4.5 / 17.5	+8.0 / 0	-4.5 / -9.5	7 / 20	+8 / 0	-7 / -12	12 / 32	+12 / 0	-12 / -20	20 / 60	+20 / 0	-20 / -40	28 / 88	+30 / -0	-28 / -58	9.85	12.41
2.5 / 12.0	+6.0 / 0	-2.5 / -6.0	5.0 / 20.0	+9.0 / 0	-5 / -11	8 / 23	+9 / 0	-8 / -14	14 / 37	+14 / 0	-14 / -23	22 / 66	+22 / 0	-22 / -44	30 / 100	+35 / -0	-30 / -65	12.41	15.75
2.8 / 12.8	+6.0 / 0	-2.8 / -6.8	5.0 / 21.0	+10.0 / 0	-5 / -11	9 / 25	+10 / 0	-9 / -15	16 / 42	+16 / 0	-16 / -26	25 / 75	+25 / 0	-25 / -50	35 / 115	+40 / -0	-35 / -75	15.75	19.69
3.0 / 16.0	+8.0 / 0	-3.0 / -8.0	6.0 / 26.0	+12.0 / 0	-6 / -14	10 / 30	+12 / 0	-10 / -18	18 / 50	+20 / 0	-18 / -30	28 / 88	+30 / 0	-28 / -58	40 / 140	+50 / -0	-40 / -90	19.69	30.09
3.5 / 19.5	+10.0 / 0	-3.5 / -9.5	7.0 / 33.0	+16.0 / 0	-7 / -17	12 / 38	+16 / 0	-12 / -22	20 / 61	+25 / 0	-20 / -36	30 / 110	+40 / 0	-30 / -70	45 / 165	+60 / -0	-45 / -105	30.09	41.49
4.0 / 24.0	+12.0 / 0	-4.0 / -12.0	8.0 / 40.0	+20.0 / 0	-8 / -20	14 / 46	+20 / 0	-14 / -26	25 / 75	+30 / 0	-25 / -45	40 / 140	+50 / 0	-40 / -90	60 / 220	+80 / 0	-60 / -140	41.49	56.19
4.5 / 30.5	+16.0 / 0	-4.5 / -14.5	9.0 / 50.0	+25.0 / 0	-9 / -25	16 / 57	+25 / 0	-16 / -32	30 / 95	+40 / 0	-30 / -55	50 / 170	+60 / 0	-50 / -110	70 / 270	+100 / 0	-70 / -170	56.19	76.39
5.0 / 37.0	+20.0 / 0	-5 / -17	10.0 / 60.0	+30.0 / 0	-10 / -30	18 / 68	+30 / 0	-18 / -38	35 / 115	+50 / 0	-35 / -65	50 / 210	+80 / 0	-50 / -130	80 / 330	+125 / 0	-80 / -205	76.39	100.9
6.0 / 47.0	+25.0 / 0	-6 / -22	12.0 / 67.0	+40.0 / 0	-12 / -27	20 / 85	+40 / 0	-20 / -45	40 / 140	+60 / 0	-40 / -80	60 / 260	+100 / 0	-60 / -160	90 / 410	+160 / 0	-90 / -250	100.9	131.9
7.0 / 57.0	+30.0 / 0	-7 / -27	14.0 / 94.0	+50.0 / 0	-14 / -44	25 / 105	+50 / 0	-25 / -55	50 / 180	+80 / 0	-50 / -100	80 / 330	+125 / 0	-80 / -205	100 / 500	+200 / 0	-100 / -300	131.9	171.9
7.0 / 72.0	+40.0 / 0	-7 / -32	14.0 / 114.0	+60.0 / 0	-14 / -54	25 / 125	+60 / 0	-25 / -65	50 / 210	+100 / 0	-50 / -110	90 / 410	+160 / 0	-90 / -250	125 / 625	+250 / 0	-125 / -375	171.9	200

Courtesy of USAS (B4.1)

Table B.1   (Continued)

Each cell lists the upper limit / lower limit. Empty cells are shown blank.

Nominal Size Range Over	To	LT 1 Fit	LT 1 Hole H7	LT 1 Shaft js6	LT 2 Fit	LT 2 Hole H8	LT 2 Shaft js7	LT 3 Fit	LT 3 Hole H7	LT 3 Shaft k6	LT 4 Fit	LT 4 Hole H8	LT 4 Shaft k7	LT 5 Fit	LT 5 Hole H7	LT 5 Shaft n6	LT 6 Fit	LT 6 Hole H7	LT 6 Shaft n7
0	0.12	-0.10 / +0.50	+0.4 / -0	+0.10 / -0.10	-0.2 / +0.8	+0.6 / -0	+0.2 / -0.2							-0.5 / +0.15	+0.4 / -0	+0.5 / +0.25	-0.65 / +0.15	+0.4 / -0	+0.65 / +0.25
0.12	0.24	-0.15 / +0.65	+0.5 / -0	+0.15 / -0.15	-0.25 / +0.95	+0.7 / -0	+0.25 / -0.25							-0.6 / +0.2	+0.5 / -0	+0.6 / +0.3	-0.8 / +0.2	+0.5 / -0	+0.8 / +0.3
0.24	0.40	-0.2 / +0.8	+0.6 / -0	+0.2 / -0.2	-0.3 / +1.2	+0.9 / -0	+0.3 / -0.3	-0.5 / +0.5	+0.6 / -0	+0.5 / +0.1	-0.7 / +0.8	+0.9 / -0	+0.7 / +0.1	-0.8 / +0.2	+0.6 / -0	+0.8 / +0.4	-1.0 / +0.2	+0.6 / -0	+1.0 / +0.4
0.40	0.71	-0.2 / +0.9	+0.7 / -0	+0.2 / -0.2	-0.35 / +1.35	+1.0 / -0	+0.35 / -0.35	-0.5 / +0.6	+0.7 / -0	+0.5 / +0.1	-0.8 / +0.9	+1.0 / -0	+0.8 / +0.1	-0.9 / +0.2	+0.7 / -0	+0.9 / +0.5	-1.2 / +0.2	+0.7 / -0	+1.2 / +0.5
0.71	1.19	-0.25 / +1.05	+0.8 / -0	+0.25 / -0.25	-0.4 / +1.6	+1.2 / -0	+0.4 / -0.4	-0.6 / +0.7	+0.8 / -0	+0.6 / +0.1	-0.9 / +1.1	+1.2 / -0	+0.9 / +0.1	-1.1 / +0.2	+0.8 / -0	+1.1 / +0.6	-1.4 / +0.2	+0.8 / -0	+1.4 / +0.6
1.19	1.97	-0.3 / +1.3	+1.0 / -0	+0.3 / -0.3	-0.5 / +2.1	+1.6 / -0	+0.5 / -0.5	-0.7 / +0.9	+1.0 / -0	+0.7 / +0.1	-1.1 / +1.5	+1.6 / -0	+1.1 / +0.1	-1.3 / +0.3	+1.0 / -0	+1.3 / +0.7	-1.7 / +0.3	+1.0 / -0	+1.7 / +0.7
1.97	3.15	-0.3 / +1.5	+1.2 / -0	+0.3 / -0.3	-0.6 / +2.4	+1.8 / -0	+0.6 / -0.6	-0.8 / +1.1	+1.2 / -0	+0.8 / +0.1	-1.3 / +1.7	+1.8 / -0	+1.3 / +0.1	-1.5 / +0.4	+1.2 / -0	+1.5 / +0.8	-2.0 / +0.4	+1.2 / -0	+2.0 / +0.8
3.15	4.73	-0.4 / +1.8	+1.4 / -0	+0.4 / -0.4	-0.7 / +2.9	+2.2 / -0	+0.7 / -0.7	-1.0 / +1.3	+1.4 / -0	+1.0 / +0.1	-1.5 / +2.1	+2.2 / -0	+1.5 / +0.1	-1.9 / +0.4	+1.4 / -0	+1.9 / +1.0	-2.4 / +0.4	+1.4 / -0	+2.4 / +1.0
4.73	7.09	-0.5 / +2.1	+1.6 / -0	+0.5 / -0.5	-0.8 / +3.3	+2.5 / -0	+0.8 / -0.8	-1.1 / +1.5	+1.6 / -0	+1.1 / +0.1	-1.7 / +2.4	+2.5 / -0	+1.7 / +0.1	-2.2 / +0.4	+1.6 / -0	+2.2 / +1.2	-2.8 / +0.4	+1.6 / -0	+2.8 / +1.2
7.09	9.85	-0.6 / +2.4	+1.8 / -0	+0.6 / -0.6	-0.9 / +3.7	+2.8 / -0	+0.9 / -0.9	-1.4 / +1.6	+1.8 / -0	+1.4 / +0.2	-2.0 / +2.6	+2.8 / -0	+2.0 / +0.2	-2.6 / +0.4	+1.8 / -0	+2.6 / +1.4	-3.2 / +0.4	+1.8 / -0	+3.2 / +1.4
9.85	12.41	-0.6 / +2.6	+2.0 / -0	+0.6 / -0.6	-1.0 / +4.0	+3.0 / -0	+1.0 / -1.0	-1.4 / +1.8	+2.0 / -0	+1.4 / +0.2	-2.2 / +2.8	+3.0 / -0	+2.2 / +0.2	-2.6 / +0.6	+2.0 / -0	+2.6 / +1.4	-3.4 / +0.6	+2.0 / -0	+3.4 / +1.4
12.41	15.75	-0.7 / +2.9	+2.2 / -0	+0.7 / -0.7	-1.0 / +4.5	+3.5 / -0	+1.0 / -1.0	-1.6 / +2.0	+2.2 / -0	+1.6 / +0.2	-2.4 / +3.3	+3.5 / -0	+2.4 / +0.2	-3.0 / +0.6	+2.2 / -0	+3.0 / +1.6	-3.8 / +0.6	+2.2 / -0	+3.8 / +1.6
15.75	19.69	-0.8 / +3.3	+2.5 / -0	+0.8 / -0.8	-1.2 / +5.2	+4.0 / -0	+1.2 / -1.2	-1.8 / +2.3	+2.5 / -0	+1.8 / +0.2	-2.7 / +3.8	+4.0 / -0	+2.7 / +0.2	-3.4 / +0.7	+2.5 / -0	+3.4 / +1.8	-4.3 / +0.7	+2.5 / -0	+4.3 / +1.8

Courtesy of USAS (B4.1)

Nominal Size Range Inches		Class LN 1			Class LN 2			Class LN 3		
		Limits of Interference	Standard Limits		Limits of Interference	Standard Limits		Limits of Interference	Standard Limits	
Over	To		Hole H6	Shaft n5		Hole H7	Shaft p6		Hole H7	Shaft r6
0 —	0.12	0 0.45	+ 0.25 − 0	+0.45 +0.25	0 0.65	+ 0.4 − 0	+ 0.65 + 0.4	0.1 0.75	+ 0.4 − 0	+ 0.75 + 0.5
0.12 —	0.24	0 0.5	+ 0.3 − 0	+0.5 +0.3	0 0.8	+ 0.5 − 0	+ 0.8 + 0.5	0.1 0.9	+ 0.5 0	+ 0.9 + 0.6
0.24 —	0.40	0 0.65	+ 0.4 − 0	+0.65 +0.4	0 1.0	+ 0.6 − 0	+ 1.0 + 0.6	0.2 1.2	+ 0.6 − 0	+ 1.2 + 0.8
0.40 —	0.71	0 0.8	+ 0.4 − 0	+0.8 +0.4	0 1.1	+ 0.7 − 0	+ 1.1 + 0.7	0.3 1.4	+ 0.7 − 0	+ 1.4 + 1.0
0.71 —	1.19	0 1.0	+ 0.5 − 0	+1.0 +0.5	0 1.3	+ 0.8 − 0	+ 1.3 + 0.8	0.4 1.7	+ 0.8 − 0	+ 1.7 + 1.2
1.19 —	1.97	0 1.1	+ 0.6 − 0	+1.1 +0.6	0 1.6	+ 1.0 − 0	+ 1.6 + 1.0	0.4 2.0	+ 1.0 − 0	+ 2.0 + 1.4
1.97 —	3.15	0.1 1.3	+0.8 −0	+1.3 + 0.7	0.2 2.1	+ 1.2 − 0	+ 2.1 + 1.4	0.4 2.3	+ 1.2 − 0	+ 2.3 + 1.6
3.15 —	4.73	0.1 1.6	+ 0.9 − 0	+1.6 +1.0	0.2 2.5	+ 1.4 − 0	+ 2.5 + 1.6	0.6 2.9	+ 1.4 − 0	+ 2.9 + 2.0
4.73 —	7.09	0.2 1.9	+ 1.0 − 0	+1.9 +1.2	0.2 2.8	+ 1.6 − 0	+ 2.8 + 1.8	0.9 3.5	+ 1.6 − 0	+ 3.5 + 2.5
7.09 —	9.85	0.2 2.2	+ 1.2 − 0	+2.2 +1.4	0.2 3.2	+ 1.8 − 0	+ 3.2 + 2.0	1.2 4.2	+ 1.8 − 0	+ 4.2 + 3.0
9.85 —	12.41	0.2 2.3	+ 1.2 − 0	+2.3 +1.4	0.2 3.4	+ 2.0 − 0	+ 3.4 + 2.2	1.5 4.7	+ 2.0 − 0	+ 4.7 + 3.5
12.41 —	15.75	0.2 2.6	+ 1.4 − 0	+2.6 +1.6	0.3 3.9	+ 2.2 − 0	+ 3.9 + 2.5	2.3 5.9	+ 2.2 − 0	+ 5.9 + 4.5
15.75 —	19.69	0.2 2.8	+ 1.6 − 0	+2.8 +1.8	0.3 4.4	+ 2.5 − 0	+ 4.4 + 2.8	2.5 6.6	+ 2.5 − 0	+ 6.6 + 5.0
19.69 —	30.09		+ 2.0 − 0		0.5 5.5	+ 3 − 0	+ 5.5 + 3.5	4 9	+ 3 − 0	+ 9 + 7
30.09 —	41.49		+ 2.5 − 0		0.5 7.0	+ 4 − 0	+ 7.0 + 4.5	5 11.5	+ 4 − 0	+11.5 + 9
41.49 —	56.19		+ 3.0 − 0		1 9	+ 5 − 0	+ 9 + 6	7 15	+ 5 − 0	+15 +12
56.19 —	76.39		+ 4.0 − 0		1 11	+ 6 − 0	+11 + 7	10 20	+ 6 − 0	+20 +16
76.39 —	100.9		+ 5.0 − 0		1 14	+ 8 − 0	+14 + 9	12 25	+ 8 − 0	+25 +20
100.9 —	131.9		+ 6.0 − 0		2 18	+10 − 0	+18 +12	15 31	+10 − 0	+31 +25
131.9 —	171.9		+ 8.0 − 0		4 24	+12 − 0	+24 +16	18 38	+12 − 0	+38 +30
171.9 —	200		+10.0 − 0		4 30	+16 − 0	+30 +20	24 50	+16 − 0	+50 +40

Courtesy of USAS (B4.1)

Table B.1 (Continued)

Nominal Size Range Inches Over	To	Class FN 1 Limits of Interference	Standard Limits Hole H6	Shaft	Class FN 2 Limits of Interference	Standard Limits Hole H7	Shaft s6	Class FN 3 Limits of Interference	Standard Limits Hole H7	Shaft t6	Class FN 4 Limits of Interference	Standard Limits Hole H7	Shaft u6	Class FN 5 Limits of Interference	Standard Limits Hole H8	Shaft x7
0	0.12	0.05	+0.25	+ 0.5	0.2	+0.4	+ 0.85				0.3	+0.4	+ 0.95	0.3	+0.6	+ 1.3
		0.5	− 0	+ 0.3	0.85	− 0	+ 0.6				0.95	− 0	+ 0.7	1.3	− 0	+ 0.9
0.12	0.24	0.1	+0.3	+ 0.6	0.2	+0.5	+ 1.0				0.4	+0.5	+ 1.2	0.5	+ 0.7	+ 1.7
		0.6	− 0	+ 0.4	1.0	− 0	+ 0.7				1.2	− 0	+ 0.9	1.7	− 0	+ 1.2
0.24	0.40	0.1	+0.4	+ 0.75	0.4	+0.6	+ 1.4				0.6	+0.6	+ 1.6	0.5	+ 0.9	+ 2.0
		0.75	− 0	+ 0.5	1.4	− 0	+ 1.0				1.6	− 0	+ 1.2	2.0	− 0	+ 1.4
0.40	0.56	0.1	+0.4	+ 0.8	0.5	+0.7	+ 1.6				0.7	+ 0.7	+ 1.8	0.6	+ 1.0	+ 2.3
		0.8	− 0	+ 0.5	1.6	− 0	+ 1.2				1.8	− 0	+ 1.4	2.3	− 0	+ 1.6
0.56	0.71	0.2	+0.4	+ 0.9	0.5	+0.7	+ 1.6				0.7	+ 0.7	+ 1.8	0.8	+ 1.0	+ 2.5
		0.9	− 0	+ 0.6	1.6	− 0	+ 1.2				1.8	− 0	+ 1.4	2.5	− 0	+ 1.8
0.71	0.95	0.2	+0.5	+ 1.1	0.6	+0.8	+ 1.9				0.8	+ 0.8	+ 2.1	1.0	+ 1.2	+ 3.0
		1.1	− 0	+ 0.7	1.9	− 0	+ 1.4				2.1	− 0	+ 1.6	3.0	− 0	+ 2.2
0.95	1.19	0.3	+0.5	+ 1.2	0.6	+0.8	+ 1.9	0.8	+0.8	+ 2.1	1.0	+0·8	+ 2.3	1.3	+ 1.2	+ 3.3
		1.2	− 0	+ 0.8	1.9	− 0	+ 1.4	2.1	− 0	+ 1.6	2.3	− 0	+ 1.8	3.3	− 0	+ 2.5
1.19	1.58	0.3	+0.6	+ 1.3	0.8	+1.0	+ 2.4	1.0	+1.0	+ 2.6	1.5	+1.0	+ 3.1	1.4	+ 1.6	+ 4.0
		1.3	− 0	+ 0.9	2.4	− 0	+ 1.8	2.6	− 0	+ 2.0	3.1	− 0	+ 2.5	4.0	− 0	+ 3.0
1.58	1.97	0.4	+0.6	+ 1.4	0.8	+1.0	+ 2.4	1.2	+1.0	+ 2.8	1.8	+1.0	+ 3.4	2.4	+ 1.6	+ 5.0
		1.4	− 0	+ 1.0	2.4	− 0	+ 1.8	2.8	− 0	+ 2.2	3.4	− 0	+ 2.8	5.0	− 0	+ 4.0
1.97	2.56	0.6	+0.7	+ 1.8	0.8	+1.2	+ 2.7	1.3	+1.2	+ 3.2	2.3	+1.2	+ 4.2	3.2	+ 1.8	+ 6.2
		1.8	− 0	+ 1.3	2.7	− 0	+ 2.0	3.2	− 0	+ 2.5	4.2	− 0	+ 3.5	6.2	− 0	+ 5.0
2.56	3.15	0.7	+0.7	+ 1.9	1.0	+1.2	+ 2.9	1.8	+1.2	+ 3.7	2.8	+1.2	+ 4.7	4.2	+ 1.8	+ 7.2
		1.9	− 0	+ 1.4	2.9	− 0	+ 2.2	3.7	− 0	+ 3.0	4.7	− 0	+ 4.0	7.2	− 0	+ 6.0
3.15	3.94	0.9	+0.9	+ 2.4	1.4	+1.4	+ 3.7	2.1	+1.4	+ 4.4	3.6	+1.4	+ 5.9	4.8	+ 2.2	+ 8.4
		2.4	− 0	+ 1.8	3.7	− 0	+ 2.8	4.4	− 0	+ 3.5	5.9	− 0	+ 5.0	8.4	− 0	+ 7.0
3.94	4.73	1.1	+0.9	+ 2.6	1.6	+1.4	+ 3.9	2.6	+1.4	+ 4.9	4.6	+1.4	+ 6.9	5.8	+ 2.2	+ 9.4
		2.6	− 0	+ 2.0	3.9	− 0	+ 3.0	4.9	− 0	+ 4.0	6.9	− 0	+ 6.0	9.4	− 0	+ 8.0
4.73	5.52	1.2	+1.0	+ 2.9	1.9	+1.6	+ 4.5	3.4	+1.6	+ 6.0	5.4	+1.6	+ 8.0	7.5	+ 2.5	+11.6
		2.9	− 0	+ 2.2	4.5	− 0	+ 3.5	6.0	− 0	+ 5.0	8.0	− 0	+ 7.0	11.6	− 0	+10.0
5.52	6.30	1.5	+1.0	+ 3.2	2.4	+1.6	+ 5.0	3.4	+1.6	+ 6.0	5.4	+1.6	+ 8.0	9.5	+ 2.5	+13.6
		3.2	− 0	+ 2.5	5.0	− 0	+ 4.0	6.0	− 0	+ 5.0	8.0	− 0	+ 7.0	13.6	− 0	+12.0
6.30	7.09	1.8	+1.0	+ 3.5	2.9	+1.6	+ 5.5	4.4	+1.6	+ 7.0	6.4	+1.6	+ 9.0	9.5	+ 2.5	+13.6
		3.5	− 0	+ 2.8	5.5	− 0	+ 4.5	7.0	− 0	+ 6.0	9.0	− 0	+ 8.0	13.6	− 0	+12.0
7.09	7.88	1.8	+1.2	+ 3.8	3.2	+1.8	+ 6.2	5.2	+1.8	+ 8.2	7.2	+1.8	+10.2	11.2	+ 2.8	+15.8
		3.8	− 0	+ 3.0	6.2	− 0	+ 5.0	8.2	− 0	+ 7.0	10.2	− 0	+ 9.0	15.8	− 0	+14.0
7.88	8.86	2.3	+1.2	+ 4.3	3.2	+1.8	+ 6.2	5.2	+1.8	+ 8.2	8.2	+1.8	+11.2	13.2	+ 2.8	+17.8
		4.3	− 0	+ 3.5	6.2	− 0	+ 5.0	8.2	− 0	+ 7.0	11.2	− 0	+10.0	17.8	− 0	+16.0
8.86	9.85	2.3	+1.2	+ 4.3	4.2	+1.8	+ 7.2	6.2	+1.8	+ 9.2	10.2	+1.8	+13.2	13.2	+ 2.8	+17.8
		4.3	− 0	+ 3.5	7.2	− 0	+ 6.0	9.2	− 0	+ 8.0	13.2	− 0	+12.0	17.8	− 0	+16.0
9.85	11.03	2.8	+1.2	+ 4.9	4.0	+2.0	+ 7.2	7.0	+2.0	+10.2	10.0	+2.0	+13.2	15.0	+ 3.0	+20.0
		4.9	− 0	+ 4.0	7.2	− 0	+ 6.0	10.2	− 0	+ 9.0	13.2	− 0	+12.0	20.0	− 0	+18.0
11.03	12.41	2.8	+1.2	+ 4.9	5.0	+2.0	+ 8.2	7.0	+2.0	+10.2	12.0	+2.0	+15.2	17.0	+ 3.0	+22.0
		4.9	− 0	+ 4.0	8.2	− 0	+ 7.0	10.2	− 0	+ 9.0	15.2	− 0	+14.0	22.0	− 0	+20.0
12.41	13.98	3.1	+1.4	+ 5.5	5.8	+2.2	+ 9.4	7.8	+2.2	+11.4	13.8	+2.2	+17.4	18.5	+ 3.5	+24.2
		5.5	− 0	+ 4.5	9.4	− 0	+ 8.0	11.4	− 0	+10.0	17.4	− 0	+16.0	24.2	+ 0	+22.0
13.98	15.75	3.6	+1.4	+ 6.1	5.8	+2.2	+ 9.4	9.8	+2.2	+13.4	15.8	+2.2	+19.4	21.5	+ 3.5	+27.2
		6.1	− 0	+ 5.0	9.4	− 0	+ 8.0	13.4	− 0	+12.0	19.4	− 0	+18.0	27.2	− 0	+25.0
15.75	17.72	4.4	+1.6	+ 7.0	6.5	+2.5	+10.6	9.5	+2.5	+13.6	17.5	+2.5	+21.6	24.0	+ 4.0	+30.5
		7.0	− 0	+ 6.0	10.6	− 0	+ 9.0	13.6	− 0	+12.0	21.6	− 0	+20.0	30.5	− 0	+28.0
17.72	19.69	4.4	+1.6	+ 7.0	7.5	+2.5	+11.6	11.5	+2.5	+ 15.6	19.5	+2.5	+23.6	26.0	+ 4.0	+32.5
		7.0	− 0	+ 6.0	11.6	− 0	+10.0	15.6	− 0	+14.0	23.6	− 0	+22.0	32.5	− 0	+30.0

Courtesy of USAS (B4.1)

Table B.2 Preferred Limits and Fits (Metric)

Dimensions in mm.

BASIC SIZE		LOOSE RUNNING			FREE RUNNING			CLOSE RUNNING			SLIDING			LOCATIONAL CLEARANCE		
		Hole H11	Shaft c11	Fit	Hole H9	Shaft d9	Fit	Hole H8	Shaft f7	Fit	Hole H7	Shaft g6	Fit	Hole H7	Shaft h6	Fit
1	MAX	1.060	0.940	0.180	1.025	0.980	0.070	1.014	0.994	0.030	1.010	0.998	0.018	1.010	1.000	0.016
	MIN	1.000	0.880	0.060	1.000	0.955	0.020	1.000	0.984	0.006	1.000	0.992	0.002	1.000	0.994	0.000
1.2	MAX	1.260	1.140	0.180	1.225	1.180	0.070	1.214	1.194	0.030	1.210	1.198	0.018	1.210	1.200	0.016
	MIN	1.200	1.080	0.060	1.200	1.155	0.020	1.200	1.184	0.006	1.200	1.192	0.002	1.200	1.194	0.000
1.6	MAX	1.660	1.540	0.180	1.625	1.580	0.070	1.614	1.594	0.030	1.610	1.598	0.018	1.610	1.600	0.016
	MIN	1.600	1.480	0.060	1.600	1.555	0.020	1.600	1.584	0.006	1.600	1.592	0.002	1.600	1.594	0.000
2	MAX	2.060	1.940	0.180	2.025	1.980	0.070	2.014	1.994	0.030	2.010	1.998	0.018	2.010	2.000	0.016
	MIN	2.000	1.880	0.060	2.000	1.955	0.020	2.000	1.984	0.006	2.000	1.992	0.002	2.000	1.994	0.000
2.5	MAX	2.560	2.440	0.180	2.525	2.480	0.070	2.514	2.494	0.030	2.510	2.498	0.018	2.510	2.500	0.016
	MIN	2.500	2.380	0.060	2.500	2.455	0.020	2.500	2.484	0.006	2.500	2.492	0.002	2.500	2.494	0.000
3	MAX	3.060	2.940	0.180	3.025	2.980	0.070	3.014	2.994	0.030	3.010	2.998	0.018	3.010	3.000	0.016
	MIN	3.000	2.880	0.060	3.000	2.955	0.020	3.000	2.984	0.006	3.000	2.992	0.002	3.000	2.994	0.000
4	MAX	4.075	3.930	0.220	4.030	3.970	0.090	4.018	3.990	0.040	4.012	3.996	0.024	4.012	4.000	0.020
	MIN	4.000	3.855	0.070	4.000	3.940	0.030	4.000	3.978	0.010	4.000	3.988	0.004	4.000	3.992	0.000
5	MAX	5.075	4.930	0.220	5.030	4.970	0.090	5.018	4.990	0.040	5.012	4.996	0.024	5.012	5.000	0.020
	MIN	5.000	4.855	0.070	5.000	4.940	0.030	5.000	4.978	0.010	5.000	4.988	0.004	5.000	4.992	0.000
6	MAX	6.075	5.930	0.220	6.030	5.970	0.090	6.018	5.990	0.040	6.012	5.996	0.024	6.012	6.000	0.020
	MIN	6.000	5.855	0.070	6.000	5.940	0.030	6.000	5.978	0.010	6.000	5.988	0.004	6.000	5.992	0.000
8	MAX	8.090	7.920	0.260	8.036	7.960	0.112	8.022	7.987	0.050	8.015	7.995	0.029	8.015	8.000	0.024
	MIN	8.000	7.830	0.080	8.000	7.924	0.040	8.000	7.972	0.013	8.000	7.986	0.005	8.000	7.991	0.000
10	MAX	10.090	9.920	0.260	10.036	9.960	0.112	10.022	9.987	0.050	10.015	9.995	0.029	10.015	10.000	0.024
	MIN	10.000	9.830	0.080	10.000	9.924	0.040	10.000	9.972	0.013	10.000	9.986	0.005	10.000	9.991	0.000
12	MAX	12.110	11.905	0.315	12.043	11.950	0.136	12.027	11.984	0.061	12.018	11.994	0.035	12.018	12.000	0.029
	MIN	12.000	11.795	0.095	12.000	11.907	0.050	12.000	11.966	0.016	12.000	11.983	0.006	12.000	11.989	0.000
16	MAX	16.110	15.905	0.315	16.043	15.950	0.136	16.027	15.984	0.061	16.018	15.994	0.035	16.018	16.000	0.029
	MIN	16.000	15.795	0.095	16.000	15.907	0.050	16.000	15.966	0.016	16.000	15.983	0.006	16.000	15.989	0.000
20	MAX	20.130	19.890	0.370	20.052	19.935	0.169	20.033	19.980	0.074	20.021	19.993	0.041	20.021	20.000	0.034
	MIN	20.000	19.760	0.110	20.000	19.883	0.065	20.000	19.959	0.020	20.000	19.980	0.007	20.000	19.987	0.000
25	MAX	25.130	24.830	0.370	25.052	24.935	0.169	25.033	24.980	0.074	25.021	24.993	0.041	25.021	25.000	0.034
	MIN	25.000	24.760	0.110	25.000	24.883	0.065	25.000	24.959	0.020	25.000	24.980	0.007	25.000	24.987	0.000
30	MAX	30.130	29.890	0.370	30.052	29.935	0.169	30.033	29.980	0.074	30.021	29.993	0.041	30.021	30.000	0.034
	MIN	30.000	29.760	0.110	30.000	29.883	0.065	30.000	29.959	0.020	30.000	29.980	0.007	30.000	29.987	0.000

Courtesy of ANSI (B4.2)

Table B.2 (*Continued*)

Dimensions in mm.

BASIC SIZE		LOOSE RUNNING Hole H11	LOOSE RUNNING Shaft c11	LOOSE RUNNING Fit	FREE RUNNING Hole H9	FREE RUNNING Shaft d9	FREE RUNNING Fit	CLOSE RUNNING Hole H8	CLOSE RUNNING Shaft f7	CLOSE RUNNING Fit	SLIDING Hole H7	SLIDING Shaft g6	SLIDING Fit	LOCATIONAL CLEARANCE Hole H7	LOCATIONAL CLEARANCE Shaft h6	LOCATIONAL CLEARANCE Fit
40	MAX	40.160	39.880	0.440	40.062	39.920	0.204	40.039	39.975	0.089	40.025	39.991	0.050	40.025	40.000	0.041
	MIN	40.000	39.720	0.120	40.000	39.858	0.080	40.000	39.950	0.025	40.000	39.975	0.009	40.000	39.984	0.000
50	MAX	50.160	49.870	0.450	50.062	49.920	0.204	50.039	49.975	0.089	50.025	49.991	0.050	50.025	50.000	0.041
	MIN	50.000	49.710	0.130	50.000	49.858	0.080	50.000	49.950	0.025	50.000	49.975	0.009	50.000	49.984	0.000
60	MAX	60.190	59.860	0.520	60.074	59.900	0.248	60.046	59.970	0.106	60.030	59.990	0.059	60.030	60.000	0.049
	MIN	60.000	59.670	0.140	60.000	59.826	0.100	60.000	59.940	0.030	60.000	59.971	0.010	60.000	59.981	0.000
80	MAX	80.190	79.850	0.530	80.074	79.900	0.248	80.046	79.970	0.106	80.030	79.990	0.059	80.030	80.000	0.049
	MIN	80.000	79.660	0.150	80.000	79.826	0.100	80.000	79.940	0.030	80.000	79.971	0.010	80.000	79.981	0.000
100	MAX	100.220	99.830	0.610	100.087	99.880	0.294	100.054	99.964	0.125	100.035	99.988	0.069	100.035	100.000	0.057
	MIN	100.000	99.610	0.170	100.000	99.793	0.120	100.000	99.929	0.036	100.000	99.966	0.012	100.000	99.978	0.000
120	MAX	120.220	119.820	0.620	120.087	119.880	0.294	120.054	119.964	0.125	120.035	119.988	0.069	120.035	120.000	0.057
	MIN	120.000	119.600	0.180	120.000	119.793	0.120	120.000	119.929	0.036	120.000	119.966	0.012	120.000	119.978	0.000
160	MAX	160.250	159.790	0.710	160.100	159.855	0.345	160.063	159.957	0.146	160.040	159.986	0.079	160.040	160.000	0.065
	MIN	160.000	159.540	0.210	160.000	159.755	0.145	160.000	159.917	0.043	160.000	159.961	0.014	160.000	159.975	0.000
200	MAX	200.290	199.760	0.820	200.115	199.830	0.400	200.072	199.950	0.168	200.046	199.985	0.090	200.046	200.000	0.075
	MIN	200.000	199.470	0.240	200.000	199.715	0.170	200.000	199.904	0.050	200.000	199.956	0.015	200.000	199.971	0.000
250	MAX	250.290	249.720	0.860	250.115	249.830	0.400	250.072	249.950	0.168	250.046	249.985	0.090	250.046	250.000	0.075
	MIN	250.000	249.430	0.280	250.000	249.715	0.170	250.000	249.904	0.050	250.000	249.956	0.015	250.000	249.971	0.000
300	MAX	300.320	299.670	0.970	300.130	299.810	0.450	300.081	299.944	0.189	300.052	299.983	0.101	300.052	300.000	0.084
	MIN	300.000	299.350	0.330	300.000	299.680	0.190	300.000	299.892	0.056	300.000	299.951	0.017	300.000	299.968	0.000
400	MAX	400.360	399.600	1.120	400.140	399.790	0.490	400.089	399.938	0.208	400.057	399.982	0.111	400.057	400.000	0.093
	MIN	400.000	399.240	0.400	400.000	399.650	0.210	400.000	399.881	0.062	400.000	399.946	0.018	400.000	399.964	0.000
500	MAX	500.400	499.520	1.280	500.155	499.770	0.540	500.097	499.932	0.228	500.063	499.980	0.123	500.063	500.000	0.103
	MIN	500.000	499.120	0.480	500.000	499.615	0.230	500.000	499.869	0.068	500.000	499.940	0.020	500.000	499.960	0.000

Dimensions in mm.

BASIC SIZE		LOCATIONAL TRANSN. Hole H7	Shaft k6	Fit	LOCATIONAL TRANSN. Hole H7	Shaft n6	Fit	LOCATIONAL INTERF. Hole H7	Shaft p6	Fit	MEDIUM DRIVE Hole H7	Shaft s6	Fit	FORCE Hole H7	Shaft u6	Fit
1	MAX	1.010	1.006	0.010	1.010	1.010	0.006	1.010	1.012	0.004	1.010	1.020	-0.004	1.010	1.024	-0.008
	MIN	1.000	1.000	-0.006	1.000	1.004	-0.010	1.000	1.006	-0.012	1.000	1.014	-0.020	1.000	1.018	-0.024
1.2	MAX	1.210	1.206	0.010	1.210	1.210	0.006	1.210	1.212	0.004	1.210	1.220	-0.004	1.210	1.224	-0.008
	MIN	1.200	1.200	-0.006	1.200	1.204	-0.010	1.200	1.206	-0.012	1.200	1.214	-0.020	1.200	1.218	-0.024
1.6	MAX	1.610	1.606	0.010	1.610	1.610	0.006	1.610	1.612	0.004	1.610	1.620	-0.004	1.610	1.624	-0.008
	MIN	1.600	1.600	-0.006	1.600	1.604	-0.010	1.600	1.606	-0.012	1.600	1.614	-0.020	1.600	1.618	-0.024
2	MAX	2.010	2.006	0.010	2.010	2.010	0.006	2.010	2.012	0.004	2.010	2.020	-0.004	2.010	2.024	-0.008
	MIN	2.000	2.000	-0.006	2.000	2.004	-0.010	2.000	2.006	-0.012	2.000	2.014	-0.020	2.000	2.018	-0.024
2.5	MAX	2.510	2.506	0.010	2.510	2.510	0.006	2.510	2.512	0.004	2.510	2.520	-0.004	2.510	2.524	-0.008
	MIN	2.500	2.500	-0.006	2.500	2.504	-0.010	2.500	2.506	-0.012	2.500	2.514	-0.020	2.500	2.518	-0.024
3	MAX	3.010	3.006	0.010	3.010	3.010	0.006	3.010	3.012	0.004	3.010	3.020	-0.004	3.010	3.024	-0.008
	MIN	3.000	3.000	-0.006	3.000	3.004	-0.010	3.000	3.006	-0.012	3.000	3.014	-0.020	3.000	3.018	-0.024
4	MAX	4.012	4.009	0.011	4.012	4.016	0.004	4.012	4.020	0.000	4.012	4.027	-0.007	4.012	4.031	-0.011
	MIN	4.000	4.001	-0.009	4.000	4.008	-0.016	4.000	4.012	-0.020	4.000	4.019	-0.027	4.000	4.023	-0.031
5	MAX	5.012	5.009	0.011	5.012	5.016	0.004	5.012	5.020	0.000	5.012	5.027	-0.007	5.012	5.031	-0.011
	MIN	5.000	5.001	-0.009	5.000	5.008	-0.016	5.000	5.012	-0.020	5.000	5.019	-0.027	5.000	5.023	-0.031
6	MAX	6.012	6.009	0.011	6.012	6.016	0.004	6.012	6.020	0.000	6.012	6.027	-0.007	6.012	6.031	-0.011
	MIN	6.000	6.001	-0.009	6.000	6.008	-0.016	6.000	6.012	-0.020	6.000	6.019	-0.027	6.000	6.023	-0.031
8	MAX	8.015	8.010	0.014	8.015	8.019	0.005	8.015	8.024	0.000	8.015	8.032	-0.008	8.015	8.037	-0.013
	MIN	8.000	8.001	-0.010	8.000	8.010	-0.019	8.000	8.015	-0.024	8.000	8.023	-0.032	8.000	8.028	-0.037
10	MAX	10.015	10.010	0.014	10.015	10.019	0.005	10.015	10.024	0.000	10.015	10.032	-0.008	10.015	10.037	-0.013
	MIN	10.000	10.001	-0.010	10.000	10.010	-0.019	10.000	10.015	-0.024	10.000	10.023	-0.032	10.000	10.028	-0.037
12	MAX	12.018	12.012	0.017	12.018	12.023	0.006	12.018	12.029	0.000	12.018	12.039	-0.010	12.018	12.044	-0.015
	MIN	12.000	12.001	-0.012	12.000	12.012	-0.023	12.000	12.018	-0.029	12.000	12.028	-0.039	12.000	12.033	-0.044
16	MAX	16.018	16.012	0.017	16.018	16.023	0.006	16.018	16.029	0.000	16.018	16.034	-0.010	16.018	16.044	-0.015
	MIN	16.000	16.001	-0.012	16.000	16.012	-0.023	16.000	16.018	-0.029	16.000	16.028	-0.039	16.000	16.033	-0.044
20	MAX	20.021	20.015	0.019	20.021	20.028	0.006	20.021	20.035	-0.001	20.021	20.048	-0.014	20.021	20.054	-0.020
	MIN	20.000	20.002	-0.015	20.000	20.015	-0.028	20.000	20.022	-0.035	20.000	20.035	-0.048	20.000	20.041	-0.054
25	MAX	25.021	25.015	0.019	25.021	25.028	0.006	25.021	25.035	-0.001	25.021	25.048	-0.014	25.021	25.061	-0.027
	MIN	25.000	25.002	-0.015	25.000	25.015	-0.028	25.000	25.022	-0.035	25.000	25.035	-0.048	25.000	25.048	-0.061
30	MAX	30.021	30.015	0.019	30.021	30.028	0.006	30.021	30.035	-0.001	30.021	30.048	-0.014	30.021	30.061	-0.027
	MIN	30.000	30.002	-0.015	30.000	30.015	-0.028	30.000	30.022	-0.035	30.000	30.035	-0.048	30.000	30.048	-0.061

Courtesy of ANSI (B4.2)

Table B.2   (Continued)

Dimensions in mm.

BASIC SIZE		LOCATIONAL TRANSN. Hole H7	Shaft k6	Fit	LOCATIONAL TRANSN. Hole H7	Shaft n6	Fit	LOCATIONAL INTERF. Hole H7	Shaft p6	Fit	MEDIUM DRIVE Hole H7	Shaft s6	Fit	FORCE Hole H7	Shaft u6	Fit
40	MAX	40.025	40.018	0.023	40.025	40.033	0.008	40.025	40.042	-0.001	40.025	40.059	-0.018	40.025	40.076	-0.035
	MIN	40.000	40.002	-0.018	40.000	40.017	-0.033	40.000	40.026	-0.042	40.000	40.043	-0.059	40.000	40.060	-0.076
50	MAX	50.025	50.018	0.023	50.025	50.033	0.008	50.025	50.042	-0.001	50.025	50.059	-0.018	50.025	50.086	-0.045
	MIN	50.000	50.002	-0.018	50.000	50.017	-0.033	50.000	50.026	-0.042	50.000	50.043	-0.059	50.000	50.070	-0.086
60	MAX	60.030	60.021	0.028	60.030	60.039	0.010	60.030	60.051	-0.002	60.030	60.072	-0.023	60.030	60.106	-0.057
	MIN	60.000	60.002	-0.021	60.000	60.020	-0.039	60.000	60.032	-0.051	60.000	60.053	-0.072	60.000	60.087	-0.106
80	MAX	80.030	80.021	0.028	80.030	80.039	0.010	80.030	80.051	-0.002	80.030	80.078	-0.029	80.030	80.121	-0.072
	MIN	80.000	80.002	-0.021	80.000	80.020	-0.039	80.000	80.032	-0.051	80.000	80.059	-0.078	80.000	80.102	-0.121
100	MAX	100.035	100.025	0.032	100.035	100.045	0.012	100.035	100.059	-0.002	100.035	100.093	-0.036	100.035	100.146	-0.089
	MIN	100.000	100.003	-0.025	100.000	100.023	-0.045	100.000	100.037	-0.059	100.000	100.071	-0.093	100.000	100.124	-0.146
120	MAX	120.035	120.025	0.032	120.035	120.045	0.012	120.035	120.059	-0.002	120.035	120.101	-0.044	120.035	120.166	-0.109
	MIN	120.000	120.003	-0.025	120.000	120.023	-0.045	120.000	120.037	-0.059	120.000	120.079	-0.101	120.000	120.144	-0.166
160	MAX	160.040	160.028	0.037	160.040	160.052	0.013	160.040	160.068	-0.003	160.040	160.125	-0.060	160.040	160.215	-0.150
	MIN	160.000	160.003	-0.028	160.000	160.027	-0.052	160.000	160.043	-0.068	160.000	160.100	-0.125	160.000	160.190	-0.215
200	MAX	200.046	200.033	0.042	200.046	200.060	0.015	200.046	200.079	-0.004	200.046	200.151	-0.076	200.046	200.265	-0.190
	MIN	200.000	200.004	-0.033	200.000	200.031	-0.060	200.000	200.050	-0.079	200.000	200.122	-0.151	200.000	200.236	-0.265
250	MAX	250.046	250.033	0.042	250.046	250.060	0.015	250.046	250.079	-0.004	250.046	250.169	-0.094	250.046	250.313	-0.238
	MIN	250.000	250.004	-0.033	250.000	250.031	-0.060	250.000	250.050	-0.079	250.000	250.140	-0.169	250.000	250.284	-0.313
300	MAX	300.052	300.036	0.048	300.052	300.066	0.018	300.052	300.088	-0.004	300.052	300.202	-0.118	300.052	300.382	-0.298
	MIN	300.000	300.004	-0.036	300.000	300.034	-0.066	300.000	300.056	-0.088	300.000	300.170	-0.202	300.000	300.350	-0.382
400	MAX	400.057	400.040	0.053	400.057	400.073	0.020	400.057	400.098	-0.005	400.057	400.244	-0.151	400.057	400.471	-0.378
	MIN	400.000	400.004	-0.040	400.000	400.037	-0.073	400.000	400.062	-0.098	400.000	400.208	-0.244	400.000	400.435	-0.471
500	MAX	500.063	500.045	0.058	500.063	500.080	0.023	500.063	500.108	-0.005	500.063	500.292	-0.189	500.063	500.580	-0.477
	MIN	500.000	500.005	-0.045	500.000	500.040	-0.080	500.000	500.068	-0.108	500.000	500.252	-0.292	500.000	500.540	-0.580

Courtesy of ANSI (B4.2)

Table B.3 American National Standard Unified Inch Screw Threads

Nominal Size		Basic Major Diameter	Series With Graded Pitches			Series With Constant Pitches								Nominal Size
Primary	Secondary		Coarse UNC	Fine UNF	Extra-Fine UNEF	4UN	6UN	8UN	12UN	16UN	20UN	28UN	32UN	
0		0.0600	—	80	—	—	—	—	—	—	—	—	—	0
	1	0.0730	64	72	—	—	—	—	—	—	—	—	—	1
2		0.0860	56	64	—	—	—	—	—	—	—	—	—	2
	3	0.0990	48	56	—	—	—	—	—	—	—	—	—	3
4		0.1120	40	48	—	—	—	—	—	—	—	—	—	4
5		0.1250	40	44	—	—	—	—	—	—	—	—	—	5
6		0.1380	32	40	—	—	—	—	—	—	—	—	UNC	6
8		0.1640	32	36	—	—	—	—	—	—	—	—	UNC	8
10		0.1900	24	32	—	—	—	—	—	—	—	—	UNF	10
	12	0.2160	24	28	32	—	—	—	—	—	—	UNF	UNEF	12
¼		0.2500	20	28	32	—	—	—	—	—	UNC	UNF	UNEF	¼
5⁄16		0.3125	18	24	32	—	—	—	—	—	20	28	UNEF	5⁄16
⅜		0.3750	16	24	32	—	—	—	—	UNC	20	28	UNEF	⅜
7⁄16		0.4375	14	20	28	—	—	—	—	16	UNF	UNEF	32	7⁄16
½		0.5000	13	20	28	—	—	—	—	16	UNF	UNEF	32	½
9⁄16		0.5625	12	18	24	—	—	—	UNC	16	20	28	32	9⁄16
⅝		0.6250	11	18	24	—	—	—	12	16	20	28	32	⅝
	11⁄16	0.6875	—	—	24	—	—	—	12	16	20	28	32	11⁄16
¾		0.7500	10	16	20	—	—	—	12	UNF	UNEF	28	32	¾
	13⁄16	0.8125	—	—	20	—	—	—	12	16	UNEF	28	32	13⁄16
⅞		0.8750	9	14	20	—	—	—	12	16	UNEF	28	32	⅞
	15⁄16	0.9375	—	—	20	—	—	—	12	16	UNEF	28	32	15⁄16
1		1.0000	8	12	20	—	—	UNC	UNF	16	UNEF	28	32	1
	1 1⁄16	1.0625	—	—	18	—	—	8	12	16	20	28	—	1 1⁄16
1 ⅛		1.1250	7	12	18	—	—	8	UNF	16	20	28	—	1 ⅛
	1 3⁄16	1.1875	—	—	18	—	—	8	12	16	20	28	—	1 3⁄16
1 ¼		1.2500	7	12	18	—	—	8	UNF	16	20	28	—	1 ¼
	1 5⁄16	1.3125	—	—	18	—	—	8	12	16	20	28	—	1 5⁄16
1 ⅜		1.3750	6	12	18	—	UNC	8	UNF	16	20	28	—	1 ⅜
	1 7⁄16	1.4375	—	—	18	—	6	8	12	16	20	28	—	1 7⁄16
1 ½		1.5000	6	12	18	—	UNC	8	UNF	16	20	28	—	1 ½
	1 9⁄16	1.5625	—	—	18	—	6	8	12	16	20	—	—	1 9⁄16
1 ⅝		1.6250	—	—	18	—	6	8	12	16	20	—	—	1 ⅝
	1 11⁄16	1.6875	—	—	18	—	6	8	12	16	20	—	—	1 11⁄16
1 ¾		1.7500	5	—	—	—	6	8	12	16	20	—	—	1 ¾
	1 13⁄16	1.8125	—	—	—	—	6	8	12	16	20	—	—	1 13⁄16
1 ⅞		1.8750	—	—	—	—	6	8	12	16	20	—	—	1 ⅞
	1 15⁄16	1.9375	—	—	—	—	6	8	12	16	20	—	—	1 15⁄16
2		2.0000	4½	—	—	—	6	8	12	16	20	—	—	2
	2 ⅛	2.1250	—	—	—	—	6	8	12	16	20	—	—	2 ⅛
2 ¼		2.2500	4½	—	—	—	6	8	12	16	20	—	—	2 ¼
	2 ⅜	2.3750	—	—	—	—	6	8	12	16	20	—	—	2 ⅜
2 ½		2.5000	4	—	—	UNC	6	8	12	16	20	—	—	2 ½
	2 ⅝	2.6250	—	—	—	4	6	8	12	16	20	—	—	2 ⅝
2 ¾		2.7500	4	—	—	UNC	6	8	12	16	20	—	—	2 ¾
	2 ⅞	2.8750	—	—	—	4	6	8	12	16	20	—	—	2 ⅞
3		3.0000	4	—	—	UNC	6	8	12	16	20	—	—	3
	3 ⅛	3.1250	—	—	—	4	6	8	12	16	—	—	—	3 ⅛
3 ¼		3.2500	4	—	—	UNC	6	8	12	16	—	—	—	3 ¼
	3 ⅜	3.3750	—	—	—	4	6	8	12	16	—	—	—	3 ⅜
3 ½		3.5000	4	—	—	UNC	6	8	12	16	—	—	—	3 ½
	3 ⅝	3.6250	—	—	—	4	6	8	12	16	—	—	—	3 ⅝
3 ¾		3.7500	4	—	—	UNC	6	8	12	16	—	—	—	3 ¾
	3 ⅞	3.8750	—	—	—	4	6	8	12	16	—	—	—	3 ⅞
4		4.0000	4	—	—	UNC	6	8	12	16	—	—	—	4
	4 ⅛	4.1250	—	—	—	4	6	8	12	16	—	—	—	4 ⅛
4 ¼		4.2500	—	—	—	4	6	8	12	16	—	—	—	4 ¼
	4 ⅜	4.3750	—	—	—	4	6	8	12	16	—	—	—	4 ⅜
4 ½		4.5000	—	—	—	4	6	8	12	16	—	—	—	4 ½
	4 ⅝	4.6250	—	—	—	4	6	8	12	16	—	—	—	4 ⅝
4 ¾		4.7500	—	—	—	4	6	8	12	16	—	—	—	4 ¾
	4 ⅞	4.8750	—	—	—	4	6	8	12	16	—	—	—	4 ⅞
5		5.0000	—	—	—	4	6	8	12	16	—	—	—	5
	5 ⅛	5.1250	—	—	—	4	6	8	12	16	—	—	—	5 ⅛
5 ¼		5.2500	—	—	—	4	6	8	12	16	—	—	—	5 ¼
	5 ⅜	5.3750	—	—	—	4	6	8	12	16	—	—	—	5 ⅜
5 ½		5.5000	—	—	—	4	6	8	12	16	—	—	—	5 ½
	5 ⅝	5.6250	—	—	—	4	6	8	12	16	—	—	—	5 ⅝
5 ¾		5.7500	—	—	—	4	6	8	12	16	—	—	—	5 ¾
	5 ⅞	5.8750	—	—	—	4	6	8	12	16	—	—	—	5 ⅞
6		6.0000	—	—	—	4	6	8	12	16	—	—	—	6

NOTE:

(1) Series designation shown indicates the UN thread form; however, the UNR thread form may be specified by substituting UNR in place of UN in all designations for external use only.

Courtesy of ANSI (B1.1)

Table B.4   Standard Coarse Pitch Metric Threads

Nominal Size	Pitch	Nominal Size	Pitch
		20	2.5
1.6	0.35	22	2.5[1]
2	0.4	24	3
2.5	0.45	27	3[1]
		30	3.5
3	0.5	36	4
3.5	0.6	42	4.5
4	0.7	48	5
5	0.8	56	5.5
6	1	64	6
8	1.25	72	6
10	1.5	80	6
12	1.75	90	6
14	2	100	6
16	2		

NOTE:
(1) For high strength structural steel fasteners only.
Courtesy of ANSI (B1.13M)

Table B.5   Standard Fine Pitch Metric Threads

Nominal Size	Pitch		Nominal Size	Pitch	
8	1	—	55	1.5	—
10	0.75	1.25	56	—	2
12	1 1.5[1]	1.25	60	1.5	—
14		1.5	64	—	2
			65	1.5	—
15	1	—	70	1.5	—
16	—	1.5	72	—	2
17	1	—	75	1.5	—
18	—	1.5	80	1.5	2
20	1	1.5	85	—	2
22	—	1.5	90	—	2
24	—	2	95	—	2
25	1.5	—	100	—	2
27	—	2	105	—	2
30	1.5	2	110	—	2
33	—	2	120	—	2
35	1.5	—	130	—	2
36	—	2	140	—	2
39	—	2	150	—	2
40	1.5	—	160	—	3
42	—	2	170	—	3
45	1.5	—	180	—	3
48	—	2	190	—	3
50	1.5	—	200	—	3

NOTE:
(1) Only for wheel studs and nuts.
Courtesy of ANSI (B1.13M)

Table B.6 Dimensions of Hex Bolts

Nominal Size or Basic Product Dia (17)		E Body Dia (7) Max	F Width Across Flats (4) Basic	F Max	F Min	G Width Across Corners Max	G Min	H Height Basic	H Max	H Min	R Radius of Fillet Max	R Min	L_T Thread Length For Bolt Lengths (11) 6 in. and shorter Basic	L_T over 6 in. Basic
1/4	0.2500	0.260	7/16	0.438	0.425	0.505	0.484	11/64	0.188	0.150	0.03	0.01	0.750	1.000
5/16	0.3125	0.324	1/2	0.500	0.484	0.577	0.552	7/32	0.235	0.195	0.03	0.01	0.875	1.125
3/8	0.3750	0.388	9/16	0.562	0.544	0.650	0.620	1/4	0.268	0.226	0.03	0.01	1.000	1.250
7/16	0.4375	0.452	5/8	0.625	0.603	0.722	0.687	19/64	0.316	0.272	0.03	0.01	1.125	1.375
1/2	0.5000	0.515	3/4	0.750	0.725	0.866	0.826	11/32	0.364	0.302	0.03	0.01	1.250	1.500
5/8	0.6250	0.642	15/16	0.928	0.906	1.083	1.033	27/64	0.444	0.378	0.06	0.02	1.500	1.750
3/4	0.7500	0.768	1 1/8	1.125	1.088	1.299	1.240	1/2	0.524	0.455	0.06	0.02	1.750	2.000
7/8	0.8750	0.895	1 5/16	1.312	1.269	1.516	1.447	37/64	0.604	0.531	0.06	0.02	2.000	2.250
1	1.0000	1.022	1 1/2	1.500	1.450	1.732	1.653	43/64	0.700	0.591	0.09	0.03	2.250	2.500
1 1/8	1.1250	1.149	1 11/16	1.688	1.631	1.949	1.859	3/4	0.780	0.658	0.09	0.03	2.500	2.750
1 1/4	1.2500	1.277	1 7/8	1.875	1.812	2.165	2.066	27/32	0.876	0.749	0.09	0.03	2.750	3.000
1 3/8	1.3750	1.404	2 1/16	2.062	1.994	2.382	2.273	29/32	0.940	0.810	0.09	0.03	3.000	3.250
1 1/2	1.5000	1.531	2 1/4	2.250	2.175	2.598	2.480	1	1.036	0.902	0.09	0.03	3.250	3.500
1 3/4	1.7500	1.785	2 5/8	2.625	2.538	3.031	2.893	1 5/32	1.196	1.054	0.12	0.04	3.750	4.000
2	2.0000	2.039	3	3.000	2.900	3.464	3.306	1 11/32	1.388	1.175	0.12	0.04	4.250	4.500
2 1/4	2.2500	2.305	3 3/8	3.375	3.262	3.897	3.719	1 1/2	1.548	1.327	0.19	0.06	4.750	5.000
2 1/2	2.5000	2.559	3 3/4	3.750	3.625	4.330	4.133	1 21/32	1.708	1.479	0.19	0.06	5.250	5.500
2 3/4	2.7500	2.827	4 1/8	4.125	3.988	4.763	4.546	1 13/16	1.869	1.632	0.19	0.06	5.750	6.000
3	3.0000	3.081	4 1/2	4.500	4.350	5.196	4.959	2	2.060	1.815	0.19	0.06	6.250	6.500
3 1/4	3.2500	3.335	4 7/8	4.875	4.712	5.629	5.372	2 3/16	2.251	1.936	0.19	0.06	6.750	7.000
3 1/2	3.5000	3.589	5 1/4	5.250	5.075	6.062	5.786	2 5/16	2.380	2.057	0.19	0.06	7.250	7.500
3 3/4	3.7500	3.858	5 5/8	5.625	5.437	6.495	6.198	2 1/2	2.572	2.241	0.19	0.06	7.750	8.000
4	4.0000	4.111	6	6.000	5.800	6.928	6.612	2 11/16	2.764	2.424	0.19	0.06	8.250	8.500

Courtesy of ANSI (B18.2.1)

Table B.7   Dimensions of Square Bolts

Nominal Size or Basic Product Dia (17)		E Body Dia (7), (14) Max	F Width Across Flats (4)			G Width Across Corners		H Height			R Radius of Fillet		L_T Thread Length For Bolt Lengths (11)	
			Basic	Max	Min	Max	Min	Basic	Max	Min	Max	Min	6 in. and shorter Basic	over 6 in. Basic
1/4	0.2500	0.260	3/8	0.375	0.362	0.530	0.498	11/64	0.188	0.156	0.03	0.01	0.750	1.000
5/16	0.3125	0.324	1/2	0.500	0.484	0.707	0.665	13/64	0.220	0.186	0.03	0.01	0.875	1.125
3/8	0.3750	0.388	9/16	0.562	0.544	0.795	0.747	1/4	0.268	0.232	0.03	0.01	1.000	1.250
7/16	0.4375	0.452	5/8	0.625	0.603	0.884	0.828	19/64	0.316	0.278	0.03	0.01	1.125	1.375
1/2	0.5000	0.515	3/4	0.750	0.725	1.061	0.995	21/64	0.348	0.308	0.03	0.01	1.250	1.500
5/8	0.6250	0.642	15/16	0.938	0.906	1.326	1.244	27/64	0.444	0.400	0.06	0.02	1.500	1.750
3/4	0.7500	0.768	1 1/8	1.125	1.088	1.591	1.494	1/2	0.524	0.476	0.06	0.02	1.750	2.000
7/8	0.8750	0.895	1 5/16	1.312	1.269	1.856	1.742	19/32	0.620	0.568	0.06	0.02	2.000	2.250
1	1.0000	1.022	1 1/2	1.500	1.450	2.121	1.991	21/32	0.684	0.628	0.09	0.03	2.250	2.500
1 1/8	1.1250	1.149	1 11/16	1.688	1.631	2.386	2.239	3/4	0.780	0.720	0.09	0.03	2.500	2.750
1 1/4	1.2500	1.277	1 7/8	1.875	1.812	2.652	2.489	27/32	0.876	0.812	0.09	0.03	2.750	3.000
1 3/8	1.3750	1.404	2 1/16	2.062	1.994	2.917	2.738	29/32	0.940	0.872	0.09	0.03	3.000	3.250
1 1/2	1.5000	1.531	2 1/4	2.250	2.175	3.182	2.986	1	1.036	0.964	0.09	0.03	3.250	3.500

Courtesy of ANSI (B18.2.1)

Bolt With Reduced Diameter Body

25° approx.

Table B.8 Dimensions of Hex Nuts and Hex Jam Nuts

Nominal Size or Basic Major Diam. of Thread	F Width Across Flats Basic	F Max.	F Min.	G Width Across Corners Max.	G Min.	H Thickness Hex Nuts Basic	H Max.	H Min.	H_1 Thickness Hex Jam Nuts Basic	H_1 Max.	H_1 Min.	Runout of Bearing Face FIM — Hex Nuts, Specified Proof Load, Up to 150,000 psi	Runout — Hex Nuts, 150,000 psi and Greater, Max.	Runout — Hex Jam Nuts, All Strength Levels
1/4 — 0.2500	7/16	0.438	0.428	0.505	0.488	7/32	0.226	0.212	5/32	0.163	0.150	0.015	0.010	0.015
5/16 — 0.3125	1/2	0.500	0.489	0.577	0.557	17/64	0.273	0.258	3/16	0.195	0.180	0.016	0.011	0.016
3/8 — 0.3750	9/16	0.562	0.551	0.650	0.628	21/64	0.337	0.320	7/32	0.227	0.210	0.017	0.012	0.017
7/16 — 0.4375	11/16	0.688	0.675	0.794	0.768	3/8	0.385	0.365	1/4	0.260	0.240	0.018	0.013	0.018
1/2 — 0.5000	3/4	0.750	0.736	0.866	0.840	7/16	0.448	0.427	5/16	0.323	0.302	0.019	0.014	0.019
9/16 — 0.5625	7/8	0.875	0.861	1.010	0.982	31/64	0.496	0.473	5/16	0.324	0.301	0.020	0.015	0.020
5/8 — 0.6250	15/16	0.938	0.922	1.083	1.051	35/64	0.559	0.535	3/8	0.387	0.363	0.021	0.016	0.021
3/4 — 0.7500	1 1/8	1.125	1.088	1.299	1.240	41/64	0.665	0.617	27/64	0.446	0.398	0.023	0.018	0.023
7/8 — 0.8750	1 5/16	1.312	1.269	1.516	1.447	3/4	0.776	0.724	31/64	0.510	0.458	0.025	0.020	0.025
1 — 1.0000	1 1/2	1.500	1.450	1.732	1.653	55/64	0.887	0.831	35/64	0.575	0.519	0.027	0.022	0.027
1 1/8 — 1.1250	1 11/16	1.688	1.631	1.949	1.859	31/32	0.999	0.939	39/64	0.639	0.579	0.030	0.025	0.030
1 1/4 — 1.2500	1 7/8	1.875	1.812	2.165	2.066	1 1/16	1.094	1.030	23/32	0.751	0.687	0.033	0.028	0.033
1 3/8 — 1.3750	2 1/16	2.062	1.994	2.382	2.273	1 11/64	1.206	1.138	25/32	0.815	0.747	0.036	0.031	0.036
1 1/2 — 1.5000	2 1/4	2.250	2.175	2.598	2.480	1 9/32	1.317	1.245	27/32	0.880	0.808	0.039	0.034	0.039
See Notes	1			2								3		

Courtesy of ANSI (B18.2.2)

Table B.9   Dimensions of Square Nuts

Nominal Size or Basic Major Diam. of Thread		F Width Across Flats			G Width Across Corners		H Thickness		
		Basic	Max.	Min.	Max.	Min.	Basic	Max.	Min.
1/4	0.2500	7/16	0.438	0.425	0.619	0.554	7/32	0.235	0.203
5/16	0.3125	9/16	0.562	0.547	0.795	0.721	17/64	0.283	0.249
3/8	0.3750	5/8	0.625	0.606	0.884	0.802	21/64	0.346	0.310
7/16	0.4375	3/4	0.750	0.728	1.061	0.970	3/8	0.394	0.356
1/2	0.5000	13/16	0.812	0.788	1.149	1.052	7/16	0.458	0.418
5/8	0.6250	1	1.000	0.969	1.414	1.300	35/64	0.569	0.525
3/4	0.7500	1 1/8	1.125	1.088	1.591	1.464	21/32	0.680	0.632
7/8	0.8750	1 5/16	1.312	1.269	1.856	1.712	49/64	0.792	0.740
1	1.0000	1 1/2	1.500	1.450	2.121	1.961	7/8	0.903	0.847
1 1/8	1.1250	1 11/16	1.688	1.631	2.386	2.209	1	1.030	0.970
1 1/4	1.2500	1 7/8	1.875	1.812	2.652	2.458	1 3/32	1.126	1.062
1 3/8	1.3750	2 1/16	2.062	1.994	2.917	2.708	1 13/64	1.237	1.169
1 1/2	1.5000	2 1/4	2.250	2.175	3.182	2.956	1 5/16	1.348	1.276
See Note		1							

Courtesy of ANSI (B18.2.2)

A NOTE 6

A NOTE 12

A NOTE 5

R See NOTE 5

See NOTE 5

-A-

See NOTE 5

30° +0 -15

(Ref)

L_T — See NOTE 10

L_G — See NOTE 10

L_B — Y (Ref)

Nominal Size or Basic Product Diameter (18)	E Body Diameter (8)		F Width Across Flats			G Width Across Corners (4)		H Height			J Wrenching Height (4)	L_T Thread Length For Screw Lengths (10)		Y Transition Thread Length (10)	Runout of Bearing Surface FIM (5)
												6 in. and Shorter	Over 6 in.		
	Max	Min	Basic	Max	Min	Max	Min	Basic	Max	Min	Min	Basic	Basic	Max	Max
1/4 0.2500	0.2500	0.2450	7/16	0.438	0.428	0.505	0.488	5/32	0.163	0.150	0.106	0.750	1.000	0.250	0.010
5/16 0.3125	0.3125	0.3065	1/2	0.500	0.489	0.577	0.557	13/64	0.211	0.195	0.140	0.875	1.125	0.278	0.011
3/8 0.3750	0.3750	0.3690	9/16	0.562	0.551	0.650	0.628	15/64	0.243	0.226	0.160	1.000	1.250	0.312	0.012
7/16 0.4375	0.4375	0.4305	5/8	0.625	0.612	0.722	0.698	9/32	0.291	0.272	0.195	1.125	1.375	0.357	0.013
1/2 0.5000	0.5000	0.4930	3/4	0.750	0.736	0.866	0.840	5/16	0.323	0.302	0.215	1.250	1.500	0.385	0.014
9/16 0.5625	0.5625	0.5545	13/16	0.812	0.798	0.938	0.910	23/64	0.371	0.348	0.250	1.375	1.625	0.417	0.015
5/8 0.6250	0.6250	0.6170	15/16	0.938	0.922	1.083	1.051	25/64	0.403	0.378	0.269	1.500	1.750	0.455	0.017
3/4 0.7500	0.7500	0.7410	1 1/8	1.125	1.100	1.299	1.254	15/32	0.483	0.455	0.324	1.750	2.000	0.500	0.020
7/8 0.8750	0.8750	0.8660	1 5/16	1.312	1.285	1.516	1.465	35/64	0.563	0.531	0.378	2.000	2.250	0.556	0.023
1 1.0000	1.0000	0.9900	1 1/2	1.500	1.469	1.732	1.675	39/64	0.627	0.591	0.416	2.250	2.500	0.625	0.026
1 1/8 1.1250	1.1250	1.1140	1 11/16	1.688	1.631	1.949	1.859	11/16	0.718	0.658	0.461	2.500	2.750	0.714	0.029
1 1/4 1.2500	1.2500	1.2390	1 7/8	1.875	1.812	2.165	2.066	25/32	0.813	0.749	0.530	2.750	3.000	0.714	0.033
1 3/8 1.3750	1.3750	1.3630	2 1/16	2.062	1.994	2.382	2.273	27/32	0.878	0.810	0.569	3.000	3.250	0.833	0.036
1 1/2 1.5000	1.5000	1.4880	2 1/4	2.230	2.175	2.598	2.480	15/16	0.974	0.902	0.640	3.250	3.500	0.833	0.039
1 3/4 1.7500	1.7500	1.7380	2 5/8	2.625	2.538	3.031	2.893	1 3/32	1.134	1.054	0.748	3.750	4.000	1.000	0.046
2 2.0000	2.0000	1.9880	3	3.000	2.900	3.464	3.306	1 7/32	1.263	1.175	0.825	4.250	4.500	1.111	0.052
2 1/4 2.2500	2.2500	2.2380	3 3/8	3.375	3.262	3.897	3.719	1 3/8	1.423	1.327	0.933	4.750	5.000	1.111	0.059
2 1/2 2.5000	2.5000	2.4880	3 3/4	3.750	3.625	4.330	4.133	1 17/32	1.583	1.479	1.042	5.250	5.500	1.250	0.065
2 3/4 2.7500	2.7500	2.7380	4 1/8	4.125	3.988	4.763	4.546	1 11/16	1.744	1.632	1.151	5.750	6.000	1.250	0.072
3 3.0000	3.0000	2.9880	4 1/2	4.500	4.350	5.196	4.959	1 7/8	1.935	1.815	1.290	6.250	6.500	1.250	0.079

Table B.11   Dimensions of Slotted Flat Countersunk Head Cap Screws

Nominal Size[1] or Basic Screw Diameter		E Body Diameter		A Head Diameter		H Head Height	J Slot Width		T Slot Depth		U Fillet Radius	F Protrusion Above Gaging Diameter		G Gaging Diameter
		Max	Min	Max, Edge Sharp	Min, Edge Rounded or Flat	Ref	Max	Min	Max	Min	Max	Max	Min	
1/4	0.2500	0.2500	0.2450	0.500	0.452	0.140	0.075	0.064	0.068	0.045	0.100	0.046	0.030	0.424
5/16	0.3125	0.3125	0.3070	0.625	0.567	0.177	0.084	0.072	0.086	0.057	0.125	0.053	0.035	0.538
3/8	0.3750	0.3750	0.3690	0.750	0.682	0.210	0.094	0.081	0.103	0.068	0.150	0.060	0.040	0.651
7/16	0.4375	0.4375	0.4310	0.812	0.736	0.210	0.094	0.081	0.103	0.068	0.175	0.065	0.044	0.703
1/2	0.5000	0.5000	0.4930	0.875	0.791	0.210	0.106	0.091	0.103	0.068	0.200	0.071	0.049	0.756
9/16	0.5625	0.5625	0.5550	1.000	0.906	0.244	0.118	0.102	0.120	0.080	0.225	0.078	0.054	0.869
5/8	0.6250	0.6250	0.6170	1.125	1.020	0.281	0.133	0.116	0.137	0.091	0.250	0.085	0.058	0.982
3/4	0.7500	0.7500	0.7420	1.375	1.251	0.352	0.149	0.131	0.171	0.115	0.300	0.099	0.068	1.208
7/8	0.8750	0.8750	0.8660	1.625	1.480	0.423	0.167	0.147	0.206	0.138	0.350	0.113	0.077	1.435
1	1.0000	1.0000	0.9900	1.875	1.711	0.494	0.188	0.166	0.240	0.162	0.400	0.127	0.087	1.661
1 1/8	1.1250	1.1250	1.1140	2.062	1.880	0.529	0.196	0.178	0.257	0.173	0.450	0.141	0.096	1.826
1 1/4	1.2500	1.2500	1.2390	2.312	2.110	0.600	0.211	0.193	0.291	0.197	0.500	0.155	0.105	2.052
1 3/8	1.3750	1.3750	1.3630	2.562	2.340	0.665	0.226	0.208	0.326	0.220	0.550	0.169	0.115	2.279
1 1/2	1.5000	1.5000	1.4880	2.812	2.570	0.742	0.258	0.240	0.360	0.244	0.600	0.183	0.124	2.505

[1]Where specifying nominal size in decimals, zeros preceding decimal and in the fourth decimal place shall be omitted.
Courtesy of ANSI (B18.6.2)

Table B.12   Dimensions of Slotted Round Head Cap Screws

Nominal Size[1] or Basic Screw Diameter		E Body Diameter		A Head Diameter		H Head Height		J Slot Width		T Slot Depth		U Fillet Radius	
		Max	Min	Max	Min	Max	Min	Max	Min	Max	Min	Max	Min
1/4	0.2500	0.2500	0.2450	0.437	0.418	0.191	0.175	0.075	0.064	0.117	0.097	0.031	0.016
5/16	0.3125	0.3125	0.3070	0.562	0.540	0.245	0.226	0.084	0.072	0.151	0.126	0.031	0.016
3/8	0.3750	0.3750	0.3690	0.625	0.603	0.273	0.252	0.094	0.081	0.168	0.138	0.031	0.016
7/16	0.4375	0.4375	0.4310	0.750	0.725	0.328	0.302	0.094	0.081	0.202	0.167	0.047	0.016
1/2	0.5000	0.5000	0.4930	0.812	0.786	0.354	0.327	0.106	0.091	0.218	0.178	0.047	0.016
9/16	0.5625	0.5625	0.5550	0.937	0.909	0.409	0.378	0.118	0.102	0.252	0.207	0.047	0.016
5/8	0.6250	0.6250	0.6170	1.000	0.970	0.437	0.405	0.133	0.116	0.270	0.220	0.062	0.031
3/4	0.7500	0.7500	0.7420	1.250	1.215	0.546	0.507	0.149	0.131	0.338	0.278	0.062	0.031

[1]Where specifying nominal size in decimals, zeros preceding decimal and in the fourth decimal place shall be omitted.
Courtesy of ANSI (B18.6.2)

Table 13 Dimensions of Slotted Fillister Head Cap Screws

Nominal Size[1] or Basic Screw Diameter		E		A		H		O		J		T		U	
		Body Diameter		Head Diameter		Head Side Height		Total Head Height		Slot Width		Slot Depth		Fillet Radius	
		Max	Min	Max	Min	Max	Min	Max	Min	Max	Min	Max	Min	Max	Min
1/4	0.2500	0.2500	0.2450	0.375	0.363	0.172	0.157	0.216	0.194	0.075	0.064	0.097	0.077	0.031	0.016
5/16	0.3125	0.3125	0.3070	0.437	0.424	0.203	0.186	0.253	0.230	0.084	0.072	0.115	0.090	0.031	0.016
3/8	0.3750	0.3750	0.3690	0.562	0.547	0.250	0.229	0.314	0.284	0.094	0.081	0.142	0.112	0.031	0.016
7/16	0.4375	0.4375	0.4310	0.625	0.608	0.297	0.274	0.368	0.336	0.094	0.081	0.168	0.133	0.047	0.016
1/2	0.5000	0.5000	0.4930	0.750	0.731	0.328	0.301	0.413	0.376	0.106	0.091	0.193	0.153	0.047	0.016
9/16	0.5625	0.5625	0.5550	0.812	0.792	0.375	0.346	0.467	0.427	0.118	0.102	0.213	0.168	0.047	0.016
5/8	0.6250	0.6250	0.6170	0.875	0.853	0.422	0.391	0.521	0.478	0.133	0.116	0.239	0.189	0.062	0.031
3/4	0.7500	0.7500	0.7420	1.000	0.976	0.500	0.466	0.612	0.566	0.149	0.131	0.283	0.223	0.062	0.031
7/8	0.8750	0.8750	0.8660	1.125	1.098	0.594	0.556	0.720	0.668	0.167	0.147	0.334	0.264	0.062	0.031
1	1.0000	1.0000	0.9900	1.312	1.282	0.656	0.612	0.803	0.743	0.188	0.166	0.371	0.291	0.062	0.031

[1] Where specifying nominal size in decimals, zeros preceding decimal and in the fourth decimal place shall be omitted.

[2] A slight rounding of the edges at periphery of head shall be permissible provided the diameter of the bearing circle is equal to no less than 90 per cent of the specified minimum head diameter.

Courtesy of ANSI (B18.6.2)

Table B.14 Dimensions of Hex Head Machine Screws

Nominal Size[1] or Basic Screw Diameter	A		W	A		W	H		J		T		F
	Regular Head[7]			Large Head[7,8]			Head Height		Slot Width		Slot Depth		Protrusion Beyond Gaging Ring
	Width Across Flats		Across Corners	Width Across Flats		Across Corners							
	Max	Min	Min	Max	Min	Min	Max	Min	Max	Min	Max	Min	Min
1 0.0730	0.125	0.120	0.134	—	—	—	0.044	0.036	—	—	—	—	0.022
2 0.0860	0.125	0.120	0.134	—	—	—	0.050	0.040	—	—	—	—	0.024
3 0.0990	0.188	0.181	0.202	—	—	—	0.055	0.044	—	—	—	—	0.026
4 0.1120	0.188	0.181	0.202	0.219	0.213	0.238	0.060	0.049	0.039	0.031	0.036	0.025	0.029
5 0.1250	0.188	0.181	0.202	0.250	0.244	0.272	0.070	0.058	0.043	0.035	0.042	0.030	0.035
6 0.1380	0.250	0.244	0.272	—	—	—	0.093	0.080	0.048	0.039	0.046	0.033	0.048
8 0.1640	0.250	0.244	0.272	0.312	0.305	0.340	0.110	0.096	0.054	0.045	0.066	0.052	0.058
10 0.1900	0.312	0.305	0.340	—	—	—	0.120	0.105	0.060	0.050	0.072	0.057	0.063
12 0.2160	0.312	0.305	0.340	0.375	0.367	0.409	0.155	0.139	0.067	0.056	0.093	0.077	0.083
1/4 0.2500	0.375	0.367	0.409	0.438	0.428	0.477	0.190	0.172	0.075	0.064	0.101	0.083	0.103
5/16 0.3125	0.500	0.489	0.545	—	—	—	0.230	0.208	0.084	0.072	0.122	0.100	0.125
3/8 0.3750	0.562	0.551	0.614	—	—	—	0.295	0.270	0.094	0.081	0.156	0.131	0.162

Courtesy of ANSI (B18.6.3)

Table B.15 Dimensions of Slotted Flat Countersunk Head Machine Screws

Nominal Size[1] or Basic Screw Diameter		L	A		H	J		T		F		G
		These Lengths or Shorter are Undercut.	Head Diameter		Head Height	Slot Width		Slot Depth		Protrusion Above Gaging Diameter		Gaging Diameter
			Max. Edge Sharp	Min. Edge Rounded or Flat	Ref	Max	Min	Max	Min	Max	Min	
0000	0.0210	—	0.043	0.037	0.011	0.008	0.004	0.007	0.003	*	*	*
000	0.0340	—	0.064	0.058	0.016	0.011	0.007	0.009	0.005	*	*	*
00	0.0470	—	0.093	0.085	0.028	0.017	0.010	0.014	0.009	*	*	*
0	0.0600	1/8	0.119	0.099	0.035	0.023	0.016	0.015	0.010	0.026	0.016	0.078
1	0.0730	1/8	0.146	0.123	0.043	0.026	0.019	0.019	0.012	0.028	0.016	0.101
2	0.0860	1/8	0.172	0.147	0.051	0.031	0.023	0.023	0.015	0.029	0.017	0.124
3	0.0990	1/8	0.199	0.171	0.059	0.035	0.027	0.027	0.017	0.031	0.018	0.148
4	0.1120	3/16	0.225	0.195	0.067	0.039	0.031	0.030	0.020	0.032	0.019	0.172
5	0.1250	3/16	0.252	0.220	0.075	0.043	0.035	0.034	0.022	0.034	0.020	0.196
6	0.1380	3/16	0.279	0.244	0.083	0.048	0.039	0.038	0.024	0.036	0.021	0.220
8	0.1640	1/4	0.332	0.292	0.100	0.054	0.045	0.045	0.029	0.039	0.023	0.267
10	0.1900	5/16	0.385	0.340	0.116	0.060	0.050	0.053	0.034	0.042	0.025	0.313
12	0.2160	3/8	0.438	0.389	0.132	0.067	0.056	0.060	0.039	0.045	0.027	0.362
1/4	0.2500	7/16	0.507	0.452	0.153	0.075	0.064	0.070	0.046	0.050	0.029	0.424
5/16	0.3125	1/2	0.635	0.568	0.191	0.084	0.072	0.088	0.058	0.057	0.034	0.539
3/8	0.3750	9/16	0.762	0.685	0.230	0.094	0.081	0.106	0.070	0.065	0.039	0.653
7/16	0.4375	5/8	0.812	0.723	0.223	0.094	0.081	0.103	0.066	0.073	0.044	0.690
1/2	0.5000	3/4	0.875	0.775	0.223	0.106	0.091	0.103	0.065	0.081	0.049	0.739
9/16	0.5625	—	1.000	0.889	0.260	0.118	0.102	0.120	0.077	0.089	0.053	0.851
5/8	0.6250	—	1.125	1.002	0.298	0.133	0.116	0.137	0.088	0.097	0.058	0.962
3/4	0.7500	—	1.375	1.230	0.372	0.149	0.131	0.171	0.111	0.112	0.067	1.186

[1] Where specifying nominal size in decimals, zeros preceding decimal and in the fourth decimal place shall be omitted.
*Not practical to gage.
Courtesy of ANSI (B18.6.3)

Table B.16   Dimensions of Slotted Fillister Head Machine Screws

Nominal Size[1] or Basic Screw Diameter		A Head Diameter		H Head Side Height		O Total Head Height		J Slot Width		T Slot Depth	
		Max	Min	Max	Min	Max	Min	Max	Min	Max	Min
0000	0.0210	0.038	0.032	0.019	0.011	0.025	0.015	0.008	0.004	0.012	0.006
000	0.0340	0.059	0.053	0.029	0.021	0.035	0.027	0.012	0.006	0.017	0.011
00	0.0470	0.082	0.072	0.037	0.028	0.047	0.039	0.017	0.010	0.022	0.015
0	0.0600	0.096	0.083	0.043	0.038	0.055	0.047	0.023	0.016	0.025	0.015
1	0.0730	0.118	0.104	0.053	0.045	0.066	0.058	0.026	0.019	0.031	0.020
2	0.0860	0.140	0.124	0.062	0.053	0.083	0.066	0.031	0.023	0.037	0.025
3	0.0990	0.161	0.145	0.070	0.061	0.095	0.077	0.035	0.027	0.043	0.030
4	0.1120	0.183	0.166	0.079	0.069	0.107	0.088	0.039	0.031	0.048	0.035
5	0.1250	0.205	0.187	0.088	0.078	0.120	0.100	0.043	0.035	0.054	0.040
6	0.1380	0.226	0.208	0.096	0.086	0.132	0.111	0.048	0.039	0.060	0.045
8	0.1640	0.270	0.250	0.113	0.102	0.156	0.133	0.054	0.045	0.071	0.054
10	0.1900	0.313	0.292	0.130	0.118	0.180	0.156	0.060	0.050	0.083	0.064
12	0.2160	0.357	0.334	0.148	0.134	0.205	0.178	0.067	0.056	0.094	0.074
1/4	0.2500	0.414	0.389	0.170	0.155	0.237	0.207	0.075	0.064	0.109	0.087
5/16	0.3125	0.518	0.490	0.211	0.194	0.295	0.262	0.084	0.072	0.137	0.110
3/8	0.3750	0.622	0.590	0.253	0.233	0.355	0.315	0.094	0.081	0.164	0.133
7/16	0.4375	0.625	0.589	0.265	0.242	0.368	0.321	0.094	0.081	0.170	0.135
1/2	0.5000	0.750	0.710	0.297	0.273	0.412	0.362	0.106	0.091	0.190	0.151
9/16	0.5625	0.812	0.768	0.336	0.308	0.466	0.410	0.118	0.102	0.214	0.172
5/8	0.6250	0.875	0.827	0.375	0.345	0.521	0.461	0.133	0.116	0.240	0.193
3/4	0.7500	1.000	0.945	0.441	0.406	0.612	0.542	0.149	0.131	0.281	0.226

[1] Where specifying nominal size in decimals, zeros preceding decimal and in the fourth decimal place shall be omitted.

[2] A slight rounding of the edges at periphery of head shall be permissible provided the diameter of the bearing circle is equal to no less than 90 per cent of the specified minimum head diameter.

Courtesy of ANSI (B18.6.3)

Table B.17 Dimensions of Square and Hex Machine Screw Nuts

Nominal Size[1] or Basic Thread Diameter		F Width Across Flats			G Width Across Corners Square		G₁ Width Across Corners Hex		H Thickness	
		Basic	Max	Min	Max	Min	Max	Min	Max	Min
0	0.0600	5/32	0.156	0.150	0.221	0.206	0.180	0.171	0.050	0.043
1	0.0730	5/32	0.156	0.150	0.221	0.206	0.180	0.171	0.050	0.043
2	0.0860	3/16	0.188	0.180	0.265	0.247	0.217	0.205	0.066	0.057
3	0.0990	3/16	0.188	0.180	0.265	0.247	0.217	0.205	0.066	0.057
4	0.1120	1/4	0.250	0.241	0.354	0.331	0.289	0.275	0.098	0.087
5	0.1250	5/16	0.312	0.302	0.442	0.415	0.361	0.344	0.114	0.102
6	0.1380	5/16	0.312	0.302	0.442	0.415	0.361	0.344	0.114	0.102
8	0.1640	11/32	0.344	0.332	0.486	0.456	0.397	0.378	0.130	0.117
10	0.1900	3/8	0.375	0.362	0.530	0.497	0.433	0.413	0.130	0.117
12	0.2160	7/16	0.438	0.423	0.619	0.581	0.505	0.482	0.161	0.148
1/4	0.2500	7/16	0.438	0.423	0.619	0.581	0.505	0.482	0.193	0.178
5/16	0.3125	9/16	0.562	0.545	0.795	0.748	0.650	0.621	0.225	0.208
3/8	0.3750	5/8	0.625	0.607	0.884	0.833	0.722	0.692	0.257	0.239

[1] Where specifying nominal size in decimals, zeros preceding decimal and in the fourth decimal place shall be omitted.

[2] Square machine screw nuts shall have tops and bottoms flat, without chamfer. The bearing surface shall be perpendicular to the axis of the threaded hole pitch cylinder within a tolerance of 4 deg.

[3] Hexagon machine screw nuts shall have tops flat with chamfered corners. Diameter of top circle shall be equal to the maximum width across flats within a tolerance of minus 15 per cent. The bearing surface shall be perpendicular to the axis of the threaded hole pitch cylinder within a tolerance of 4 deg.

[4] Bottoms of hexagon machine screw nuts are normally flat, but for special purposes may be chamfered, if so specified by purchaser.

Courtesy of ANSI (B18.6.3)

Table B.18   Dimensions of Slotted Headless Set Screws

FLAT POINT   DOG POINT   HALF DOG POINT

CUP POINT   OVAL POINT   CONE POINT

Nominal Size[1] or Basic Screw Diameter		I Crown Radius Basic	J Slot Width		T Slot Depth		C Cup and Flat Point Diameters		P Dog Point Diameters		Q Point Length				R Oval Point Radius	Y Cone Point Angle 90° ±2° For These Nominal Lengths or Longer; 118° ±2° For Shorter Screws
											Dog		Half Dog			
		Basic	Max	Min	Max	Min	Max	Min	Max	Min	Max	Min	Max	Min	Basic	
0	0.0600	0.060	0.014	0.010	0.020	0.016	0.033	0.027	0.040	0.037	0.032	0.028	0.017	0.013	0.045	5/64
1	0.0730	0.073	0.016	0.012	0.020	0.016	0.040	0.033	0.049	0.045	0.040	0.036	0.021	0.017	0.055	3/32
2	0.0860	0.086	0.018	0.014	0.025	0.019	0.047	0.039	0.057	0.053	0.046	0.042	0.024	0.020	0.064	7/64
3	0.0990	0.099	0.020	0.016	0.028	0.022	0.054	0.045	0.066	0.062	0.052	0.048	0.027	0.023	0.074	1/8
4	0.1120	0.112	0.024	0.018	0.031	0.025	0.061	0.051	0.075	0.070	0.058	0.054	0.030	0.026	0.084	5/32
5	0.1250	0.125	0.026	0.020	0.036	0.026	0.067	0.057	0.083	0.078	0.063	0.057	0.033	0.027	0.094	3/16
6	0.1380	0.138	0.028	0.022	0.040	0.030	0.074	0.064	0.092	0.087	0.073	0.067	0.038	0.032	0.104	3/16
8	0.1640	0.164	0.032	0.026	0.046	0.036	0.087	0.076	0.109	0.103	0.083	0.077	0.043	0.037	0.123	1/4
10	0.1900	0.190	0.035	0.029	0.053	0.043	0.102	0.088	0.127	0.120	0.095	0.085	0.050	0.040	0.142	1/4
12	0.2160	0.216	0.042	0.035	0.061	0.051	0.115	0.101	0.144	0.137	0.115	0.105	0.060	0.050	0.162	5/16
1/4	0.2500	0.250	0.049	0.041	0.068	0.058	0.132	0.118	0.156	0.149	0.130	0.120	0.068	0.058	0.188	5/16
5/16	0.3125	0.312	0.055	0.047	0.083	0.073	0.172	0.156	0.203	0.195	0.161	0.151	0.083	0.073	0.234	3/8
3/8	0.3750	0.375	0.068	0.060	0.099	0.089	0.212	0.194	0.250	0.241	0.193	0.183	0.099	0.089	0.281	7/16
7/16	0.4375	0.438	0.076	0.068	0.114	0.104	0.252	0.232	0.297	0.287	0.224	0.214	0.114	0.104	0.328	1/2
1/2	0.5000	0.500	0.086	0.076	0.130	0.120	0.291	0.270	0.344	0.334	0.255	0.245	0.130	0.120	0.375	9/16
9/16	0.5625	0.562	0.096	0.086	0.146	0.136	0.332	0.309	0.391	0.379	0.287	0.275	0.146	0.134	0.422	5/8
5/8	0.6250	0.625	0.107	0.097	0.161	0.151	0.371	0.347	0.469	0.456	0.321	0.305	0.164	0.148	0.469	3/4
3/4	0.7500	0.750	0.134	0.124	0.193	0.183	0.450	0.425	0.562	0.549	0.383	0.367	0.196	0.180	0.562	7/8

[1] Where specifying nominal size in decimals, zeros preceding decimal and in the fourth decimal place shall be omitted.

[2] Tolerance on radius for nominal sizes up to and including 5 (0.125 in.) shall be plus 0.015 in. and minus 0.000, and for larger sizes, plus 0.031 in. and minus 0.000. Slotted ends on screws may be flat at option of manufacturer.

[3] Point angle X shall be 45° plus 5°, minus 0°, for screws of nominal lengths equal to or longer than those listed in Column Y, and 30° minimum for screws of shorter nominal lengths.

[4] The extent of rounding or flat at apex of cone point shall not exceed an amount equivalent to 10 per cent of the basic screw diameter.

Courtesy of ANSI (B18.6.2)

Table B.19 Dimensions of Square Head Set Screws

OPTIONAL HEAD CONSTRUCTIONS

Nominal Size[1] or Basic Screw Diameter		F Width Across Flats		G Width Across Corners		H Head Height		K Neck Relief Diameter		S Neck Relief Fillet Radius	U Neck Relief Width	W Head Radius
		Max	Min	Max	Min	Max	Min	Max	Min	Max	Min	Min
10	0.1900	0.188	0.180	0.265	0.247	0.148	0.134	0.145	0.140	0.027	0.083	0.48
1/4	0.2500	0.250	0.241	0.354	0.331	0.196	0.178	0.185	0.170	0.032	0.100	0.62
5/16	0.3125	0.312	0.302	0.442	0.415	0.245	0.224	0.240	0.225	0.036	0.111	0.78
3/8	0.3750	0.375	0.362	0.530	0.497	0.293	0.270	0.294	0.279	0.041	0.125	0.94
7/16	0.4375	0.438	0.423	0.619	0.581	0.341	0.315	0.345	0.330	0.046	0.143	1.09
1/2	0.5000	0.500	0.484	0.707	0.665	0.389	0.361	0.400	0.385	0.050	0.154	1.25
9/16	0.5625	0.562	0.545	0.795	0.748	0.437	0.407	0.454	0.439	0.054	0.167	1.41
5/8	0.6250	0.625	0.606	0.884	0.833	0.485	0.452	0.507	0.492	0.059	0.182	1.56
3/4	0.7500	0.750	0.729	1.060	1.001	0.582	0.544	0.620	0.605	0.065	0.200	1.88
7/8	0.8750	0.875	0.852	1.237	1.170	0.678	0.635	0.731	0.716	0.072	0.222	2.19
1	1.0000	1.000	0.974	1.414	1.337	0.774	0.726	0.838	0.823	0.081	0.250	2.50
1 1/8	1.1250	1.125	1.096	1.591	1.505	0.870	0.817	0.939	0.914	0.092	0.283	2.81
1 1/4	1.2500	1.250	1.219	1.768	1.674	0.966	0.908	1.064	1.039	0.092	0.283	3.12
1 3/8	1.3750	1.375	1.342	1.945	1.843	1.063	1.000	1.159	1.134	0.109	0.333	3.44
1 1/2	1.5000	1.500	1.464	2.121	2.010	1.159	1.091	1.284	1.259	0.109	0.333	3.75

[1] Where specifying nominal size in decimals, zeros preceding decimal and in the fourth decimal place shall be omitted.
Courtesy of ANSI (B18.6.2)

Table B.20

FRACTION–DECIMAL
Conversion Table

Fraction			2-Place Decimal	3-Place Decimal
		1/32	.03	.031
	1/16		.06	.062
		3/32	.09	.094
1/8			.12	.125
		5/32	.16	.156
	3/16		.19	.188
		7/32	.22	.219
1/4			.25	.250
		9/32	.28	.281
	5/16		.31	.312
		11/32	.34	.344
3/8			.38	.375
		13/32	.41	.406
	7/16		.44	.438
		15/32	.47	.469
1/2			.50	.500
		17/32	.53	.531
	9/16		.56	.562
		19/32	.59	.594
5/8			.62	.625
		21/32	.66	.656
	11/16		.69	.688
		23/32	.72	.719
3/4			.75	.750
		25/32	.78	.781
	13/16		.81	.812
		27/32	.84	.844
7/8			.88	.875
		29/32	.91	.906
	15/16		.94	.938
		31/32	.97	.969
1			1.00	1.000

Table B.21 Dimensions of Preferred Sizes of Type A Plain Washers

Nominal Washer Size			Inside Diameter A			Outside Diameter B			Thickness C		
			Basic	Plus	Minus	Basic	Plus	Minus	Basic	Max	Min
—	—		0.078	0.000	0.005	0.188	0.000	0.005	0.020	0.025	0.016
—	—		0.094	0.000	0.005	0.250	0.000	0.005	0.020	0.025	0.016
—	—		0.125	0.008	0.005	0.312	0.008	0.005	0.032	0.040	0.025
No. 6	0.138		0.156	0.008	0.005	0.375	0.015	0.005	0.049	0.065	0.036
No. 8	0.164		0.188	0.008	0.005	0.438	0.015	0.005	0.049	0.065	0.036
No. 10	0.190		0.219	0.008	0.005	0.500	0.015	0.005	0.049	0.065	0.036
$\frac{3}{16}$	0.188		0.250	0.015	0.005	0.562	0.015	0.005	0.049	0.065	0.036
No. 12	0.216		0.250	0.015	0.005	0.562	0.015	0.005	0.065	0.080	0.051
$\frac{1}{4}$	0.250	N	0.281	0.015	0.005	0.625	0.015	0.005	0.065	0.080	0.051
$\frac{1}{4}$	0.250	W	0.312	0.015	0.005	0.734	0.015	0.007	0.065	0.080	0.051
$\frac{5}{16}$	0.312	N	0.344	0.015	0.005	0.688	0.015	0.007	0.065	0.080	0.051
$\frac{5}{16}$	0.312	W	0.375	0.015	0.005	0.875	0.030	0.007	0.083	0.104	0.064
$\frac{3}{8}$	0.375	N	0.406	0.015	0.005	0.812	0.015	0.007	0.065	0.080	0.051
$\frac{3}{8}$	0.375	W	0.438	0.015	0.005	1.000	0.030	0.007	0.083	0.104	0.064
$\frac{7}{16}$	0.438	N	0.469	0.015	0.005	0.922	0.015	0.007	0.065	0.080	0.051
$\frac{7}{16}$	0.438	W	0.500	0.015	0.005	1.250	0.030	0.007	0.083	0.104	0.064
$\frac{1}{2}$	0.500	N	0.531	0.015	0.005	1.062	0.030	0.007	0.095	0.121	0.074
$\frac{1}{2}$	0.500	W	0.562	0.015	0.005	1.375	0.030	0.007	0.109	0.132	0.086
$\frac{9}{16}$	0.562	N	0.594	0.015	0.005	1.156	0.030	0.007	0.095	0.121	0.074
$\frac{9}{16}$	0.562	W	0.625	0.015	0.005	1.469	0.030	0.007	0.109	0.132	0.086
$\frac{5}{8}$	0.625	N	0.656	0.030	0.007	1.312	0.030	0.007	0.095	0.121	0.074
$\frac{5}{8}$	0.625	W	0.688	0.030	0.007	1.750	0.030	0.007	0.134	0.160	0.108
$\frac{3}{4}$	0.750	N	0.812	0.030	0.007	1.469	0.030	0.007	0.134	0.160	0.108
$\frac{3}{4}$	0.750	W	0.812	0.030	0.007	2.000	0.030	0.007	0.148	0.177	0.122
$\frac{7}{8}$	0.875	N	0.938	0.030	0.007	1.750	0.030	0.007	0.134	0.160	0.108
$\frac{7}{8}$	0.875	W	0.938	0.030	0.007	2.250	0.030	0.007	0.165	0.192	0.136
1	1.000	N	1.062	0.030	0.007	2.000	0.030	0.007	0.134	0.160	0.108
1	1.000	W	1.062	0.030	0.007	2.500	0.030	0.007	0.165	0.192	0.136
$1\frac{1}{8}$	1.125	N	1.250	0.030	0.007	2.250	0.030	0.007	0.134	0.160	0.108
$1\frac{1}{8}$	1.125	W	1.250	0.030	0.007	2.750	0.030	0.007	0.165	0.192	0.136
$1\frac{1}{4}$	1.250	N	1.375	0.030	0.007	2.500	0.030	0.007	0.165	0.192	0.136
$1\frac{1}{4}$	1.250	W	1.375	0.030	0.007	3.000	0.030	0.007	0.165	0.192	0.136
$1\frac{3}{8}$	1.375	N	1.500	0.030	0.007	2.750	0.030	0.007	0.165	0.192	0.136
$1\frac{3}{8}$	1.375	W	1.500	0.045	0.010	3.250	0.045	0.010	0.180	0.213	0.153
$1\frac{1}{2}$	1.500	N	1.625	0.030	0.007	3.000	0.030	0.007	0.165	0.192	0.136
$1\frac{1}{2}$	1.500	W	1.625	0.045	0.010	3.500	0.045	0.010	0.180	0.213	0.153
$1\frac{5}{8}$	1.625		1.750	0.045	0.010	3.750	0.045	0.010	0.180	0.213	0.153
$1\frac{3}{4}$	1.750		1.875	0.045	0.010	4.000	0.045	0.010	0.180	0.213	0.153
$1\frac{7}{8}$	1.875		2.000	0.045	0.010	4.250	0.045	0.010	0.180	0.213	0.153
2	2.000		2.125	0.045	0.010	4.500	0.045	0.010	0.180	0.213	0.153
$2\frac{1}{4}$	2.250		2.375	0.045	0.010	4.750	0.045	0.010	0.220	0.248	0.193
$2\frac{1}{2}$	2.500		2.625	0.045	0.010	5.000	0.045	0.010	0.238	0.280	0.210
$2\frac{3}{4}$	2.750		2.875	0.065	0.010	5.250	0.065	0.010	0.259	0.310	0.228
3	3.000		3.125	0.065	0.010	5.500	0.065	0.010	0.284	0.327	0.249

Courtesy of ANSI (B18.22.1)

Table B.22   Dimensions of Woodruff Keys

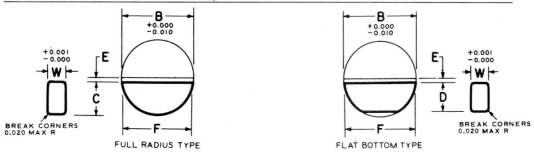

FULL RADIUS TYPE   FLAT BOTTOM TYPE

Key No.	Nominal Key Size W × B	Actual Length F +0.000-0.010	Height of Key				Distance Below Center E
			C		D		
			Max	Min	Max	Min	
202	¹⁄₁₆ × ¼	0.248	0.109	0.104	0.109	0.104	¹⁄₆₄
202.5	¹⁄₁₆ × ⁵⁄₁₆	0.311	0.140	0.135	0.140	0.135	¹⁄₆₄
302.5	³⁄₃₂ × ⁵⁄₁₆	0.311	0.140	0.135	0.140	0.135	¹⁄₆₄
203	¹⁄₁₆ × ³⁄₈	0.374	0.172	0.167	0.172	0.167	¹⁄₆₄
303	³⁄₃₂ × ³⁄₈	0.374	0.172	0.167	0.172	0.167	¹⁄₆₄
403	⅛ × ³⁄₈	0.374	0.172	0.167	0.172	0.167	¹⁄₆₄
204	¹⁄₁₆ × ½	0.491	0.203	0.198	0.194	0.188	³⁄₆₄
304	³⁄₃₂ × ½	0.491	0.203	0.198	0.194	0.188	³⁄₆₄
404	⅛ × ½	0.491	0.203	0.198	0.194	0.188	³⁄₆₄
305	³⁄₃₂ × ⅝	0.612	0.250	0.245	0.240	0.234	¹⁄₁₆
405	⅛ × ⅝	0.612	0.250	0.245	0.240	0.234	¹⁄₁₆
505	⁵⁄₃₂ × ⅝	0.612	0.250	0.245	0.240	0.234	¹⁄₁₆
605	³⁄₁₆ × ⅝	0.612	0.250	0.245	0.240	0.234	¹⁄₁₆
406	⅛ × ¾	0.740	0.313	0.308	0.303	0.297	¹⁄₁₆
506	⁵⁄₃₂ × ¾	0.740	0.313	0.308	0.303	0.297	¹⁄₁₆
606	³⁄₁₆ × ¾	0.740	0.313	0.308	0.303	0.297	¹⁄₁₆
806	¼ × ¾	0.740	0.313	0.308	0.303	0.297	¹⁄₁₆
507	⁵⁄₃₂ × ⅞	0.866	0.375	0.370	0.365	0.359	¹⁄₁₆
607	³⁄₁₆ × ⅞	0.866	0.375	0.370	0.365	0.359	¹⁄₁₆
707	⁷⁄₃₂ × ⅞	0.866	0.375	0.370	0.365	0.359	¹⁄₁₆
807	¼ × ⅞	0.866	0.375	0.370	0.365	0.359	¹⁄₁₆
608	³⁄₁₆ × 1	0.992	0.438	0.433	0.428	0.422	¹⁄₁₆
708	⁷⁄₃₂ × 1	0.992	0.438	0.433	0.428	0.422	¹⁄₁₆
808	¼ × 1	0.992	0.438	0.433	0.428	0.422	¹⁄₁₆
1008	⁵⁄₁₆ × 1	0.992	0.438	0.433	0.428	0.422	¹⁄₁₆
1208	⅜ × 1	0.992	0.438	0.433	0.428	0.422	¹⁄₁₆
609	³⁄₁₆ × 1⅛	1.114	0.484	0.479	0.475	0.469	⁵⁄₆₄
709	⁷⁄₃₂ × 1⅛	1.114	0.484	0.479	0.475	0.469	⁵⁄₆₄
809	¼ × 1⅛	1.114	0.484	0.479	0.475	0.469	⁵⁄₆₄
1009	⁵⁄₁₆ × 1⅛	1.114	0.484	0.479	0.475	0.469	⁵⁄₆₄

Courtesy of USAS (B17.2)

Table B.23 Standard Welding Symbols

Typical Welding Symbols

Double-Fillet Welding Symbol	Chain Intermittent Fillet Welding Symbol	Staggered Intermittent Fillet Welding Symbol
Plug Welding Symbol	**Back Welding Symbol**	**Backing Welding Symbol**
Spot Welding Symbol	**Stud Welding Symbol**	**Seam Welding Symbol**
Square-Groove Welding Symbol	**Single-V Groove Welding Symbol**	**Double-Bevel-Groove Welding Symbol**
Symbol with Back Gouging	**Flare-V Groove Welding Symbol**	**Flare-Bevel-Groove Welding Symbol**
Multiple Reference Lines	**Complete Penetration**	**Edge Flange Welding Symbol**
Flash or Upset Welding Symbol	**Melt-Thru Symbol**	**Joint with Backing**
Joint with Spacer	**Flush Contour Symbol**	**Convex Contour Symbol**

Courtesy of American Welding Society (A2.1-86)

'It should be understood that these charts are intended only as shop aids. The only complete and official presentation of the standard welding symbols is in A2.4.

INDEX

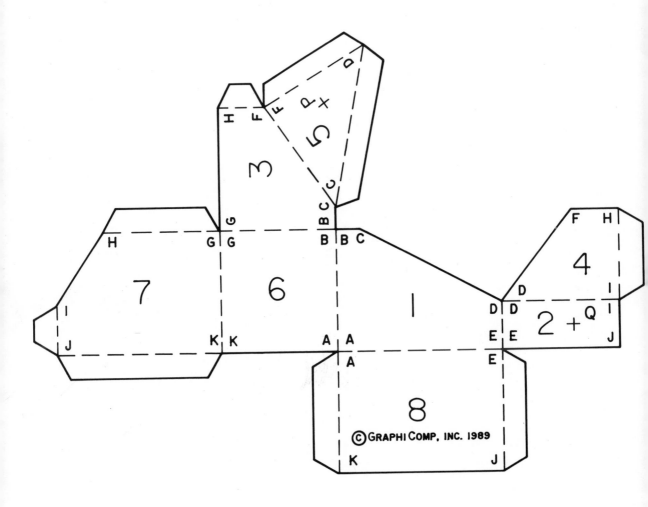

© GraphiComp, Inc. 1989